# Smart Nanomaterials Technology

Nanotechnology is a rapidly growing scientific field and has attracted a great interest over the last few years because of its abundant applications in different fields like biology, physics and chemistry. This science deals with the production of minute particles called nanomaterials having dimensions between 1 and 100 nm which may serve as building blocks for various physical and biological systems. On the other hand, there is the class of smart materials where the material that can stimuli by external factors and results a new kind of functional properties. The combination of these two classes forms a new class of smart nanomaterials, which produces unique functional material properties and a great opportunity to larger span of application. Smart nanomaterials have been employed by researchers to use it effectively in agricultural production, soil improvement, disease management, energy and environment, medical science, pharmaceuticals, engineering, food, animal husbandry and forestry sectors.

This book series in Smart Nanomaterials Technology aims to comprehensively cover topics in the fabrication, synthesis and application of these materials for applications in the following fields:

- Energy Systems—Renewable energy, energy storage (supercapacitors and electrochemical cells), hydrogen storage, photocatalytic water splitting for hydrogen production
- Biomedical—controlled release of drugs, treatment of various diseases, biosensors,
- Agricultural—agricultural production, soil improvement, disease management, animal feed, egg, milk and meat production/processing,
- Forestry—wood preservation, protection, disease management
- Environment—wastewater treatment, separation of hazardous contaminants from wastewater, indoor air filters.

Petar Antov · Muhammad Adly Rahandi Lubis ·
Seng Hua Lee · Hamid R. Taghiyari
Editors

# Nanomaterials Additives in Bioadhesives for Wood Composites

Springer

*Editors*
Petar Antov
Mechanical Wood Technology
University of Forestry
Sofia, Bulgaria

Muhammad Adly Rahandi Lubis
Research Center for Biomass
and Bioproducts
National Research and Innovation Agency
Cibinong, Indonesia

Seng Hua Lee
Department of Wood Industry, Faculty
of Applied Sciences
Universiti Teknologi MARA Pahang
Branch
Pahang, Jengka, Malaysia

Hamid R. Taghiyari
Faculty of Materials Engineering and New
Technologies
Shahid Rajaee Teacher Training University
Tehran, Iran

ISSN 3004-8273 ISSN 3004-8281 (electronic)
Smart Nanomaterials Technology
ISBN 978-981-96-3399-9 ISBN 978-981-96-3400-2 (eBook)
https://doi.org/10.1007/978-981-96-3400-2



This Springer imprint is published by the registered company Springer Nature Singapore Pte Ltd.
The registered company address is: 152 Beach Road, #21-01/04 Gateway East, Singapore 189721, Singapore

# Contents

**Bio-based Wood Adhesives: Recent Developments** .................... 1
Antonio Pizzi and Petar Antov

**Potential of Nanotechnology in Wood-Based Materials** ................ 27
Jakub Kawalerczyk, Dorota Dziurka, Dorota Dukarska,
Joanna Walkiewicz, and Radosław Mirski

**Nanocellulose Applications in Bio-based Wood Adhesives** ............. 67
Mohd Ezwan Bin Selamat and Lum Wei Chen

**Nanolignin in Bioadhesives for Wood Composites** .................... 83
Widya Fatriasari, Asma Sohail, Muhammad Rasyidur Ridho,
Elvara Windra Madyaratri, Swetha Gillela, Sumit Manohar Yadav,
Tuan Sherwyn Hamidon, and Mohd. Hazwan Hussin

**Sustainable and Eco-Friendly Starch-Based Adhesives for High-Performance Wood-Based Composites** ...................... 119
Muhammad Adly Rahandi Lubis, Apri Heri Iswanto,
and Muhammad Iqbal Maulana

**Multifunctional Nanotechnology Application for Wood Properties Enhancement: Adhesive and Coating** ............................... 135
Ria Aniza, Anelie Petrissans, and Mathieu Petrissans

**Potential of Nanomaterials as Formaldehyde Scavengers in Wood-Based Composites** ......................................... 151
Petar Antov, Jakub Kawalerczyk, Seng Hua Lee,
Muhammad Adly Rahandi Lubis, and Viktor Savov

**Fire Retarding Treatments of Wood and Wood-Based Materials** ....... 183
Linda Makovicka Osvaldova and Widya Fatriasari

**Wood Protection: Nanomaterials to Improve Durability in Wood & Wood-Based Composites** ........ 207
Hamid R. Taghiyari, Mahdi Arabi, Dorina Camelia Ilies, Elham Nadali, Zharas Berdenov, and Vasile Grama

**Life-Cycle Assessment of Nanoparticle in Wood Adhesive and Coating: The State-of-The Art** ........ 231
Ria Aniza, Anelie Petrissans, and Mathieu Petrissans

**Economic Aspects of Nanomaterials Application in Bioadhesives and Wood-Based Composites** ........ 249
Nikolay Neykov, Viktor Savov, and Aureliu-Florin Halalisan

# Bio-based Wood Adhesives: Recent Developments

**Antonio Pizzi and Petar Antov**

**Abstract** This review is aimed at presenting only the most recent concepts and advances of a great majority bio-sourced, sustainable, green, and high-performance wood adhesives derived from carbohydrates, lignin, and tannins that are actively researched as substitutes for synthetic oil-derived adhesives for wood panels. For all these bio-sourced adhesives the review exposes the considerable progresses which have occurred relatively recently, with a host of new approaches and ideas having been proposed and tested, some even implemented, but even with many more already appearing on the horizon.

**Keywords** Wood bioadhesives · Environment-friendly · Wood-based panels · Carbohydrates · Tannin · Lignin

## 1 Introduction

Research on bio-based adhesives is greatly fashionable at present, with an increasing number of groups doing research on all sorts of bio-based materials to prepare thermosetting adhesives suitable for wood. It must be clearly pointed out that this review concerns only the most recent and interesting developments in this field, as excellent, extensive, and in-depth reviews on bio-adhesives already exist [5, 6, 36, 37, 50, 85–87, 91–93]. It must also be pointed out that this review focuses mainly on bio-adhesives that are all or in the far greater majority bio-sourced.

There are several bio-based adhesives based on renewable natural materials that are at the forefront of new developments. Some of these are already industrial, sometime for many years, such as tannin adhesives and some soy adhesives, while others

A. Pizzi (✉)
LERMAB-ENSTIB, University of Lorraine, Epinal, France
e-mail: antonio.pizzi@univ-lorraine.fr

P. Antov
Faculty of Forest Industry, University of Forestry, Sofia, Bulgaria

 
P. Antov et al. (eds.), *Nanomaterials Additives in Bioadhesives for Wood Composites*, Smart Nanomaterials Technology, https://doi.org/10.1007/978-981-96-3400-2_1

are on the way of industrialization, and many others are still at the experimental stage.

A definition of bio-based wood adhesives, or renewable natural materials sourced adhesives, of no oil origin, is actually due. The need of such a definition is dictated by the fact that in broader terms even resins such as urea–formaldehyde can be considered biosourced as both its two components are from non-oil derived raw materials. This is untrue. Bio-based adhesives are now being considered within a specific narrow meaning. Only natural, non-oil derived, non-mineral origin materials, usable unaltered or slightly modified and capable of matching the performance and adaptability to the tasks as the synthetic adhesives are included in the definition. Thus, the variety of materials which one can include within the narrowest sense of such a definition is rather limited. Only lignocellulosic origin polyphenolics such as lignins and hydrolyzable and condensed tannins, carbohydrate oligomers, proteins and their hydrolysates, unsaturated vegetable oils, isosorbide, cardanol, dissolved wood and wood welding self-adhesion wood welding can be the ones included.

## 2 Carbohydrate Adhesives

The greatest advances have possibly been made recently in this area, such advances influencing also other fields like tannins, lignin, and even protein adhesives such as soy flour. (a) to modify presently used commercial synthetic adhesives such as urea-formaldehyde (UF) and phenol-formaldehyde (PF). This is a traditional approach and the literature on this subject is already very large [27, 28, 68–72]. (b) By degrading carbohydrates yielding simpler compounds usable to build adhesives by synthesis and (c) by using the biosourced materials themselves as wood panel binders. And (d) by coreaction with other biosourced materials. Route (b) above leads to the formation of furan resins. While biosourced materials from the acid treatment of carbohydrate agricultural waste such as furfuryl alcohol and furfural are the building blocks of furanic resins, these latter are at present considered as just synthetic resins [10]. They are instead true bioresins from natural materials already used extensively as binders other than for wood. Reviews on them already exist in the literature [10]. Nonetheless, it must be pointed out that the relatively high cost of the building blocks of furanic resins and the dark color of such resins render unsuitable at present their use as wood panel adhesives. Such resins have important niche applications, however, in industrial applications where both these drawbacks are not a disadvantage. However, they can very well be used as binders for wood panels notwithstanding that the furfuryl alcohol has a relatively higher toxicity before polymerization that will need to be addressed. Successful application of these furanic-aldehyde resins as binders for wood panels by coupling non-volatile, non-toxic biosourced aldehydes and furfuryl alcohol has recently also been reported [132].

The use of liquefied wood, thus derived from cellulosic materials, has been reported by a number of research groups showing more than adequate wood adhesive performances [2, 3, 98, 106, 138]. Liquefaction by sulfuric acid at atmospheric pressure coupled with ethylene glycol or phenol of lignocellulosic and cellulosic materials was the approach used.

Specific oxidation of carbohydrate by sodium periodate is the most encouraging recent approach taken. The periodate cleaves the carbon bonds between vicinal 1,2–glycols and oxidizes the two hydroxyl groups yielding two aldehyde groups. This reaction is well known and widely used in carbohydrate chemistry [13]. The mild conditions under which this reaction occurs in water are particularly well suited for use with water-soluble carbohydrates and presents a high level of selectivity [26, 39, 40, 47]. The periodate reaction on carbohydrates is well known for monosaccharides such as glucose [13, 130, 131] and disaccharides such as sucrose [130, 131] (Fig. 1). Also for higher molecular weight carbohydrate oligomers even up to cellulose [13, 26, 39, 40, 47, 130, 131] (Fig. 1). For long carbohydrate oligomers, such as cellulose and hemicelluloses and their oligomeric fragments, the specific oxidation reaction by periodate leads to the condensation with other oligomeric carbohydrate chains of the aldehydes generated leading to crosslinking, and for cellulose to hard panels (Fig. 1) [26].

By increasing the level of oxidation with further periodate, the aldehyde groups so formed can react with species such as a soy protein [39, 40], a flavonoid tannin [133] or other reactive species, including lignin [21], cross-linking can also occur, leading to feasible wood adhesives [21, 26, 39, 40, 130, 131].

Several different aldehydes have been identified when sodium periodate is used to treat glucose and sucrose for 1 h at 120 °C. These are generated by a number of different mechanisms from both simple and even from some insoluble oligomeric carbohydrates such as those in soy flour to give soy flour wood binders of good performance [39, 40], as well as yielding good adhesives also from the fragments of hemicelluloses in commercial tannin extracts [133].

Larger carbohydrate oligomers also yielded aldehydes of much higher molecular weight [39] where several reactive aldehyde groups were generated on the carbohydrate chains. Such an approach also yielded good tannin-based and lignin-based wood panel adhesives [21, 133].

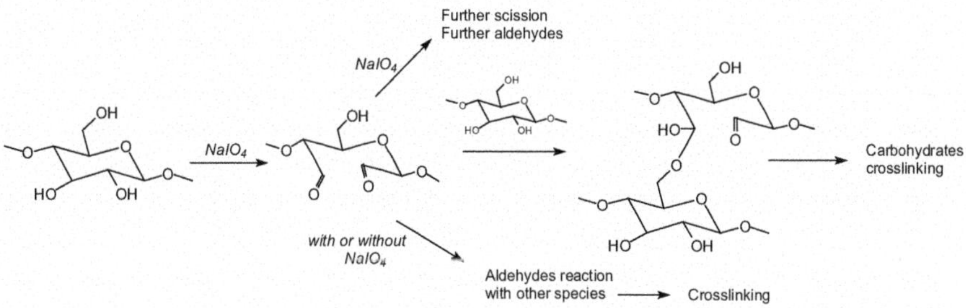


**Fig. 1** Specific oxidation of carbohydrates by sodium periodate leading to non-toxic and non-volatile bio-sourced aldehydes and to different cross-linked products according to the conditions used

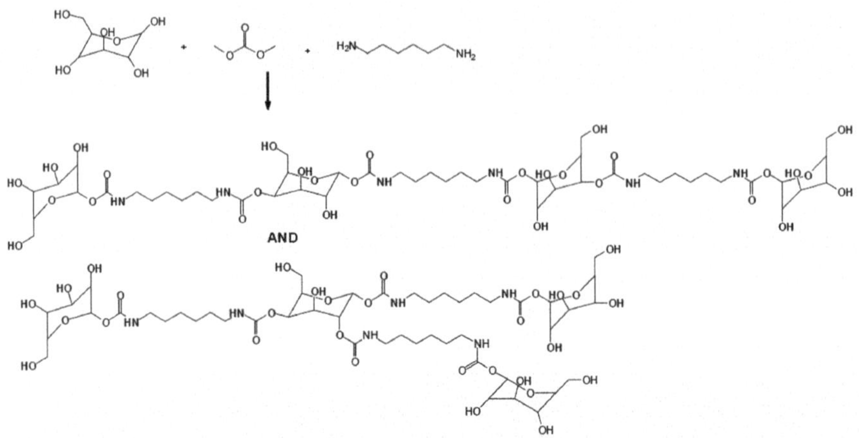


**Fig. 2** Schematic representation of the formation of NIPU oligomers from monosaccharides reacted with dimethyl carbonate and hexamethylene diamine. The two tetramers shown have been identified and are among the multitude of other NIPU oligomers identified

A different and more recent approach to the preparation of wood panel adhesives from glucose and sucrose passed through the preparation of non-isocyanate polyurethanes (NIPU). Glucose and sucrose were reacted with dimethyl carbonate and a diamine to yield NIPU resins [130] (Fig. 2). This brought these biosourced NIPU resins to be tested as thermosetting adhesives for wood panels [136]. The results were very encouraging.

Reacting glucose with maleic anhydride and a diamine is also an equally new approach yielding a polyamide-based bioadhesive, rather than making NIPUs, also yielding good performance wood particleboard adhesives [134].

Possibly the most revolutionary recent approach has been to use citric acid as a wood panels binder. Different systems of this approach have been experimented with carbohydrates. Japanese researchers were the first to advance such a citric acid binder approach for wood panels reporting good bonding results for particleboard [122, 123]. Flat veneers direct citric acid bonding, as in LVL (Laminated Veneer Lumber) and plywood, were then successfully tested [31] (Fig. 3) and applied to weldline improved waterproofing in wood welding [4]. The work reported by Del Menezzi et al. [31] carried out an in-depth analysis of the reasons why citric acid bonded wood. It showed that citric acid esterified the hydroxyl groups on both the wood carbohydrate constituents and lignin, and this even bridging two joint veneers [31] thus showing why it was a real plywood adhesive of some sort. Again, sugar such as sucrose + citric acid was demonstrated as well to be a very viable adhesive for wood composite panels [114, 124, 127, 146]. Their work reported also the formation of furfural and hydroxymethyl furfural by the acid action on the sugar attesting of the very low pH reached on the glueline under the conditions they used (Fig. 4). This further confirmed the same cross-linking manner with wood carbohydrates, glucose and with lignin by the action of citric acid directly on wood already found by other authors [31].

The work with Citric acid + carbohydrates has been extended to glucose [59] sucrose [124] and sorbitol [55, 60], and also polymeric carbohydrates such as citric acid/starch [56, 137], citric acid/cellulose [4, 31, 122, 123], and others [58]. While all these approaches do give good bonding results a grave and main aspect of the problem has been ignored in all these works. This is the extremely acid pH of the citric acid-based adhesives at the interface with the wood causing slow but sure degradation by acid hydrolysis of the wood carbohydrates at the interface with the adhesive. In fact, very acid pHs are not allowed by any standards in traditional wood adhesives. This is because they lead to the medium to long-term hydrolysis of the wood surface polymeric carbohydrates at the interface between the substrate and the hardened adhesive. This results in disastrous failure of the bonded wood joint in the medium to long term. The damage is done at two levels: (i) by the residual acid in the hardened glueline and even worse (ii) by the contact of the liquid catalyzed resin with the wood before curing. The greatest drawback is that traditional accelerated tests are not able to detect such a problem, as only medium to long-term tests might be able to detect it.

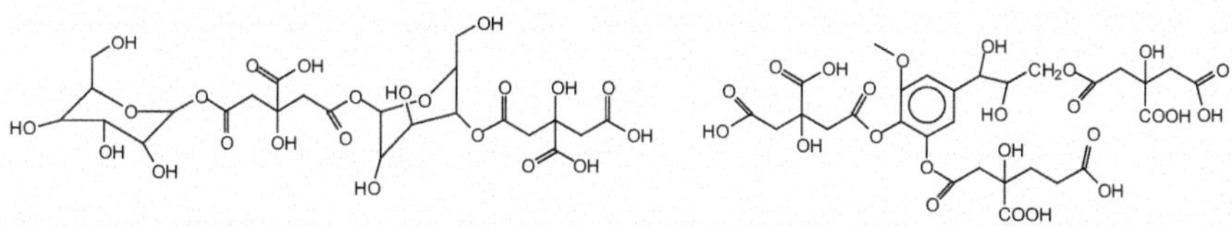

**Fig. 3** Example of citric acid bridges linking glucoses or holocellulose chains, and of reaction products of citric acid with lignin. The compounds shown have been identified among several others of the same type showing that both holocellulose and lignin are involved in wood bonding by citric acid alone [31]

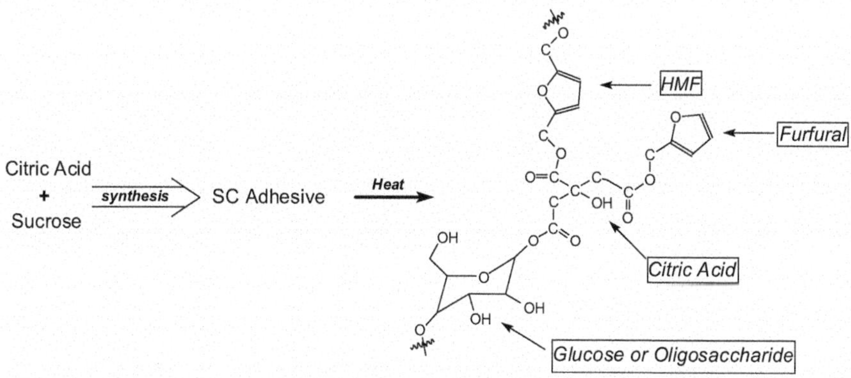


**Fig. 4** Proposed bridges formed by the reaction of citric acid with sucrose and with the furfural and hydroxymethyl furfural generated from sucrose by the citric acid treatment in sugar + citric acid adhesives [114, 146]

However, no wood adhesives, traditional or new, are allowed to be used commercially for the acid hydrolysis problem at the adhesive/wood interface [19, 73, 74, 96, 112]. For example, the unfortunate and disastrous failure after 2 years in service of the McMillan-Bloedel industrial beams joined with acid-setting PF adhesives rather than alkaline phenol-resorcinol-formaldehyde adhesives in the 1970s–1980s is still well present in the memory of the Canadian wood industry. This example is reported to indicate that excessive acidity of the adhesive is a problem not to be underrated in wood bonding with any new adhesive system.

Recently even stronger acids than citric acid have been used together with glucose to prepare wood adhesives. For example, a sucrose-p-toluenesulfonic acid adhesive has been proposed with good bonding results being reported [100]. The pH of this adhesive was reported to vary according to the proportions of acid, between pH 0.18 and pH 0.84, thus extremely acid. Equally the more researched glucose-citric acid adhesives show also a too low a pH in the range from 0.8 to 1.3 when used as wood adhesives [58], 2.2–2.9 (under different conditions) [29]. Even the sucrose-citric acid system in its application as an adhesive had pH ranges also reported to be lower than 1 [124]. In all these cases the low to extremely low pHs show clearly that these adhesives, under the conditions they function, are unsuitable for bonding wood.

Acid setting adhesive systems might be used if at a pH lower than pH 4 during their curing a quick neutralization of the glueline occurs thus avoiding any residual free acid strong enough to damage the wood substrate. The literature reports that several approaches have been tried in the far past to overcome the well-known problem of very acid adhesives [49, 74, 119]. Of these the only older systems showing some promise were (i) to use fillers or extenders impregnated with $Mg(OH)_2$ and (ii) to add to the adhesive a mix of aluminum powder and Reax [74]. These systems gave limited results with formaldehyde-based adhesives. However, one self-neutralizing system has been developed to effectively counteract the effect of strong acids of acid curing phenol-formaldehyde adhesives on the wood glueline interface [94] (Fig. 5). This is based on adding to the adhesive before using a complex of a base, namely morpholine, with a weak acid, this being moreover easy to prepare by forming it at around 60 °C. After the complex the strong hardening acid catalyst is added and the wood panel is heated rapidly for curing in the hot press. As the temperature rises at around 60 °C the complex hydrolyses separating the weak acid from the base, the strong acid then starting to form again the complex preferentially with the morpholine base and complexing with it as the temperature on cooling the wood panels falls under 60 °C. Thus, the strong acid catalyst has well-initiated curing of the adhesive and the only free acid left in the glueline is the weak acid, the weaker the better. When the system was developed for PF adhesives, the final pH of the glueline varied between 3 and 6 according to the weak acid added and the proportion of the complex used on the adhesive [94].

This system has very recently also been successfully applied to the adhesives based on citric acid/carbohydrate polymers, in particular to the citric acid/starch adhesive system [42, 137] but not yet adapted for mono and di-saccharides.

A different approach to try to solve the problem of the excessive glueline acidity, when citric acid is used, is to couple it with resins that are intrinsically so reactive

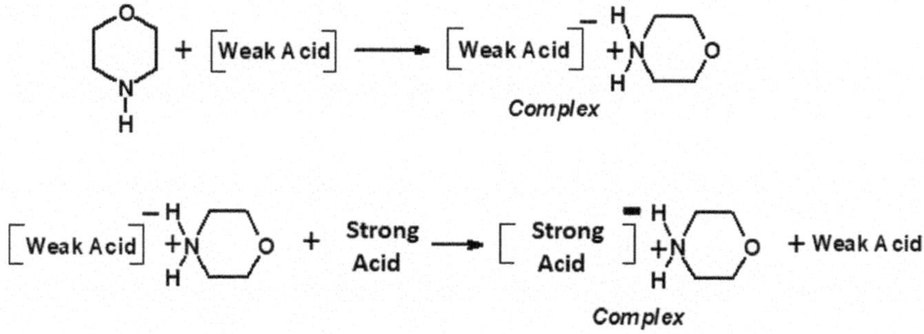


**Fig. 5** Acid-setting wood adhesives self-neutralizing system developed for PF resins and recently also applied to citric acid/polymeric carbohydrate adhesive systems

that even if buffered or self-buffered hardening occurs without the pH becoming too low, and hence without any damage to the wood-adhesive interface. An example of this are melamine-formaldehyde resins, melamine-urea-formaldehyde resins, and even melamine-alternate aldehydes resins such as for example melamine-dialdehyde starch resins, when a specifically-effective strong buffer is added [32, 53, 143, 144]. This indicates that the use of citric acid is of interest and utility also when added to more traditional resins when due precautions to control the pH at the glueline interface are taken.

Zhao et al. [145] have presented a no formaldehyde and no aldehyde emission eco-friendly urea-oxidized starch adhesive. This was obtained by condensing urea and oxidized starch, the system with addition of nano-titanium dioxide (nano-$TiO_2$). Formaldehyde was replaced by oxidized starch to avoid formaldehyde emission from the adhesive [147]. The scheme of synthesis of this adhesive is shown in Fig. 6.

Adhesives prepared by this method or its variation were also presented [79]. The same approach but for an adhesive with urea cross-linking $H_2O_2$-oxidized starch and glyoxalated lignin has also been reported [24] (Fig. 7) as well as resins obtained by reaction of melamine with oxidized starch [144] (Fig. 8).

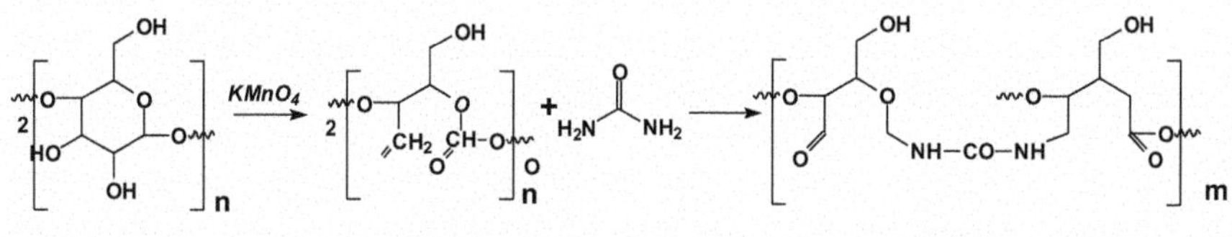


**Fig. 6** Reaction of oxidized starch (called dialdehyde starch) oxidized by permanganate, cross-linked by urea without any formaldehyde [145]

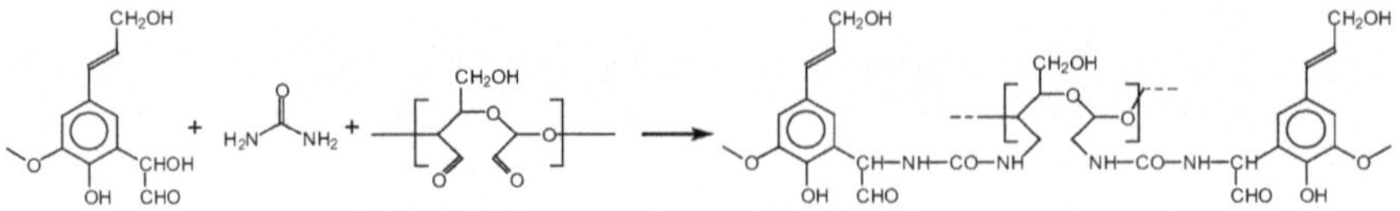


**Fig. 7** Reaction of oxidized starch (called dialdehyde starch) and glyoxalated lignin cross-linked by urea without any formaldehyde [24]

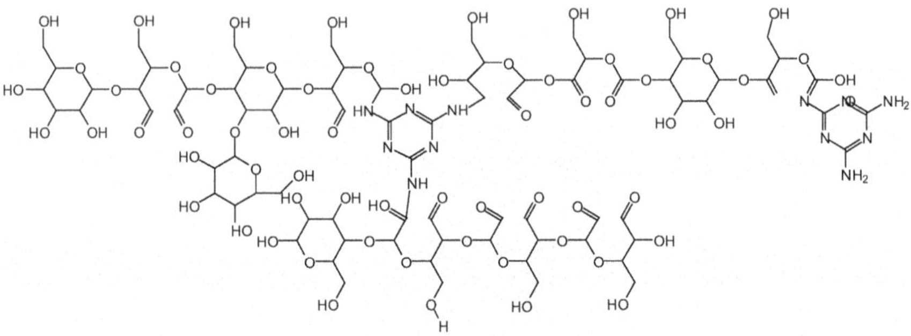

**Fig. 8** An oligomer of dialdehyde starch crosslinked by melamine [144]

# 3 Tannin Adhesives

Tannins are mainly extracted from the bark or wood of trees, but also from other sources such as the residues of grape pressings in the wine industry [83, 84]. Tannin-based adhesives have been in industrial use since the early 1970s in a few countries of the southern hemisphere [85–88, 91–93, 125]. Several in-depth reviews already exist on the use of tannins for wood adhesives to which the reader is referred [37, 38, 85–88, 91–93]. In this review only the recent new advances are presented.

The relative scarcity of tannins produced in the world, compared to the tonnage of synthetic adhesives used in the panel industry, has led to extending the tannin resource with other abundant biomaterials, with the preparation of bio-adhesives formed by copolymerization of tannins and lignin [66, 75] (cf. Lignin adhesives), of soy flour proteins and tannin [41–44] (cf. Protein adhesives), of tannin-furfuryl alcohol adhesives, as furfuryl alcohol is biobased too [1, 77] and very recently of tannins co-reacted with furanic and non-furanic humins [20, 22] to form NIPU adhesives, humins being also bio-sourced materials.

Predominantly non-furanic commercial humins have been used to prepare humins-based non-isocyanate polyurethanes (NIPU) resins for wood panel adhesives. Pure humin-based NIPU resins and tannin-humins NIPU resins have been prepared, the latter to upgrade the humins performance. Species in the raw humins and species formed in the NIPU resins were identified [20, 22]. Humins fulvic acid and derivatives, humic acid and its fragments, some lignans present and furanic

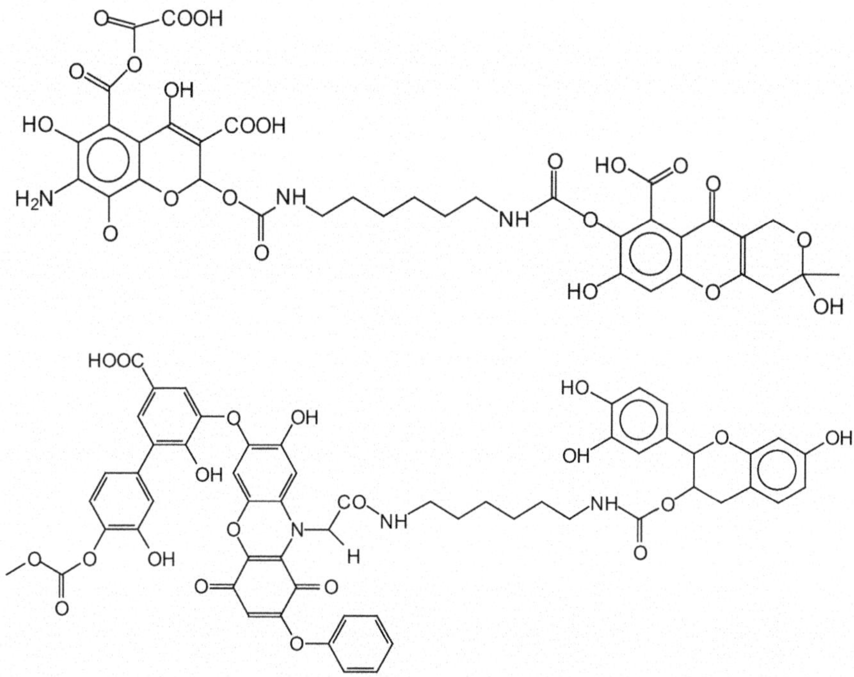


**Fig. 9** Examples of NIPU oligomer species of humins and of humins-flavonoids [20, 22]

oligomers present formed NIPU linkages among themselves and with the flavonoid units of the tannin. Some examples are shown in Fig. 9.

Two temperature curing peaks occur in these resins, one at about 130 °C and the other at about 220 °C. The 130 °C one belongs to the linear growth and to the consequent formation of a physical entanglement network of the oligomers formed. As these entanglements are only physical, not chemical, as the temperature increases the oligomer chains increased movement causes disentanglement to eventually arrive to form a chemically cross-linked network at the second peak temperature. The second peak is much more pronounced in NIPUs resins derived from the coreaction of tannins and humins. All these resins were used to bond and test laboratory particleboard. Both the ones glued with just non-furanic humins NIPU and humins-tannin NIPU adhesives presented internal bond strengths well within the range specified by the standards for particleboards for interior applications. However, the humins-tannin NIPU resin proved to be clearly a better binder than the humins NIPU without tannin [22].

A few approaches have been recently followed for the total elimination of formaldehyde by (1) substituting it with aldehydes, less or non-toxic, and non-volatile [7, 16, 46, 103, 121], such as glyoxal, glutaraldehyde, vanillin, and vanillin derivate aldehydes, these two latter ones yielding fully bio-based tannin adhesives [103]. An example of the reactions involved is shown in Fig. 10.

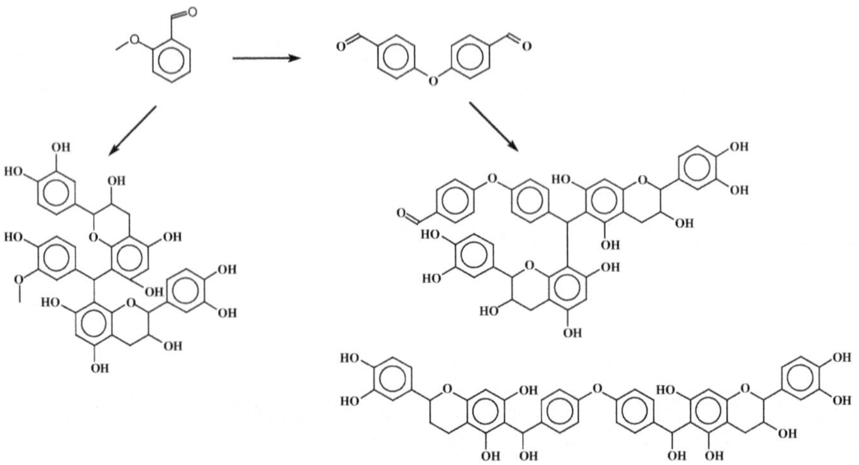

**Fig. 10** Tannin-vanillin and tannin-vanillin derivates resins for adhesives

(2) Very recently, also fully bio-based, some carbohydrate extracts from African trees have shown to release both hydroxymethyl furfural and furan 2,5-dialdehyde as hardeners [76] during hot-pressing. An example of the reactions involved is shown in Fig. 11.

This approach does also fit in the adhesives based on the reaction of the very reactive procyanidins of pine bark tannin with hydroxymethyl furfural [104] which also gives acceptable wood bonding results.

(3) Recent publications presented promising results achieved by two newer approaches. One of these was the same as the carbohydrates specific oxidation by sodium periodate and was applied for soy flour adhesives (cf. under protein adhesives) [39, 134]. Tannin extract contains 8–10% of carbohydrate monomers and oligomers, and the periodate cleaved these fractions to yield oligomers of different molecular

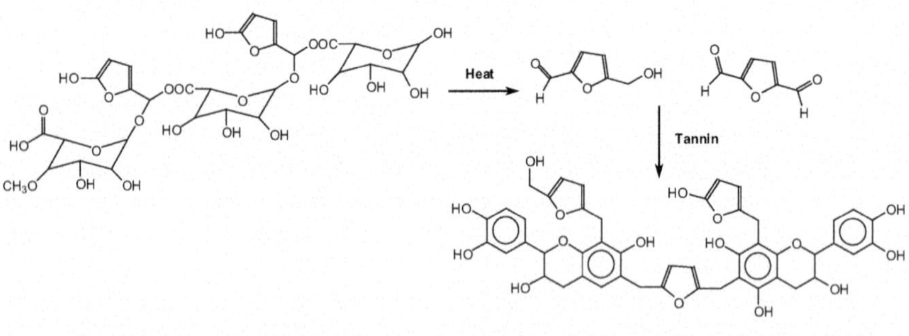


**Fig. 11** Fully bio-based adhesive compounds obtained by cross-linking flavonoid tannins with tropical wood carbohydrate exudates generating reactive furanic compounds linked or not linked to the original carbohydrate oligomer [76]

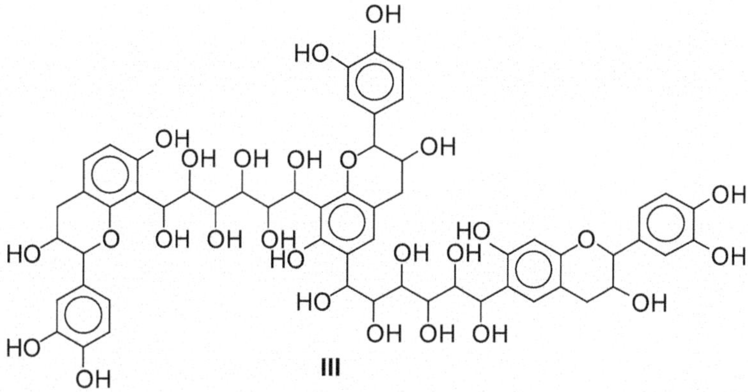


**Fig. 12** Example of a tannin oligomer obtained by oxidation of hemicelluloses fragments present in a condensed tannin extract [133]

weights generated presenting on them a number of aldehyde groups generated by the oxidizing treatment. Addition of ulterior carbohydrates, such as glucose, further increased the number of aldehyde groups generated. These aldehydes being of high but different molecular weight have been shown to be both non-toxic and non-volatile [133]. These reacted with the tannin extract polyphenolic tannin fraction yielding good particleboard bonding results [133]. Figure 12 presents one example of an oligomer derived by reacting several tannin flavonoid units with the tannin extract carbohydrates bioaldehydes (namely fragments of hemicelluloses).

The second newest approach is on the use of tannin non-isocyanate polyurethanes (NIPUs) adhesives. Tannin NIPU resins for coatings have been prepared before [116–118]. The Tannin NIPU adhesives work well on plywood, but to satisfy the standards wet test requirement they need a small contribution from a biosourced glycerol diglycidyl ether to pass wet shear strength standard requirements [25].

Polycondensation and cross-linking by triethyl phosphate (TEP) of mimosa bark tannin extract also yields binders presenting at higher temperatures a particularly good thermal stability, as high as 410 °C [8, 97]. Chemical spectroscopy by CP-MAS $^{13}$C NMR, $^{31}$P NMR, and MALDI-ToF have indicated that the main reaction involves the alcoholic behavior –OH on the heterocycle of the flavonoid units C3 as well as the phenolic –OH groups of the flavonoid units B-ring. The analysis appeared to indicate the A-ring phenolic hydroxyls do not participate in the reaction with TEP. A difference in the relative proportions of these two reaction sites for tannin and catechin monomer has been noticed. The reaction is temperature dependant. An example of the compounds formed in the hardened adhesive structure is shown in Fig. 13.

Such approach also led to adhesives based on kraft lignin and TEP, with equally good results [9]. In the same line adhesives based on tannin-sucrose-diammonium phosphate based on earlier findings on adhesives based on sucrose-diammonium

**Fig. 13** Example of compound generated by cross-linking a condensed tannin with triethyl phosphate [8]

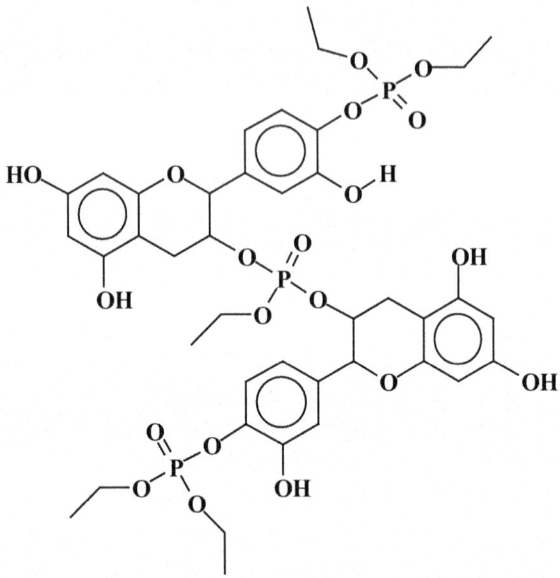


phosphate [145] were also developed indicating that phosphates inorganic salts [145] and organic salts [9] have also potential that can be developed in wood adhesives.

## 4 Lignin-Based Adhesives

Formulation technologies proposed for lignin-based adhesives have been extensively reviewed many times and the reader is earnestly encouraged to refer to them [85–87, 91–93]. Among natural polymers, after cellulose, lignin is the second most abundant one. It is a polyphenol that for its structure has always focused researcher's interest for use in wood adhesives. Notwithstanding such interest and the extensive research carried out on this polymer, several adhesives based on just lignin have not been able to be adopted industrially/commercially without the addition of synthetic resins. While many industrial trials were carried out, for a variety of reasons such as inadequately long hot press times or equipment corrosion, commercial success has eluded them. Methylolation of either lignin or lignosulphonates and their addition to phenol-formaldehyde (PF) adhesive resins has been one of the most used approaches as well as copolymerizing phenol, lignin, and formaldehyde to prepare LPF plywood adhesives [6, 17, 18, 126, 138, 140]. Recent advances in this latter approach have reported the commercialization of a 50/50 by weight lignin/phenol LPF plywood adhesive [126]. A further encouraging improvement of this LPF resin has been obtained by treating lignin with ionic liquids in the preparation of both LPF and LPGlyoxal adhesives [140]. However, recently, new encouraging approaches of promising technologies for completely lignin-based bio-adhesives

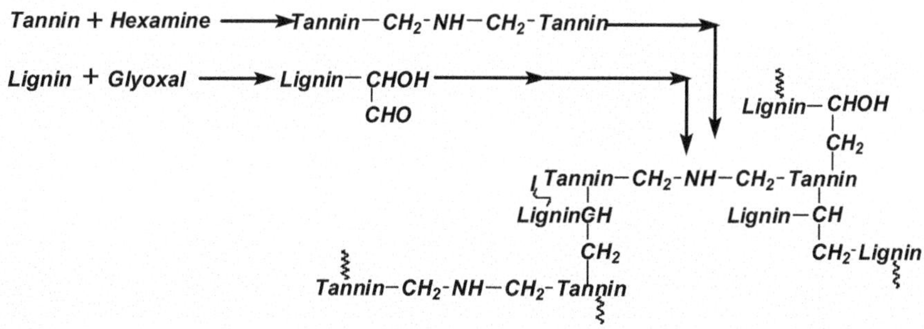


**Fig. 14** Schematic representation of the adhesive formed by reacting glyoxalated lignin with a tannin-hexamine copolymer

have also been presented. Among these one counts (a) adhesives for particleboard, thin hardboard, and other agglomerate wood panels bonded with glyoxalated lignin copolymerized with tannin/hexamine [66, 75] (Fig. 14), and (b) equivalently formulated high-performance resins of high solids content for agriculture-generated fiber composites [11, 107, 148].

The polycondensation reaction leading to triethyl phosphate (TEP) and kraft desulphurized lignin hardened, cross-linked networks has also been investigated in parallel to the same reaction for the tannins discussed above. Solid-state CP-MAS $^{13}C$ NMR and MALDI ToF mass spectroscopy confirmed that the reaction of TEP takes place on both the lignin phenolic and aliphatic hydroxyl groups with a variety of structures being formed and identified. Higher temperatures and the addition of ammonia appear to favor the condensation reaction. Initial wood adhesion tests gave very encouraging results, in particular a good hydrophobic behavior of the TEP-lignin resin treated wood surface [9].

The most recent approach was issued from the carbohydrates specific oxidation by sodium periodate in the research program on soy flour adhesives (cf. later) [39, 40]. Thus, previously demethylated lignin was oxidized by sodium periodate. This treatment cleaved a number of the lignin aliphatic chains and Beta-O-4 bonds with vicinal −OH groups generating several different non-volatile and non-toxic aldehyde groups on the lignin. Such aldehyde groups on the lignin structure then proceeded to react with the lignin aromatic rings yielding a lignin-based adhesive for plywood of good performance [21].

Last but not least, NIPU adhesives based on organosolv lignin were also prepared to bond wood particleboard with good Internal Bond (IB) strength results [105]. However, a pressing temperature of 230 °C was needed to obtain good results. Such a temperature is not unusual in modern continuous presses in the particleboard industry. The pressing temperature could be further decreased down to 180 °C by using an epoxy-silane additive at a level of 10–15% on resin solids while still presenting good bonding, yielding structures as shown in Fig. 15.

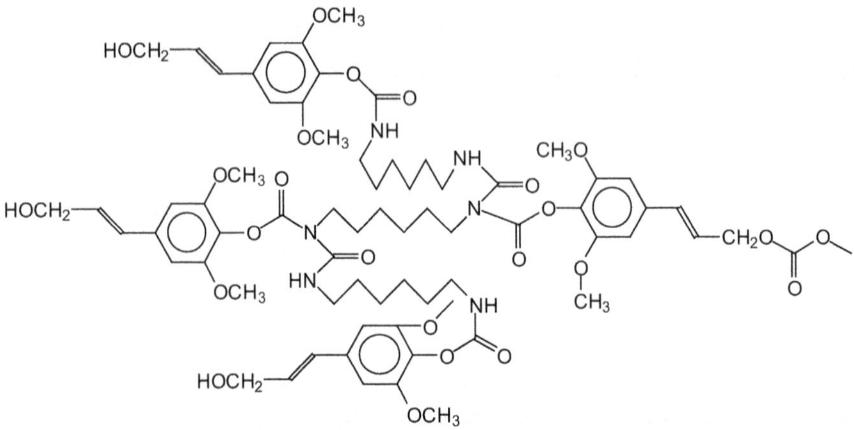


**Fig. 15** Example of an organosolv lignin-based NIPU oligomer

# 5 Protein Adhesives

This research field is dominated by extensive research on soy protein isolates/hydrolysates and soy flour adhesives. A considerable number of different approaches have been tried with encouraging results. The more unusual and interesting non-traditional system is based on the pre-reaction of the soy protein hydrolysate with maleic anhydride to form an adduct that is then reacted in the panel with polyethylene imine [61, 62]. This system works well, but seems to be expensive. Equally, work on proteins-glutaraldehyde adhesives has been published [135].

Several very new approaches to yield totally bio-based soy flour and soy protein wood adhesives have very recently been published. The first is based on the reaction of a hydrolysable tannin (tannic acid) or a condensed tannin with soy protein isolate and/or soy flour. This gave rather encouraging results for plywood, particleboard, and hardboard [41–44]. The results showed that the addition of tannic acid or condensed tannins to soy-based adhesive decreased soy-based adhesive viscosity and its pH and improved soy flour adhesion properties. The glass transition temperature and denaturation temperature of soy-based adhesives decreased by adding tannic acid. Fourier-transform infrared (FTIR) spectroscopy results also showed that soy flour amino acids appeared to react well with tannic acid. Furthermore, delamination and shear strength test results showed good water resistance of plywood bonded with soy-based tannic acid-modified adhesive. The linkage between the soy protein and the tannin appeared to be of three kinds. When mixed at ambient temperature a mix of ionic and covalent linkages between soy protein amino acids and tannin were identified [41]. While these contributed to cross-linking especially after plywood hot pressing the secondary forces complexes between tannin and protein, as in leathers, are likely to still dominate the interaction between the two materials. Later a much more interesting reaction to form covalent bonds between soy protein and condensed

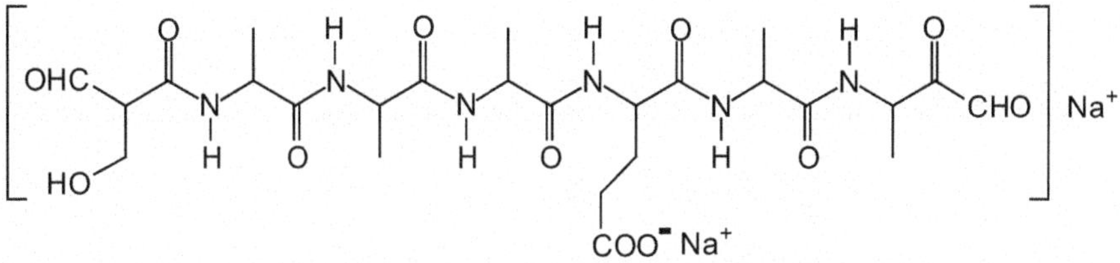


**Fig. 16** Example of soy protein multi-aminoacids sequence fragment in soy flour in specific oxidation adhesives [39]

tannins without any hardeners was found allowing the preparation and application of a totally bio-based adhesive for wood panels [41].

The second is the formation of soy protein isolate-based polyamides formed by reaction of soy protein isolate with maleic anhydride with hexamethylene diamine [132, 133]. This being a totally new resin for such an application. The third and more revolutionary one to obtain totally biobased soy adhesives is based on the specific oxidation of carbohydrates by sodium periodate. The bonding results using soy flour gave good bonding results for plywood [40]. The periodate treatment was found to cause cleavage of monomeric and polymeric carbohydrates in Soy flour forming a variety of non-toxic and non-volatile monomeric and oligomeric aldehydes. These reacted with the protein part of the soy flour to give good bonding results [39]. Even more interesting was that periodate oxidation formed aldehyde groups on the soy protein itself which also contributed to good bonding results [39]. Thus, aldehydes on both linear and branched peptide chains were also generated from the specific oxidation cleavage of the protein, explaining why good adhesive results were also obtained by soy protein hydrolysate not containing any carbohydrates [39]. An example of one of these structures has been identified, for example, a serine-[alanine]$_5$-glutamic acid dialdehyde + Na+ of structure as shown in Fig. 16.

More recently, non-isocyanate polyurethane (NIPU) wood panel adhesives were prepared by reacting soy protein isolates (SPI) with dimethyl carbonate (DMC) and hexamethylene diamine [23]. The branched and linear oligomers detected proceeded then to cross-link to a hardened network structure. Some unusual structures were detected, such as urethane bridges derived from carbamic acid coupled with lactam structures. A two steps adhesive curing process was observed. Linear oligomers first started and continued to form in the lower temperature range (maximum 130 °C), yielding a physically entangled network. As the temperature increased this physical network started to progressively but partially disentangle. In a second step chemical cross-linking then ensued leading to the formation of a covalently hardened network, its peak being at around 220 °C. Plywood panels bonded with the SPI-NIPU adhesives bonded plywood panels passed well the dry strength requirements of relevant standards, but adding 15% of biosourced glycerol diglycidyl ether was necessary to better the panel's wet strength.

## 6 Other Wood Panels Bio-Adhesives

Several alternative biomaterials have been used to prepare bio-adhesives. Up to the present none of the resins described in this section have reached the stage of industrial trials. However, they present an interest for future developments due to their unusual concepts and research on them is thus still worth pursuing.

Cardanol as the main component of raw cashew nut shell liquid has also been used to formulate adhesives. Cardanol possess an aromatic nucleus based on resorcinol linked to a long unsaturated fatty acid chain, both these features rendering it a suitable renewable material to prepare waterproof thermosetting adhesives. Both its resorcinol ring and its carbon–carbon double bonds in its fatty acid chain can lead to yield hardened adhesive resins. Ozonolysis is also an alternate technique that has been developed to yield suitable reactive groups, such as aldehydes, on its structure. By this latter route cardanol hydroperoxide is the main product obtained in a first step that by being then reduced by addition of glucose or by zinc/acetic acid in a second reaction step gives a high proportion of cardanol-derived aldehyde groups. These latter ones then react and cross-link with the cardanol aromatic resorcinol moieties leading to self-condensation to a hardened network [120]. A good review on cardanol thermoset polymers other than for wood composites does exist [52].

Vegetable oils have also been investigated as wood adhesives with a number of techniques being reported [115]. Acrylics Epoxidised Soy Oil (AESO) coupled with urea-formaldehyde (UF) resins was used to bond wheat straw particleboard at the weight resin solid contents level of 13% and 8% of each material respectively. These AESO-bonded panels showed a better performance than the UF-bonded controls, particularly for IB strength and thickness swelling [115]. Bioresins based on Soy bean oil and other oils based bioresins have also been reported in the literature [128]. They were developed using animal and plant triglycerides functionalized with a variety of chemical groups (e.g., epoxy, carboxyl, hydroxyl, vinyl, amine, etc.) rendering them capable of polymerizing. The report claims also the making of renewable biofibers composites by using hemp, straw, flax, wood particles, and fibers but with no results being reported. Epoxidized oil resins were also proposed and tested as composite wood panel binders that could be tightly controlled by an adequate triglycerides and polycarboxylic anhydrides selection [67]. This work claims to control the variation of cross-linking conditions by using different specialized catalysts with panels prepared at a range of different temperatures (120°–180 °C) showing good water tolerance even at high temperatures. Their main problem appears to be that the panels bonded with them present a far too slow hot pressing time and the relatively high cost of the resins.

Another newer and rather unusual approach is the use of bonding by fungal mycelium. This has been used to bond wood composites. The mycelium-bonded composites (MBB) were composed of hemp shives, thus an agriculture waste, and wood chips which were bonded together by natural growth fungal mycelium [149]. As a result, Mycocomposites from hemp (HMC) and wood (WMC) were prepared and tested for their mechanical properties, for water absorption and biodegradation properties. WMC compression tests indicated that its strength was 60% better than that of HMC composites. However, the WMC panel's thickness swelling and water absorption were relatively high. These composites were biodegradable, a drawback for some uses but an advantage for others.

Lastly, wood friction welding with no adhesives can be counted in this category too. Wood welding occurs when by applying under pressure rapid friction to the two wood samples to be joined the wood intercellular material, namely xylans hemicelluloses and lignin, does melt and flow. A composite having a welded joint of much higher density than the joined wood samples is thus generated. The much higher interfacial density has thus a much higher strength than the underlying wood substrate in the two joined surfaces, the higher density interface being composed by entangled wood fibers held together by a matrix of molten intercellular material [45, 54, 57, 64, 89]. Wood welding gives good results both when the wood is welded by linear friction and by rotational welding of dowels. Its water resistance that is originally limited has been vastly improved using different approaches [90, 95]. While linear friction welding is really impressive to see it is also a somewhat limited technique because (i) relatively expensive machinery is necessary and (ii) only short pieces up to about one meter can be welded with existing equipment. Conversely, rotational dowel welding is very flexible, needs only inexpensive but good hand or fixed drills, and has no application limitations. Several small workshops have already adopted rotational dowel welding to prepare wood furniture with neither adhesives, nor nails, nor staples nor screws, for shops areas refitting and for small wood pieces assembly, such as for example wooden toys [14, 15, 80–82, 99, 108–111]. Plywood by using high temperature for welding together the veneers has also been made but has shown not to be of practical use as its hot press time is far too long to be of any industrial significance [30, 65].

Binders using several other natural, renewable biomaterials, oligomers, and polymers have also been prepared but for application in other fields than for wood composites. However, these too should be tried and tested for wood composites. Thus, isosorbide, urushiol, humins, etc. [12, 48, 51, 101, 102] used for adhesives for other materials should also be tried for wood composite binders.

## 7 Some Reputed Synthetic Resins Can also be Bio

For the non-traditional synthetic adhesives the, at first shocking, concept is that urea–formaldehyde can be classed as a bio-adhesive too. Both urea and formaldehyde can and are derived from natural materials, with urea obtained industrially from the nitrogen of the air, while formaldehyde, also a natural material, is instead manufactured today by a non-bio route. Consequently, the focus bears on the elimination of formaldehyde and its substitution by non-toxic, or at least less toxic, and non-volatile alternative materials thus eliminating the drawback and problem of formaldehyde emission. A first approach to eliminate formaldehyde was the preparation of urea-glyoxal (UG) resins for wood composite adhesives [33, 34]. Resins of this type, but of lesser complexity, and much less condensed are already used in the textile industry. It was necessary then to markedly change both their preparation process and their formulation to be able to apply such resins to wood composites. The first drawback of UG resins was their considerably slower cure than UF resins during the bonded wood panels hot pressing. This was determined to be due to their much higher activation energy of curing. To solve this problem an initial approach consisted to rather use the more reactive melamine in a melamine-glyoxal resin (MG). This led to a limited improvement in the sense that the MG resins prepared were usable for paper impregnation as the impregnated paper is directly in contact with the hot press platen, thus at higher effective temperature [35], faster curing but not as yet fast curing enough to bond wood panels well. An in-depth chemical, structural, and mechanism study of this system has also been recently published [63]. The problem has been finally solved by introducing ionic liquids (IL) as UG hardeners. Ionic liquids were shown to very much decrease the UG adhesives curing activation energy. The IL UG resins were then used to bond laboratory wood particleboard at pressing times as fast as those of UF adhesives with good results [141, 142]. This breakthrough allowed them to use ionic liquids to accelerate the curing of melamine-glyoxal resins (IL MG) and of melamine-glyoxal-glutaraldehyde (IL MGG') adhesives that yielded excellent results and a good performance as particleboard binders [131].

A renewed interest in research on urea-furfural and urea-furfuryl alcohol adhesives has also come to the fore notwithstanding that such resins are based on old technologies for other substrates other than wood [78, 113, 132]. However, the interest in hydroxymethyl furfural to prepare urea-hydroxymethyl furfural adhesives and tannin-hydroxymethyl furfural is new [104]. The drawback of these excellent resins is the very dark color they develop on curing. While this is not a drawback for already dark resins such as tannin-based ones, it is instead a serious commercial handicap for the traditionally white or transparent urea-based adhesives. Also furfuryl alcohol + an aldehyde adhesive have been prepared and tested with very promising performances in bonding plywood, these being furfuryl-alcohol formaldehyde, furfuryl alcohol-glyoxal, and furfuryl alcohol-glutaraldehyde [85].

Even more different approaches to mainly synthetic thermoset adhesives have been tried and presented. For example, glucose or sucrose reacted with glycerol triacetate and a diamine yielded a variety of mixed oligomers (Fig. 17) and a very acceptable wood adhesive for plywood [129]. The system was found by pure chance, but it worked, and the oligomers and polymers formed were identified and characterized.

## 8 Conclusion

The field of wood bio-adhesives is a rapidly growing field of research, of future economic importance, fast moving both in the conception of newer approaches and the use of newer biomaterials but even more in the conception and development of newer and sometimes even revolutionary types of binders. In the strongly growing field of bio-based adhesives, there have been almost incredible progresses caused

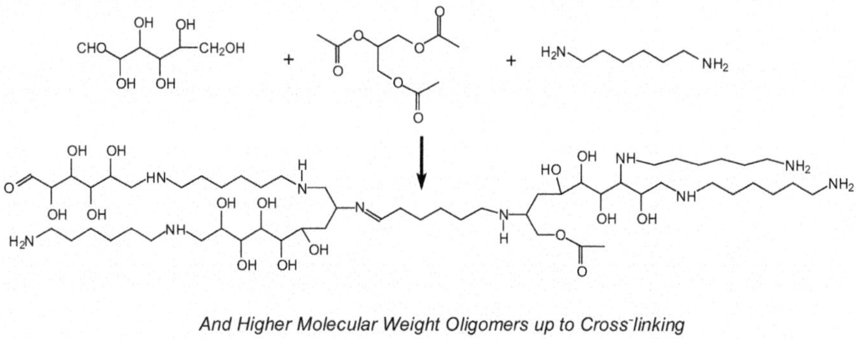


**Fig. 17** Example of a mixed oligomer species leading to adhesive cross-linking formed, among others, in the new adhesive system derived by the reaction of glucose or sucrose with triacetin and hexamethylene diamine [129]

among others by outside constraints often imposed by new government's regulations to reduce or eliminate formaldehyde and its emissions as well as other volatiles toxic materials emissions in service or in manufacturing. Consumers awareness and consequent pressure on industry to use more environmentally acceptable materials has also played a role. The industry's wish to eliminate or at least decrease petrochemicals dependence also has played a secondary role due to real or imagined fears of oil reserves future decreases with its related raw materials costs increases for the traditional synthetic wood binders now used. This last trend has been up to now rather secondary as the requirement by the wood industry that new bio-adhesives present equal or better performance than the presently used ones but possibly even at lower costs shows that the pressure of the other trends reported above is still not sufficiently strong to avoid the sometime just lip-service in interest in bio-adhesives. Notwithstanding this, the use of bio-adhesives for wood in the world has been steadily but slowly progressing. This review presents only a brief overview of what has occurred and is still occurring in this field. Lesser or more difficult to industrially implement approaches have not being mentioned or not enough, as this review has been intentionally focused on adhesives industrially applicable. However, the progress in the bio-adhesives field which is of future primary economic importance has been accelerating with the number of new ideas, approaches, and new proposed binder systems continuously increasing mirroring an exciting and interesting research future.

**Acknowledgements** This research was supported by the project "Development, Exploitation Properties and Application of Eco-Friendly Wood-Based Composites from Alternative Lignocellulosic Raw Materials", project No. НИС-Б-1290/19.10.2023, carried out at the University of Forestry, Sofia, Bulgaria.

## References


1. Abdullah UHB, Pizzi A (2013) Tannin-Furfuryl alcohol wood panel adhesives without formaldehyde. Eur J Wood Prod 71:131–132
2. Alma MH, Yoshioka M, Yao Y, Shiraishi N (1998) Preparation of sulfuric acid-catalyzed phenolated wood resin. Wood Sci Techno 32:297–308
3. Alma MH, Yoshioka M, Yao Y, Shiraishi N (1996) The preparation and flow properties of HCl catalyzed phenolated wood and its blends with commercial novolak resin. Holzforschung 50:85–90
4. Amirou S, Pizzi A, Delmotte L (2017) Citric acid as waterproofing additive in butt joints linear wood welding. Eur J Wood Prod 75:651–654
5. Antov P, Savov V, Mantanis GI, Neykov N (2020) Medium-density fibreboards bonded with phenol-formaldehyde resin and calcium lignosulfonate as an eco-friendly additive. Wood Mat Sci Eng 16(1):42–48
6. Antov P, Savov V, Neykov N (2020) Sustainable bio-based adhesives for eco-friendly wood composites a review. Wood Res. 65:51–62
7. Ballerini A, Despres A, Pizzi A (2005) Non-toxic, zero-emission tannin-glyoxal adhesives for wood panels. Holz Roh Werkst 63:477–478
8. Basso MC, Pizzi A, Polesel-Maris J, Delmotte L, Colin B, Rogaume Y (2017) MALDI-TOF and 13C NMR analysis of the cross-linking reaction of condensed tannins by triethyl phosphate. Ind Crops Prod 95:621–631
9. Basso MC, Pizzi A, Delmotte L, Abdalla S (2017) MALDI-TOF and $^{13}$C NMR analysis of the cross-linking reaction of lignin by triethyl phosphate. Polymers 9:206
10. Belgacem MN, Gandini A (2003) Furan-based adhesives. In: Pizzi A, Mittal KL (eds) Handbook of adhesive technology, 2nd ed. Marcel Dekker, New York, pp 615–634
11. Bertaud F, Tapin-Lingua S, Pizzi A, Navarrete P, Petit-Conil M (2012) Development of green adhesives for fibreboards manufacturing, using tannins and lignin from pulp mill residues. Cellul Chem Technol 46:449–455
12. Bobade SK, Paluvai NR, Mohanty S, Nayaka SK (2016) Bio-based thermosetting resins for future generation: a review. Polymer-Plastics Technol Eng 55:1863–1896
13. Bobbitt JM (1956) Periodate oxidation of carbohydrates. Adv Carbohydr Chem 11:1–41
14. Bocquet JF, Pizzi A, Resch L (2007) Full-scale (Industrial) wood floor assembly and structures by welded through dowels Holz Roh Werkst 65:149–156
15. Bocquet JF, Pizzi A, Despres A, Mansouri HR, Resch L, Michel D, Letort F (2007) Wood joints and laminated wood beams assembled by mechanically welded wood dowels. J Adhes Sci Technol 21:301–317
16. Böhm R, Hauptmann M, Pizzi A, Friederich C, Laborie M-P (2016) The chemical, kinetic and mechanical characterization of Tannin-based adhesives with different crosslinking systems. Int J Adhesion Adhesives 68:1–8
17. Calvé LR (1990) Phenolic-lignin wood adhesives for plywood. In: Proceedings of the 19th IUFRO world congress, Montreal, QC, Canada, 6–10 August
18. Calvé LR (1992) Fast cure and pre-cure resistant cross-linked phenol-formaldehyde adhesives and methods of making same. US Patent 5,173,527
19. Cameron FA, Pizzi A (1984) Acid-setting cold-setting phenolic adhesives for glulam: a controversial issue. J Appl Polym Sc, Proceedings of Appl Polym Symposium Nr. 40: 229–234. Wiley
20. Chen X, Guigo N, Pizzi A, Sbirrazzuoli N, Li B, Fredon E, Gerardin C (2020) Ambient temperature self-blowing tannin-humins biofoams. Polymers 12:2732
21. Chen X, Xi X, Pizzi A, Fredon E, Du G, Gerardin C (2021) Oxidized demethylated lignin as a bio-based adhesive for wood bonding. J Adhesion 97(9):873–890
22. Chen X, Pizzi A, Essawy H, Fredon E, Gerardin C, Guigo N, Sbirrazzuoli N (2021) Non-furanic humins-based non-isocyanate polyurethane (NIPU) thermoset wood adhesives. Polymers 13:372

23. Chen X, Pizzi A, Xi X, Zhou X, Fredon E, Gerardin C (2021) Soy protein isolate non-isocyanates polyurethanes (NIPU) wood adhesives. J Renew Mater 9(6):1045–1057
24. Chen X, Pizzi A, Fredon F, Gerardin C, Li J, Zhou X, Du G (2022) Preparation and properties of a novel type of tannin-based wood adhesive. J Adhesion 98(7):871–888
25. Chen X, Pizzi A, Zhang B, Zhou X, Fredon E, Gerardin C, Du G (2022) Particleboard bioadhesive by glyoxalated lignin and oxidized dialdehyde starch cross-linked by urea. Wood Sci Technol 56:63–85
26. Codou A, Guigo N, Heux L, Sbirrazzuoli N (2015) Partial periodate oxidation and thermal cross-linking for the processing of thermoset all-cellulose composites. Compos Sci Technol 117:54–61
27. Conner AH, River BH, Lorenz LF (1986) Carbohydrates-modified PF resins for wood panels. J Wood Chem Technol 6:591–596
28. Conner AH, Lorenz LF, River BH (1989) Carbohydrate-modified PF resins formulated at neutral conditions. ACS Symp Ser 385:355–369
29. Crisostomo M, Del Menezzi C, Militz H, Kurkowiak K, Mayer A, Carvalho L, Martins J (2023) Effect of the citric acid on the properties of sapwood of pinus sylvestris submitted to thermomechanical treatment. Forests 14(9):1839
30. Cristescu C, Karlsson O (2013) Changes in content of furfurals and phenols in self-bonded laminated boards. BioResources 8(3):4056–4071
31. Del Menezzi C, Amirou S, Pizzi A, Xi X, Delmotte L (2018) Reactions with wood carbohydrates and lignin of citric acid as a bond promoter of wood veneer panels. Polymers 10:833
32. Menezzi D, Pizzi A, Amirou S, Xi X (2023) MUF resins improved by citric acid as adhesives for wood veneer panels. J Renew Mater 11(2):539–553
33. Deng S, Du G, Li X, Pizzi A (2014) Performance and reaction mechanism of zero formaldehyde-emission urea-glyoxal (UG) resins. J Taiwan Inst Chem Engineers 45(4):2029–2038
34. Deng S, Pizzi A, Du G, Zhang J, Zhang J (2014) Synthesis and chemical structure of a novel glyoxal-urea-formaldehyde (GUF) co-condensed resins with different MMU/G molar ratios by $^{13}$CNMR and MALDI-TOF-MS. J Appl Polym Sci 131(21):1–10
35. Deng S, Pizzi A, Du G, Lagel MC, Delmotte L, Abdalla S (2018) Synthesis, structure characterization and application of melamine-glyoxal adhesive resins. Eur J Wood Prod 76:283–296
36. Dunky M (2020) Wood adhesives based on natural resources: a critical review. Part I. Protein-based adhesives. Rev Adhesion Adhesives 8(3):199–332
37. Dunky M (2020) Wood adhesives based on natural resources: a critical review. Part II. Carbohydrate-based adhesives. Rev Adhesion Adhesives 8(3):334–378
38. Dunky M (2020) Wood adhesives based on natural resources: a critical review. Part III. Tannin- and lignin-based adhesives. Rev Adhesion Adhesives 8(4):379–392
39. Frihart CR, Pizzi A, Xi X, Lorenz L (2019) Reactions of soy flour and soy protein by non-volatile aldehydes generation by specific oxidation. Polymers 11:1478
40. Frihart CR, Lorenz L (2019) Specific oxidants improve the wood bonding strength of soy and other plant flours. J Polym Sci A: Polym Chem 57:1017–1023
41. Ghahri S, Pizzi A, Hajihassani A (2022) A study of concept to prepare totally biosourced wood adhesives from only soy protein and tannin. Polymers 14:1150
42. Ghahri S, Pizzi A, Mohebby B, Mirshoktaie A, Mansouri HR (2018) Soy-based, tannin-modified plywood adhesives. J Adhesion 94:218–237
43. Ghahri S, Pizzi A, Mohebby B, Mirshoktaie A, Mansouri HR (2018) Improving water resistance of soy-based adhesive by vegetable tannin. J Polym Environm 26:1881–1890
44. Ghahri S, Pizzi A (2018) Improving soy-based adhesives for wood particleboard by tannins addition. Wood Sci Technol 52:261–279
45. Gfeller B, Zanetti M, Properzi M, Pizzi A, Pichelin F, Lehmann M, Delmotte L (2003) Wood bonding by vibrational welding. J Adhes Sci Technol 17:1425–1590

46. Grigsby WJ, McIntosh CD, Warnes JM, Suckling ID, Anderson CR (2008) Adhesives, US patent 7,319,115-B2
47. Guigo N, Mazeau K, Putaux JL, Heux L (2014) Surface modification of cellulose microfibrils by periodate oxidation and subsequent reductive amination with benzylamine: a topochemical study. Cellulose 21:4119–4133
48. Hammami N, Jarroux N, Robitzer M, Majdoub M, Habas J-P (2016) Optimized synthesis according to one-step process of a biobased thermoplastic polyacetal derived from isosorbide. Polymers 8(8):294
49. Higuchi M, Sakata I (1979) Studies on the improvement of urea-formaldehyde resins adhesives.-1: a possibility of water resistance improvement. Mokuzai Gakkaishi 25:496–502
50. Hussin MH, Abd Latif NH, Hamidon TS, Idris NN, Hashim R, Appaturi JN, Brosse N, Ziegler-Devin I, Chrusiel L, Fatriasari W, Syamani FA, Iswanto AH, Hua LS, Azry Osman Al Edrus SS, Chen LW, Antov P, Savov V, Lubis MAR, Kristak L, Reh R, Sedliačik J (2022) Latest advancements in high-performance bio-based wood adhesives: A critical review. J Mater Res Technol 21:3909–3946
51. Je H, Won J (2021) Natural urushiol as a novel under-water adhesive. Chem Eng J 404:126424
52. Jial P, Song F, Li Q, Xia H, Shu X, Zhou Y (2019) Recent development of cardanol based polymer materials: a review. J Renew Mater 7:601–619
53. Kamoun C, Pizzi A, Zanetti M (2003) Upgrading of MUF resins by buffering additives—Part 1: hexamine sulphate effect and its limits. J Appl Polym Sci 90(1):203–214
54. Kanazawa F, Pizzi A, Properzi M, Delmotte L, Pichelin F (2005) Influence parameters in wood dowels welding by high speed rotation. J Adhes Sci Technol 19:1025–1038
55. Kurkowiak K, Hentges D, Dumarçay S, Géradin P, Militz H (2022) Understanding the mode of action of sorbitol and citric acid (SorCA) in wood. In: Proceedings of the tenth european conference on wood modification, April 25–26, Nancy, France
56. Kusumah SS, Jayadi D, Wibowo T, Pramasari DA, Widyaningrum BA, Darmawan T, Ismadi T, Dwianto W, Umemura K (2019) Investigation of eco-friendly plywood bonded with citric acid—starch based adhesive. In: IOP conference series: earth and environmental science, volume 460, International conference of biomass and bioenergy 19–20 August, West Java, Indonesia. https://doi.org/10.1088/1755-134
57. Leban JM, Pizzi A, Properzi M, Pichelin F, Gelhaye P, Rose C (2005) Wood Welding. A challenging alternative to conventional wood gluing. Scandinavian J Forest Res 20:534–538
58. Li C, Hou D, Xi X, Lei H, Tondi G, Shi J, Du, G (2024) Eco-friendly bio-based adhesives prepared from different polyester-type glucose with high cross-linked structure. Unpublished work
59. Li C, Lei H, Wu Z, Xi X, Du G, Pizzi A (2022) Fully biobased adhesive from glucose and citric acid for plywood with high performance. ACS Appl Mater Interfaces 14(20):23859–23867
60. Lin C-F, Karlsson O, Jones D, Sandberg D (2022) Bio-based adhesive derived from citric acid and sorbitol. Wood Mater Sci Eng 17(5):397–399
61. Liu Y, Li K (2002) Chemical modification of soy protein for wood adhesives. Macromol Rapid Comm 23:739–742
62. Liu K, Li K (2007) Development and characterization of adhesives from soy protein for bonding wood. Int J Adhesion Adhesives 27:59–67
63. Liu, H, Pizzi A, Qing Z, Li X, Du G, Deng S (2024) A novel glyoxal-based resin with highly cross-linked networks: reaction mechanism, characterizations and performance measurements. Int J Adhesion Adhesives 130:103650
64. Mansouri H, Leban J-M, Pizzi A (2010) End-grain butt joints obtained by friction welding of high density eucalyptus wood. Wood Sci Technol 44:399–406
65. Mansouri H, Leban J-M, Pizzi A (2010) High density panels by wood veneers welding without any adhesives. J Adhes Sci Technol 24:1529–1534
66. Mansouri HR, Navarrete P, Pizzi A, Tapin-Lingua S, Benjelloun-Mlayah B, Rigolet S (2011) Synthetic-resin-free wood panel adhesives from low molecular mass lignin and tannin. Eur J Wood Prod 69:221–229

67. Miller R, Shonfeld U (2002) Company literature; Esbacher Weg 15; Preform Raumgliederungssysteme GmBH, Feuchtwangen, Germany
68. Moubarik A, Pizzi A, Charrier F, Allal A, Badia M-A, Mansouri HR, Charrier B (2013) Mechanical characterization of industrial particleboard panels glued with cornstarch-mimosa tannin-urea formaldehyde resins. J Adhes Sci Technol 27:423–429
69. Moubarik A, Mansouri HR, Pizzi A, Charrier F, Allal A, Charrier B (2013) Corn flour-mimosa tannin-based adhesives without formaldehyde for interior particleboard production. Wood Sci Technol 47:1–9
70. Moubarik A, Allal A, Pizzi A, Charrier F, Charrier B (2010) Characterization of a formaldehyde-free cornstarch-tannin wood adhesives for interior plywood. Eur J Wood Prod 68:427–433
71. Moubarik A, Charrier B, Allal A, Charrier F, Pizzi A (2010) Development and optimisation of a new formaldehyde-free cornstarch and tannin adhesive. Eur J Wood Prod 68:167–177
72. Moubarik A, Pizzi A, Allal A, Charrier F, Charrier B (2009) Cornstarch and tannin in phenol-formaldehyde resins for plywood production. Ind Crops Prod 30:188–193
73. Müller A (1953) Kaltleime auf Phenolharzbasis unr die Gefhar der Holzschädigung (Cold-setting phenolic resin adhesives and the risk of damage to wood). Holz Roh Werkst 11:429–435
74. Myers GE (1983) Use of acid scavengers to improve durability of acid-catalyzed adhesive wood bonds. Forest Prod J 33:49–57
75. Navarrete P, Mansouri HR, Pizzi A, Tapin-Lingua S, Benjelloun-Mlayah B, Rigolet S (2010) Synthetic-resin-free wood panel adhesives from low molecular mass lignin and tannin. J Adhesion Sci Technol 24:1597–1610
76. Ndiwe B, Pizzi A, Tibi B, Danwe R, Konai N, Amirou S (2019) African tree bark exudate extracts as biohardeners of fully biosourced thermoset tannin adhesives for wood panels. Ind Crops Prod 132:253–268
77. Nicollin A, Li X, Girods P, Pizzi A, Rogaume Y (2013) Fast pressing composite using tannin-furfuryl alcohol resin and vegetal fibers reinforcement. J Renew Mater 1:311–316
78. Novotny EE, Johnson WW (1931) Urea-Furfural Resins and Process of Making Them. U.S. patent 1,827,824
79. Oktay S, Kızılcan N, Bengu B (2021) Oxidized cornstarch—Urea wood adhesive for interior particleboard production. Int J Adhesion Adhesives 110:102947
80. O'Loising C, Oudjene M, Shotton E, Pizzi A, Fanning P (2012) Mechanical behaviour and 3D stress analysis of multilayered wooden beams made with welded-through wood dowels. Compos Struct 94:313–321
81. O'Loising C, Oudjene M, Ait-Adler H, Fanning P, Pizzi A, Shotton E, Meghlat E-M (2012) Experimental study of timber-to-timber composite beam using welded-through wood dowels. Construct Build Mater 36:245–250
82. Oudjene M, Khalifa M, Segovia C, Pizzi A (2010) Application of numerical modelling to dowel-welded wood joints. J Adhes Sci Technol 24:359–370
83. Ping L, Pizzi A, Guo ZD, Brosse N (2012) Condensed tannins from grape pomace: characterization by FTIR and MALDI TOF and production of environment friendly wood adhesive. Ind Crops Prod 40:13–20
84. Ping L, Gambier F, Pizzi A, Guo ZD, Brosse N (2012) Wood adhesives from agricultural by-products: lignins and tannins for the elaboration of particleboards. Cellulose Chem Techn 46(7–8):457–462
85. Pizzi A (1983) Wood adhesives chemistry and technology. Marcel Dekker, New York
86. Pizzi A (1994) Advanced wood adhesives technology. Marcel Dekker, New York
87. Pizzi A (2006) Recent developments in biobased adhesives for wood bonding: opportunities and issues. J Adhesion Sci Technol 20(8):829–846
88. Pizzi A (2023) Little secrets for the successful industrial use of tannin adhesives: a review. J Renew Mater 11(9):3403–3415
89. Pizzi A, Leban J-M, Kanazawa F, Properzi M, Pichelin F (2004) Wood dowels bonding by high speed rotation welding. J Adhes Sci Technol 18:1263–1278

90. Pizzi A, Mansouri HR, Leban JM, Delmotte L, Omrani P, Pichelin F (2011) Enhancing the exterior performance of wood linear and rotational welding. J Adhes Sci Technol 25:2717–2730
91. Pizzi A, Mittal KL (2003) Handbook of adhesive technology, 2nd edn. Marcel Dekker, New York
92. Pizzi A, Mittal KL (2017) Handbook of adhesive technology, 3rd edn. CDC, Boca Raton
93. Pizzi A, Papadopoulos A, Policardi F (2020) Wood composites and their polymer binders. Polymers 12(5):1115
94. Pizzi A, Vosloo R, Cameron FA, Orovan E (1986) Self-neutralizing acid-set PF wood adhesives. Holz Roh Werkst 44:229–234
95. Pizzi A, Zhou X, Navarrete P, Segovia C, Mansouri HR, Placentia Pena MI, Pichelin F (2013) Enhancing water resistance of welded dowel wood joints by acetylated lignin. J Adhes Sci Technol 27:252–262
96. Plath E (1953) Die Prüfung von PF Harzen aus Säureschadigungen (testing of PF resins for acid damages). Holz Roh Werkst 15:466–471
97. Polesel-Maris J, Jutang I (2017) Anti-adhesive coatings based on condensed tannins. Patent PCT WO2017/037393 A1
98. Pu H, Han K, Dai R, Shan Z (2021) Semi-liquefied bamboo modified urea-formaldehyde resin to synthesize composite adhesives. Int J Adhes Adhes, 103061
99. Renaud A (2009) Minimalist Z-chair assembly by rotational dowel welding. Eur J Wood Prod 67:111–112
100. Sakai S, Chen S, Matsuo-Ueda M, Umemura K (2023) Curing behavior of sucrose with p-toluenesulfonic acid. Polymers 15(23):4592
101. Sangregorio A, Guigo N, van der Waal JC, Sbirrazzuoli N (2019) All 'green' composites comprising flax fibres and humins resins. Compos Sci Technol 171:70–77
102. Sangregorio A, Muralidhara A, Guigo N, Marlair G, Angelici C, Thygesen G, de Jong E, Sbirrazzuoli N (2020) Humins based resin for wood modification and property improvement. Green Chem 22(9):2786–2798
103. Santiago-Medina FJ, Foyer G, Pizzi A, Calliol S, Delmotte L (2016) Lignin-derived non-toxic aldehydes for ecofriendly tannin adhesives for wood panels. Int J Adhesion Adhesives 70:239–248
104. Santiago-Medina FJ, Pizzi A, Abdalla S (2017) Hydroxymethylfurfural hardening of pine tannin wood adhesives. J Renew Mater 5(5):435–447
105. Sarazin J, Pizzi A, Amirou S, Schmidt D, Sernek M (2021) Organosolv lignin non-isocyanate based polyurethanes (NIPU) as wood adhesive. J Renew Mater 9(5):881–907
106. Saud AS, Maniam GP, Rahim MHA (2021) Introduction of eco-friendly adhesives: source, types, chemistry and characterization. In: Jawaid M, Khan TA, Nasir M, Asim M (eds) Eco-friendly adhesives for wood and natural fiber composites. Composites science and technology. Springer, Singapore. https://doi.org/10.1007/978-981-33-4749-6_1
107. Sauget A, Nicollin A, Pizzi A (2013) Fabrication and mechanical analysis of mimosa tannin and commercial flax fibers biocomposites. J Adhesion Sci Technol 27(20):2204–2218
108. Segovia C, Pizzi A (2009) Performance of dowel-welded wood furniture linear joints. J Adhes Sci Technol 23:1293–1301
109. Segovia C, Pizzi A (2009) Performance of dowel-welded T-joints for wood furniture. J Adhes Sci Technol 23:2073–2084
110. Segovia C, Renaud A, Pizzi A (2010) Performance of dowel-welded L-joints for wood furniture. J Adhes Sci Technol 25:1829–1837
111. Segovia C, Zhou X, Pizzi A (2013) Wood blockboards for construction by wood welding with pre-oiled dowels. J Adhes Sci Technol 27:577–585
112. Sodhi S (1957) über die Säure-Schädigung von Holz bei Der Verleimung mit Phenolharzleimen (On the acid corrosion of wood by gluing with phenol resin adhesives). Holz Roh Werkst 15:261–263
113. Stierli, RF, Newton NY (1949) Urea-formaldehyde-furfuryl alcohol resins. U.S. Patent 2,487,394

114. Sun S, Zhao Z, Umemura K (2019) Further exploration of sucrose-citric acid adhesive: synthesis and application on plywood. Polymers 11:1875
115. Tasooji M, Tabarsa T, Khazaeian A, Wool RP (2010) Acrylated epoxidised soy oil as an alternative to urea-formaldehyde in making wheat straw particleboard. J Adhes Sci Technol 24(8–10):1717–1727
116. Thebault M, Pizzi A, Dumarcay S, Gerardin P, Fredon E, Delmotte L (2014) Polyurethanes from hydrolydsable tannins obtained without using isocyanates. Ind Crops Prod 59:329–336
117. Thebault M, Pizzi A, Essawy H, Baroum A, van Assche G (2015) Isocyanate free condensed tannin-based polyurethanes. Eur Polym J 67(6):513–523
118. Thebault M, Pizzi A, Santiago-Medina FJ, Al-Marzouki FM, Abdalla S (2017) Isocyanate-free polyurethanes by coreaction of condensed tannins with aminated tannins. J Renew Mater 5(1):21–29
119. Tiedeman GT, Gillern MF (1975) Adhesives containing metal powders. US patent 3,872,051
120. Tomkinson J (2002) Adhesives based on natural resources. In: Dunky M, Pizzi A, Van Leemput M (eds) Wood adhesion and glued products: wood adhesives. European Commission, Directorate General for Research: Brussels, Belgium, pp 46–65
121. Trosa A, Pizzi A (2001) A no-aldehyde emission hardener for tannin-based wood adhesives. Holz Roh Werkst 59:266–271
122. Umemura K, Ueda T, Munawar S, Kawai S (2012) Application of citric acid as natural adhesive for wood. J Appl Polym Sci 123:1991–1996
123. Umemura K, Kawai S (2015) Development of wood-based materials bonded with citric acid. Forest Prod J 65:38–42
124. Umemura K, Sugihara O, Kawai S (2013) Investigation of a new natural adhesive composed of citric acid and sucrose for particleboard. J Wood Sci 59:203–208
125. Valenzuela J, von Leyser E, Pizzi A, Westermeyer C, Gorrini B (2012) Industrial production of pine tannin-bonded particleboard and MDF. Eur J Wood Prod 70(5):735–740
126. Valkonen S (2017) Lignin-based binders: an industrial reality, latest developments. In: Proceedings of the international conference on wood adhesives, Atlanta, GA, USA, 25–27 October
127. Widyorini R, Nugraha PA, Rahman MZA, Prayitno TA (2016) Bonding ability of a new adhesive composed of citric acid-sucrose for particleboard. BioResources 11:4526–4535
128. Wool RP (1998) Proceedings of the second European Panel Products Symposium; Academic Press, Bangor, UK
129. Xi X, Pizzi A (2020) No-aldehydes glucose/sucrose-triacetin-diamine wood adhesives for particleboard. J Renew Mater 8(7):715–725
130. Xi X, Pizzi A, Delmotte L (2018) Isocyanate-free polyurethane coatings and adhesives from mono- and di-saccharides. Polymers 10:402
131. Xi X, Pizzi A, Amirou S (2018) Melamine-glyoxal-glutaraldehyde wood panel adhesives without formaldehyde. Polymers 10:22
132. Xi X, Wu Z, Pizzi A, Gerardin C, Lei H, Du G (2020) Furfuryl alcohol-aldehyde plywood adhesive resins. J Adhesion 96(9):814–838
133. Xi X, Pizzi A, Frihart CR, Lorenz L, Gerardin C (2020) Tannin plywood adhesives by non-volatile aldehydes generation from specific oxidation of mono- and disaccharides. Int J Adhesion Adhesives 98:102499
134. Xi X, Pizzi A, Gerardin C, Chen X, Amirou S (2020) Soy protein isolate based polyamides as wood adhesives. Wood Sci Technol 54(1):89–102
135. Xi X, Pizzi A, Gerardin C, Liao J, Amirou S, Abdalla S (2021) Glutaradehyde-wheat protein-based adhesives for plywood. J Adhesion 97(1):88–100
136. Xi X, Wu Z, Pizzi A, Gerardin C, Lei H, Zhang B, Du G (2019) Non-isocyanate polyurethane adhesive from sucrose used for particleboard. Wood Sci Technol 53:393–405
137. Xu D, Li C, Pizzi A, Xi X, Wang Z, Du G, Lei H (2024) Self-neutralizing citric acid–glucose and citric acid–starch wood adhesives, in preparation
138. Yang W, Han Y, Zhang W, Zhang D (2021) High-strength and low-cost biobased polyurethane foam composites enhanced by poplar wood powder liquefaction. Polymers 13:2999

139. Younesi-Kordkheili H, Pizzi A (2018) Properties of plywood panels bonded with ionic liquid-modified lignin–phenol–formaldehyde resin. J Adhesion 94:143–154
140. Younesi-Kordkheili H, Pizzi A (2019) Some properties of particleboard panels bonded with phenol-lignin-glyoxal resin. J Adhesion 96(16):1385–1395
141. Younesi-Kordkheili H, Pizzi A (2016) Acid ionic liquids as a new hardener in urea-glyoxal adhesive resins. Polymers 8(3):57–71
142. Younesi-Kordkheili H, Pizzi A (2017) Ionic liquids as enhancers of urea-glyoxal panel adhesives as substitutes of urea-formaldehyde resins. Eur J Wood Prod 75(3):481–483
143. Zanetti M, Pizzi A (2023) Upgrading of MUF resins by buffering additives—Part 2: hexamine sulphate mechanisms and alternate buffers. J Appl Polym Sci 90(1):215–226
144. Zhang B, Pizzi A, Petrissans M, Petrissans A, Colin B (2023) A melamine-dialdehyde starch wood surface finish without formaldehyde. J Renew Mater 11(11):3867–3889
145. Zhao X, Peng L, Wang H, Wang Y, Zhang (2018) Environment-friendly urea-oxidized starch adhesive with zero formaldehyde-emission. Carbohydrate Polym 181:1112–1118
146. Zhao Z, Sakai S, Chen Z, Zhu N, Huang C, Sun S, Zhang M, Umemura K, Yong X (2019) Further exploration of sucrose-citric acid adhesive: investigation of optimal hot-pressing conditions for plywood and curing behavior. Polymers 11:1996
147. Zhao Z, Hayashi S, Xu W, Wu Z, Tanaka S, Sun S, Zhang M, Kanayama K, Umemura K (2018) A novel eco-friendly wood adhesive composed by sucrose and ammonium dihydrogen phosphate. Polymers 10(11):1251
148. Zhu J, Abhyanker H, Njuguna J, Perreux D, Thiebaud F, Chapelle D, Pizzi A, Sauget A, Nicollin A (2013) Tannin-based flax fibre-reinforced bio-composites for structural application in Superlight Electric Vehicles. Ind Crops Prod 50:68–76
149. Zimele Z, Irbe I, Grinins J, Bikovens O, Verovkins A, Bajare D (2020) Novel mycelium-based biocomposites (MBB) as building materials. J Renew Mater 8:1067–1076

# Potential of Nanotechnology in Wood-Based Materials

**Jakub Kawalerczyk, Dorota Dziurka, Dorota Dukarska, Joanna Walkiewicz, and Radosław Mirski**

**Abstract** With a tremendous progress observed over the last decade, nanoscience already has become a frontier of technology by enabling almost endless possibilities in many fields. It creates the opportunity to work on a matter representing near-atomic scale in order to invent the novel, so far unknown solutions. Due to the fact that some of these discoveries are implemented into a development of innovative, high-performance materials, this field can be considered as an indicator of the progress in modern times. Nanotechnology offers the strong potential also for a wood industry sector which might lead to numerous advances in, for example, wood-based materials manufacturing. As indicated by the indexes of consumption of wood-containing composites around the world, it is a dynamically developing branch of industry. Therefore, it attracts a lot of attention of scientists who are still looking for innovative ways to improve these materials. Nanoparticles owe their reinforcing nature to a number of excellent properties such as, for example, tremendous surface area and high thermal conductivity which are rather rare in the case of their bulk counterparts. This is why they were found to effectively improve characteristics of wood-based composites bonded with synthetic or natural adhesives. It is particularly important because the implementation of such modifications may contribute to the extension of their use. Therefore, the purpose of this chapter is to sum up the current knowledge on the potential of nanomaterials such as mineral nanoparticles, nano-oxides, nano-silver, carbon-based nanoparticles, nanocellulose, and nanolignin in the enhancement of both strength and water-related properties of wood-based materials bonded with synthetic and bio-based adhesives (based on starch, tannins, and lignin).

**Keywords** Nanotechnology · Wood-based materials · Nanoparticles · Perspectives

---

J. Kawalerczyk (✉) · D. Dziurka · D. Dukarska · J. Walkiewicz · R. Mirski
Faculty of Forestry and Wood Technology, Department of Mechanical Wood Technology, Poznań University of Life Sciences, Poznań, Poland
e-mail: jakub.kawalerczyk@up.poznan.pl

 
P. Antov et al. (eds.), *Nanomaterials Additives in Bioadhesives for Wood Composites*, Smart Nanomaterials Technology, https://doi.org/10.1007/978-981-96-3400-2_2

# 1 Introduction

Wood-based materials in general can be defined as the type of materials obtained by processing the raw wood into a composite with the use of synthetic or natural adhesives [71]. Based on the form of wood used for their production, they can be divided into the materials manufactured from: (i) veneers such as laminated veneer lumber (LVL) or plywood; (ii) wood particles, such as particleboard; and (iii) wood fibers such as fiberboard. Despite the fact that wood, due to a number of positive, features resulting from its porous, hierarchical, and macromolecular structure, has accompanied humanity for centuries, the development in the area of wood-based materials during the transitional period from nineteenth and twentieth century provided new opportunities for applications [6, 177]. It is assumed that the production of softwood plywood began in the early 1900s, the fiberboard production technology was developed in the 1920s, while particleboards around 1950s and since then they are used all over the world until today [23, 187]. Compared to solid wood, the wood-based materials retain many advantageous features such as their reproducibility, reduced isotropy, and much greater homogeneity [188, 214]. Furthermore, in comparison with steel or concrete, they are characterized by numerous favorable properties including lower energy consumption, lower thermal expansion, lower $CO_2$ emission, lighter weight, and higher strength-to-density ratio [96, 152, 221]. Popularity of wood-based panels is shown by the indicators of production, export, and import quantities, which overall show constant growth on a global scale. Moreover, based on theoretical predictions they are expected to keep growing [27, 77]. For example, the analysis of data regarding the indexes of particleboard manufacturing shows that between 2010 and 2016 there was a total increase of about 10% in the annual production. Furthermore, in 2020 it reached the value of over 96 million $m^3$, of which 40.5% was European production [172, 183]. These tendencies result from the simultaneous development of the industry branches that use these materials such as furniture manufacturing and wooden construction and make them very important from an economic standpoint [16, 83]. The continuous growth of the industry means that large amounts of wood are processed annually. As stated by Pędzik et al. [171] based on Statistical Yearbook of Forestry [175], around 42 million $m^3$ of wood was harvested by Polish State Forests in 2019 and almost half of it was classified as a medium-sized wood, applied mainly for particleboard production. Additionally, the production of wood-based materials is also associated with a high consumption of various types of binding agents [232]. These are most often synthetic thermosetting polymers such as urea-formaldehyde (UF), melamine-urea-formaldehyde (MUF), and phenol-formaldehyde (PF) resins. However, appearing concerns about the formaldehyde emissions, increased environmental consciousness, and stricter environmental legislations are the examples of driving forces behind the progress which can be observed in developments concerning the synthesis of formaldehyde-free bio-based adhesives containing, for example, lignin, tannins, and starch [12, 90, 111].

According to Fridley [71], the considerable increase both in the consumption of wood-based materials and in the advancement of this area results from two aspects.

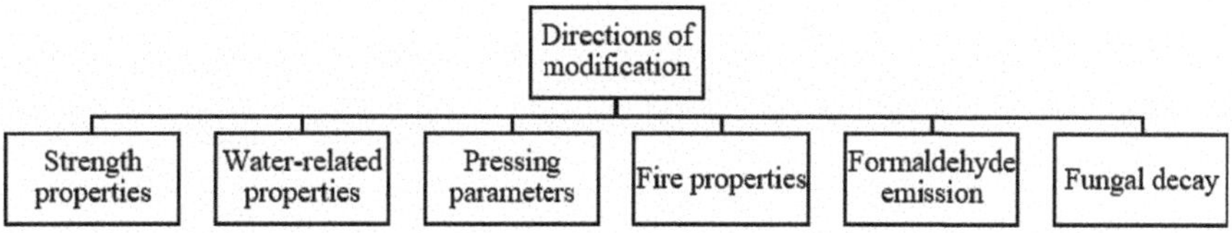


**Fig. 1** Examples of wood-based materials modification directions

First is related to ever-changing wood resources which manifests itself in the fluctuating availability, price, and competitiveness with other industrial plants. Secondly, there are an increased performance demands resulting from numerous stress factors that wood-based materials are exposed to. Among them, long-term mechanical loads, stresses caused by moisture changes, exposure to fire, and fungal decay can be listed as the examples [156]. Reduction of the negative impact caused by these factors can contribute to the increase in number of the potential application possibilities for wood-based materials even more. Consequently, this kind of research attracts major attention of scientists around the world [169, 180]. In order to enhance the performance of wood-containing composites, two different approaches, excluding surface treatments, can be implemented [225]. First one is the modification of fragmented wood (regardless of its form) and the second one is an introduction of modifiers to the adhesives. Exemplary directions of the modification described in literature are summarized in Fig. 1.

Promising results in the modification of both wood particles and adhesives intended for the production of wood-based materials are constantly being developed using the solutions from the field of nanotechnology (NT). National Nanotechnology Initiative (NNI) defined this science field as the "understanding and control of matter at the nanoscale, at dimensions between 1 and 100 nm, where unique phenomena enable novel applications" [20, 194]. It is assumed that the rapid progress of nanoscience began in 1959, when American physicist Richard Feynman, winner of the Nobel Prize, stated that "there is a plenty room at the bottom" during the American Physical Society Meeting. By making this statement, he introduced the concept of not only creating the nanomaterials (NMs) but also manipulating them and operating on them in a controlled manner which makes him a father of modern nanotechnology [89, 213]. In the twenty-first century, years after Feynman's lecture, the achievements in this field have become an indicator of the overall technological progress. It aroused so much interest that nanoscience was recognized in the legislation signed in 2003 by George Bush, former president of the United States, which made it a state priority and resulted in the accelerated growth of NNI [89, 192]. Since then, humanity has witnessed many remarkable inventions and achievements of modern nanotechnology (Fig. 2).

However, although nanoscience is often considered a new field of science that emerged at the end of the nineteenth century, it is assumed that nanoparticles of natural origin have always accompanied the life on earth [46]. The examples of nanomaterials applied without the knowledge of their nanostructured characteristics

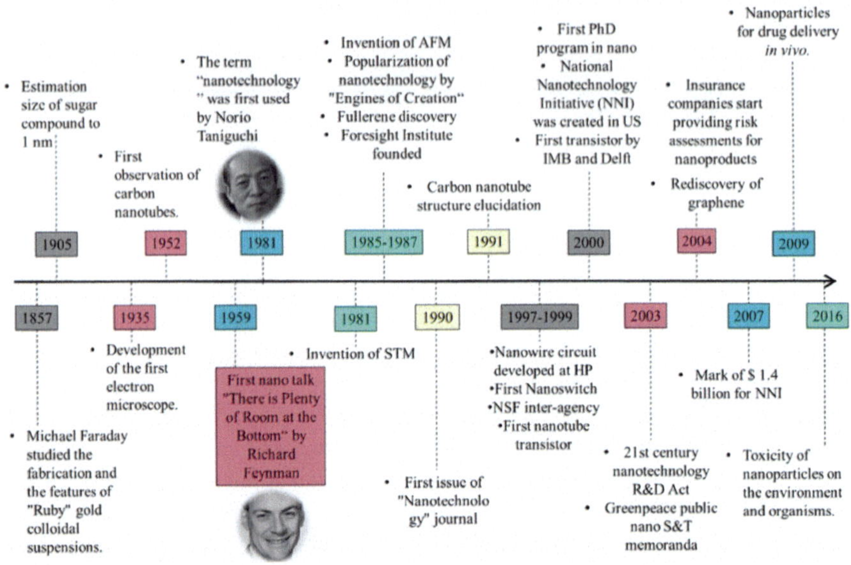


**Fig. 2** Timeline presenting the progress in modern nanotechnology [24] (Open access)

can be found by analyzing the materials representing even prehistoric or ancient times. The ancestors used a mixture of nanomaterials (such as fullerenes, graphene and carbon nanotubes) and waxes for cave paintings in prehistoric times [18, 235]. Furthermore, interesting examples of applied nanomaterials can also be found in the ancient civilizations such as carbon nanoparticles-containing soot of oil lamps used for writing on papyrus, dyes filled with nanostructured clay, asbestos nanofibers-reinforced ceramics or copper nanoparticles-enhanced glass and enamels [14, 24, 32]. Moreover, some of the exhibits from medieval times being the examples of the use of nanoparticles such as the Lycurgus cup, Damascus sabers, or ceramic plates can be admired to this day [25, 63, 173]. The timeline presenting the use of nanoparticles in over the times is shown in Fig. 3.

Such major popularity of nanomaterials in various fields of science and industry results from the fact that they are characterized by unique properties. They have excellent mechanical features when compared to their bulk counterparts. It results from the surface, volume, and quantum effects, however, many variables such as, for example, grain size, grain boundary structure, occurrence of defects can significantly affect their final characteristics [147, 227]. Moreover, NMs also show exceptional physical properties in terms of their structure, linear optical and nonlinear optical features, electrical and magnetic characters, etc., which are rather uncommon in case of macroscopic materials [129]. Additionally, Kumar and Dixit [115] stated that nano-sized particles are distinguished by superior chemical qualities such as their reactivity, stability, dispersibility, and compatibility. Taking into account all of the above-mentioned characteristics, it can be concluded that NMs have a great potential

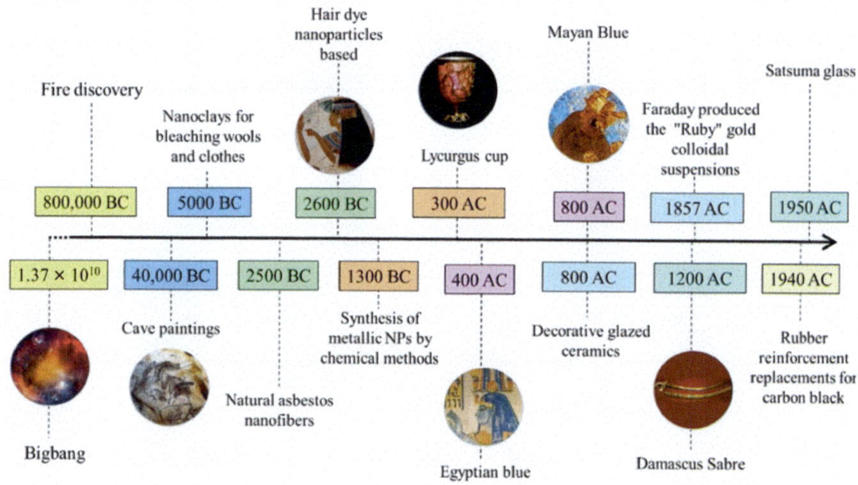


**Fig. 3** Timeline presenting the use of nanotechnology by ancient civilizations [24] (Open access)

in the application as a modifier of both wood and adhesives for the production of wood-based materials.

Despite the development of new technologies and materials, wood-based materials do not lose their importance, moreover, their global consumption is constantly growing. In addition, implementing the solutions offered by nanostructural approach, combining the achievements of many fields such as advanced materials, biology, chemistry, physics, and others, can lead to finding new ways of their application. Therefore, the aim of this work was to summarize the potential and current knowledge on the effect of nanoparticles addition on the properties of manufactured wood-based composites with the emphasis on improvements in their strength properties and water-related properties.

## 2 Characteristic of Selected Nanoparticles

The world of nanoparticles is very diverse and they can be classified according to many different criteria, for example, whether they are natural or synthetic, according to their chemical composition or according to their dimensionality as it is shown in Fig. 4 [93, 113, 182]. However, along with the dynamic development of the discipline and new discoveries in the field of nanotechnology, the new ways of classifying and categorizing nanomaterials are constantly emerging.

Recently, an increasing attention is being paid to mineral nanoparticles (MN), also known as nanoclays. They consist of the structural units characterized by a layered morphology which form a clay crystallites due to the van der Waals forces [81]. Moreover, it is the arrangement of these adjacent layers that allows the different

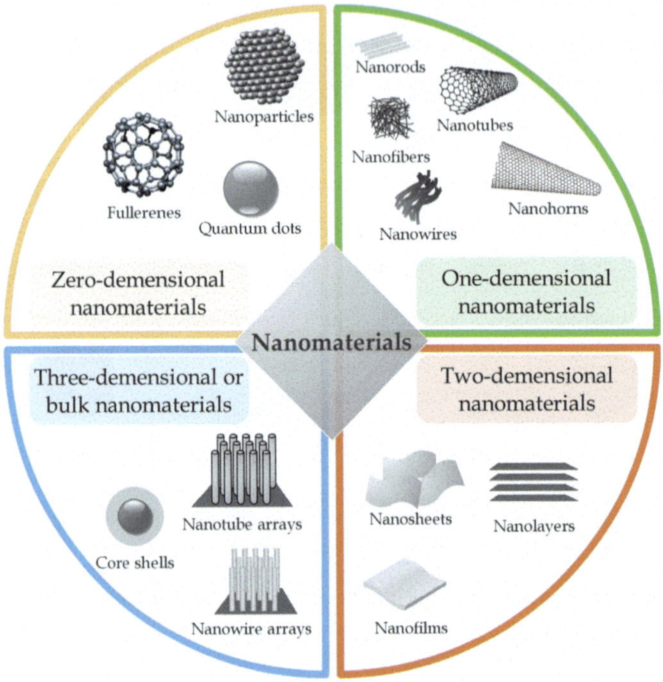


**Fig. 4** Classification of NMs based on their dimensionality [113] (Open access)

types of nanoclays to be defined and distinguished [151]. Their advantages include low cost, low environmental impact, and practically unlimited availability since they can be obtained by synthesis and be found in nature [81]. An interesting example is the montmorillonite (MMT) which is built of the two silica tetrahedral sheets and the alumina sheet between them. These aluminosilicates are usually characterized by the very low diameter and a very high aspect ratio [5, 7, 226, 247]. The appearance of MMT nanoparticles is shown in Fig. 5.

Another interesting example of NMs is nano-sepiolite (SEP). It is a microcrystalline hydrous magnesium silicate with a formula of $Mg_4Si_6O_{15}(OH)_2.6H_2O$ and a needle-like morphology (Fig. 6). It is composed of two layers of silica in tetrahedron and a central layer of octahedral magnesium-oxide hydroxide [40]. Due to the nanometer tunnel structure, which is characteristic for phyllosilicates, it has a superior specific surface area around 200–300 $m^2/g$. Moreover, when compared to platelet-like MMT morphology, SEP nanoparticles are characterized by the smaller interface surface which contributes to a better dispersion within the polymer matrix [2, 29, 108, 163, 196, 240, 253]. In addition to the high surface area and tunnel structure the following favorable features of SEP can be listed: porosity, fiber structure, surface activity, crystal morphology, and the ability to form stable suspensions characterized by a high viscosity despite low concentrations [35].

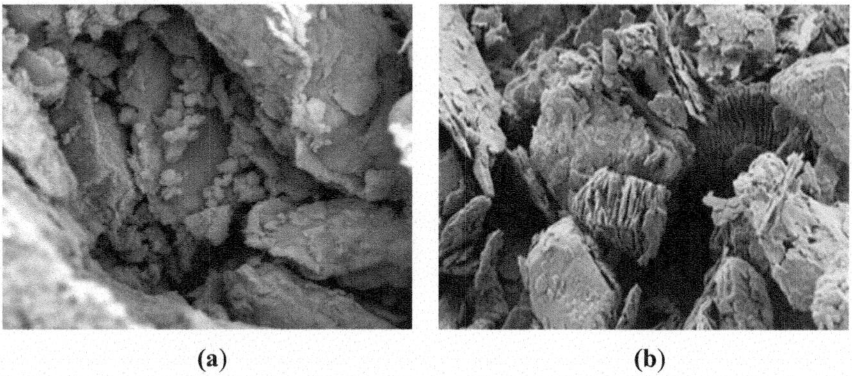

**Fig. 5** SEM photographs of MMT made by: **a** Jozefaciuk et al. [98], **b** Zhang et al. [250] (Open access)

**Fig. 6** SEM photographs of SEP made by: **a** Yao et al. [234], **b** Savun-Hekimoğlu et al. [186] (Open access)

The attention has also recently turned to the application of nano-wollastonite (NW) (Fig. 7). It is a naturally-occurring calcium metasilicate ($CaSiO_3$) mineral distributed by nature, composed of the silicon oxides and calcium. Furthermore, NW is characterized by a needle-like morphology and many favorable features which, as reported by Tolinski [212], can be similar to glass fibers in certain ways. High melting point, thermal stability, aspect ratio, hardness, specific gravity, and low thermal expansion can be considered as particularly beneficial and these attributes consequently have contributed to the growing popularity of NW in recent years [134, 137, 144, 210, 211].

Another group of nanoparticles that attracts much attention of scientists are nano-oxides (NO). They demonstrate an exceptional chemical and physical qualities comparing to their bulk counterparts which makes them one of the most intensively studied types of nanomaterials in general [191, 251]. One of them is a nano-silica

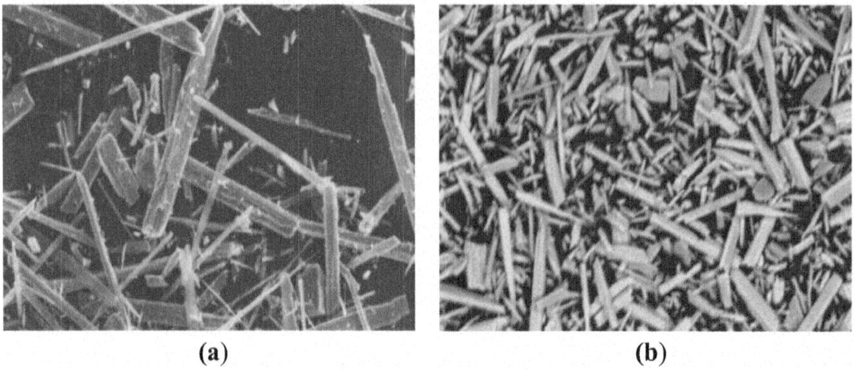


**Fig. 7** SEM photographs of NW made by: **a** Di Giuseppe et al. [51], **b** Panin et al. [168] (Open access)

(nano-$SiO_2$), sometimes also called as silica dust or quartz dust, which is widely studied as an additive for the production of multifunctional, high-performance materials [73]. Nano-silica can be extracted from natural sources; however, this type is not preferred for industrial and scientific applications due to a high content of metal impurities. This is why the synthetic forms (colloidal silica, pyrogenic silica, precipitated silica, etc.) are much more popular and appreciated [179, 217]. Although its properties are very diverse and they depend on many factors, in general it is assumed that nano-$SiO_2$ is characterized by a high reactivity (among others due to the presence of silanol groups), good chemical compatibility, large surface energy, strong surface adsorption, good dispersion, and great mechanical properties [254]. The appearance of spherical nano-silica is presented in Fig. 8.

Metal-oxide nanoparticles (MONPs) also exhibit a wide range of positive features and therefore, they are considered as the most often used NMs [65]. Nanoparticles such as nano-$TiO_2$ (Fig. 9a) and nano-ZnO (Fig. 9b) show many advantageous

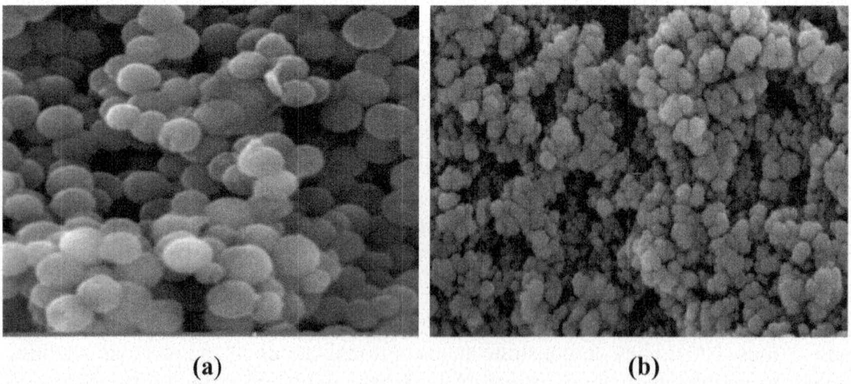


**Fig. 8** SEM photographs of nano-$SiO_2$ made by: **a** Alvarado et al. [8], **b** Da'na [47] (Open access)

**Fig. 9** SEM photographs of: **a** nano-$TiO_2$ made by Hu et al. [88], **b** nano-ZnO made by Nageswara Rao et al. [148], **c** nano-silver made by Fozia et al. [69] (Open access)

features such as great optical properties, large surface area, thermal and chemical stability, mechanical properties, high crystallinity, anti-corrosiveness, and microbial nature [140, 178]. Their exceptional properties comparing to bulk counterparts result from a quantum confinement and the increased availability of surface atoms. MONPs can be synthesized by many methods including physical methods (ball milling, laser ablation, sputtering, etc.) and chemical methods (sol–gel or hydrothermal synthesis, chemical vapor deposition, etc.) [170]. What is important, chosen method of the nanoparticles synthesis has a crucial effect on their properties, and consequently also on the potential ways of application [165]. Another example of metal widely studied in recent years by scientists representing the field of nanotechnology is a silver at the nanoscale. Silver nanoparticles (AgNPs), presented in Fig. 9c, have been exploited for various purposes due to, for example, their high surface-to-volume ratio which in turn results in a long list of beneficial properties such as, e.g., high stability, conductivity, reactivity and other physical, chemical and biological characteristics [150, 248]. Moreover, the large part of the studied nano-silver applications is related to the fact that it demonstrates the strong antibacterial effect [4].

According to Gogotsi and Presser [76], carbon-based nanoparticles are the most important and promising nanostructures. For almost three decades, they have been gaining a lot of attention which resulted in the implementation in many industrial areas such as automotive industry, high-frequency devices production, advanced sensors manufacturing, and medicine, just to name a few [41, 118, 231]. This is confirmed by the number of patents and publications appearing each year, which according to Choudhary et al. [43] is constantly growing since 2004. This growth is caused by the number of unique properties such as thermal and electric conductivity, hydrophobicity, chemical resistance, mechanical strength and stiffness [79, 131, 132, 157]. The best-known example of carbon nano-derivative are carbon nanotubes (CNT) which have been discovered in 1991 (Fig. 10a) [92]. They are large, cylindrical molecules with a hexagonal arrangement of carbon atoms which are created by rolling up the sheets of graphene [85]. Based on the number of graphene sheets, they can be divided into single-walled carbon nanotubes composed of one sheet of graphene and multiwalled carbon nanotubes composed of the multiple sheets of

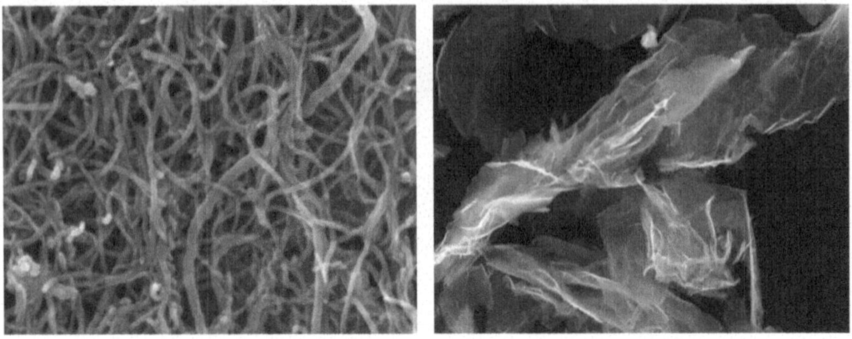

(a) (b)

**Fig. 10** SEM photographs of: **a** carbon nanotubes made by Jung et al. [99], **b** graphene nanoplatelets made by Le and Huang [120] (Open access)

graphene [68]. The progress in the field of carbon nanotubes stimulated the development of the entire nanotechnology field. In 2004 graphene nanoplatelets (GNP) have been discovered [160]. These are two-dimensional platelet-like structures being a single layer or multiple layers of graphite (Fig. 10b) [112]. Most often GNPs are obtained by applying the liquid phase exfoliation technique, however, many other methods such as microwave radiation, ball milling, or wet-jet milling are also effective in this matter [36, 239]. Regardless of the method of production, due to the unique atomic structure they exhibit an excellent properties [135].

The most abundant polymer found in biosphere is cellulose. According to Klemm et al. [110], its resources are comparable to the reserves of the main fossil and mineral stocks. Each year the industrial cellulose is produced in the amount of approx. 7.5 tons [149]. Moreover, as predicted by U.S. Department of Agriculture that market of nanocellulose will reach the level of 35 million tons per year in 2050 [153]. Sources of nanocellulose include plants, microorganisms (algae, bacteria, fungi), and even animals such as tunicates [49, 55]. However, plants remain the main source of obtaining nanocellulose due to the fact that it constitutes around 40–50% of their total mass [195]. In addition to wood, its source can also be various types of biomass such as, e.g., fast-growing grasses, agricultural residues, garden wastes [21, 166]. When analyzing the chemical structure of cellulose, it can be seen that it consists of both crystalline and amorphous regions. Because of the strong intermolecular hydrogen bonds a chain is very difficult to break, especially a well-ordered crystalline parts [155, 243]. Thus, there are several chemical methods of obtaining nanocellulose which have to be included, for example, enzymatic hydrolysis, acid hydrolysis, or hydrolysis with the use of ionic liquids [22]. Nevertheless, the intensity of research in this area is progressing very quickly and consequently new, highly efficient methods are constantly being developed. Nanocellulose can be obtained in the form of cellulose nanocrystals (NCC) (Fig. 11a) or cellulose nanofibers so-called nanofibrillated cellulose (CNF) (Fig. 11b). Regardless of the cellulosic derivative, the nano-sized

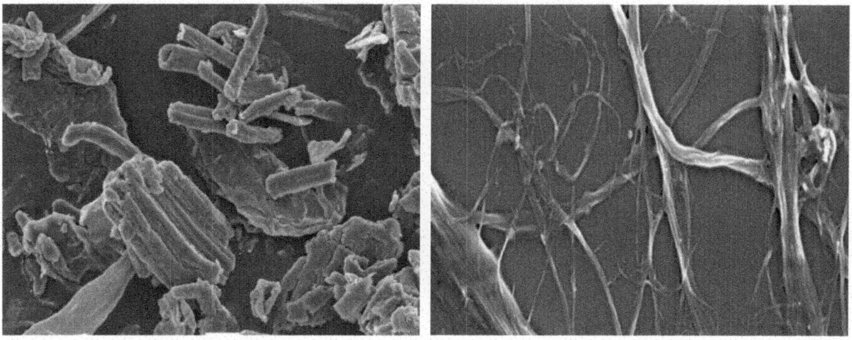


**Fig. 11** SEM photographs of: **a** NCC made by Fitriani et al. [67], **b** CNF made by Noonan et al. [158] (Open access)

cellulose is characterized by high-energy conversion capacity, chemical compatibility, high reactivity, great mechanical strength, and exceptional surface area [3, 256].

Another example of a biopolymer widely distributed through nature is lignin. It is estimated that its average share in lignocellulosic biomass is ranging from 10 to 20%, which makes lignin the second most abundant natural polymer in general [224]. Moreover, according to Hussin et al. [91] 50–70 million tons of lignin is produced by the pulp and paper industry each year and about 40% of it is used as fuel for energy purposes in the form of black liquor. Besides availability, it shows a irregularly-branched complex structure composed of aromatic moieties and containing numerous polyphenols [26, 34]. Lignin nanoparticles (LNPs) presented in Fig. 12 can be obtained by various methods such as solvent exchange, flash precipitation, $CO_2$ saturation, and dialysis [215]. LNPs are characterized by many positive features such as antioxidative and antibacterial properties resulting from its surface chemistry and shape, environmental friendliness, high porosity, and a very high surface area [127, 208, 216]. However, a strong hydrogen bonding between the molecules can cause major difficulties in a size reduction and dispersion of LNPs. Therefore, in recent years there have been numerous studies on their chemical modification, which effectively improves both the reactivity and surface properties [244].

Based on the above characteristics, it can be concluded that nanoparticles in general are characterized by the wide range of positive properties. Many of them can be highly beneficial in terms of the potential applications as modifiers of wood and adhesives used in the production of wood-based materials and consequently, they can significantly improve some of their properties.

**Fig. 12** SEM photographs of nano-lignin made by **a** Del Buono et al. [48] and **b** Luo et al. [133] (Open access)

# 3 Improvement of the Properties of Wood-Based Materials Bonded with Formaldehyde-Based Adhesives

## *3.1 Improvement of Strength Properties*

The strength properties of the wood-based panels are the basic criterion determining their applications and thus, mineral nanoparticles have been widely used to improve them. Nanoclays have been extensively studied as a filler for particleboards manufacturing. Hosseyni et al. [87] introduced MMT in the amount up to 6% to the UF adhesive which resulted in a significant improvement in the strength properties of boards. They observed an increase in the modulus of rupture (MOR) of the boards by up to 39%. Similar research was performed by Salari et al. [184]. Authors enhanced UF resin with organo-modified MMT (0–5%) and studied the effect on OSB boards made from paulownia wood. It was found that all mechanical properties such as MOR, modulus of elasticity (MOE), and internal bond (IB) were increased as a result of the modification. According to authors the acceleration in resin curing and a decrease in its viscosity could be the reason behind observed effects. Furthermore, the results of experiments conducted by Laosman et al. [119] also showed an increase in MOR, MOE, and IB of particleboard made from wood wastes and the best results were obtained by adding up to 1% of nanoclay. On the contrary, the outcomes of some studies also showed a very minor, barely noticeable effect or no effect at all on the strength of manufactured particleboard [139, 228]. According to available literature data, the enhancement in mechanical characteristic of medium density fiberboard (MDF) by the addition of MMT can be achieved in two ways: by the addition of MMT to the adhesive and by mixing the nanoparticles with resinated mat. The reinforcement of UF adhesive considerably improved properties of MDF made from both wood fibers and the mixture of wood fibers with chicken feathers [86, 164]. Furthermore, Ashori and Nourbakhsh [17] found out that mixing the mat with nanoclay increased both MOR and MOE and the optimal load of nanoparticles was 6%. In case of plywood, the introduction of small amounts of MMT (0–8%) to UF adhesive resulted in the increase of wet shear strength of the resultant panels. Based on the results of differential scanning calorimetry (DSC), it was found that implemented modification had an accelerating effect on resin curing [121]. Similar effect was observed by Yadav et al. [230] researching the possibility to use pristine nanoclay which also led to the improvement in plywood shear strength. Nanoclays can be also applied as secondary fillers for the production of plywood. Studies have shown that using the formulation of UF resin including wheat flour and 5% of MMT or the formulation of PF resin containing both tannin extender and 3% of nanoparticles can contribute to the significant enhancement in the properties of resultant panels [54, 60]. According to the authors, the reason could be the ability of nano-sized materials to ease the stress transfer due to their exceptional surface area [60].

Compared to MMT, the number of publications on the use of nano-sepiolite for wood bonding is scarce. Taghiyari et al. [199] investigated the effect of SEP-reinforced UF adhesive on the properties of oriented strand lumber (OSL). OSL is a structural composite made from wood strands with the primary orientation along the length of the board. It was developed mainly for the production of beams and columns [138]. The outcomes of performed studies showed the significant increase in the hardness of manufactured panels. According to authors the observed effect resulted from the increase in thermal conductivity coefficient which acceleration of the heat transfer to the core section of the mat [206]. Moreover, it was also found that the increase in hardness was possible due to the even dispersion of nanoparticles (Fig. 13) [202].

**Fig. 13** SEM photographs of SEP nanoparticles applied for UF resin reinforcement by Taghiyari et al. [199] (Open access)

Although none of the studies analyzed the effect of the addition of nano-SEP on the properties of wood-based materials such as particleboard or plywood, some studies indicate great potential of these nanoparticles. The incorporation of macroscopic tunnel-structured sepiolite to UF and MUF resin resulted in a considerable improvement of plywood wet shear strength due to formation of toughened fraction of cured adhesive [123, 124]. Therefore, even better effect can be expected in the case of using particles in the dimensional range of nanotechnology.

Some studies also examined how the addition of nano-wollastonite influences mechanical properties of the obtained wood-based panels. Yadav [229] applied ball-milled wollastonite nanoparticles as a filler for UF adhesive and determined the properties of particleboards. Results have shown the significant improvement in MOR, MOE, and IB due to the introduction of NW in the amount of 0.5–2%. According to authors, the reinforcing effect resulted from both the increase in a crosslink density of the adhesive and the increase in wood-adhesive bonding. Research performed by Taghiyari et al. [200] showed a very similar effect consisting of the enhancement in mechanical properties of manufactured particleboard caused by the implementation of NW internally as a modifier of UF adhesive, and on the surface of the board by mixing the nanoparticles with water-based paint. It was concluded that regardless of the application method, the introduced nano-sized wollastonite contributed to the improvement in mechanical properties due to the increase in bond formation between the wood particles. Furthermore, Taghiyari et al. [198] investigated the effect of UF adhesive modification with NW (2–8%) on the properties of MDF boards. Studies have shown the increase in all strength properties such as MOR, MOE, and IB. Based on the outcomes, the authors recommended the addition of nano-modifier in the amount of 6% as the optimal variant. Moreover, the continuation of research concerned the use of water-based paint formulation containing NW and it also increased the mechanical strength of manufactured fiberboard [204]. Moreover, the positive impact on the strength properties could also be noticed in case of composite boards made from the mixture of both rice straw with wood fibers and camel thorn with wood fibers, bonded with NW-reinforced UF adhesive [201, 209]. According to authors, the applied modification was effective due to the increase in heat transfer during the hot pressing caused by the presence of NW. In addition, findings of another study conducted by Taghiyari et al. [199] on the use of micron-sized wollastonite indicate that it can be also successfully implemented to the production of plywood.

Many studies examining the effect of nano-silica on the properties of wood-based materials have appeared in recent years. Roumeli et al. [181] found out that the use of UF/$SiO_2$ adhesive led to the improvement in the strength characteristics of particleboard. A few loadings were investigated; however, the best results were observed in the variant assuming 1% of nanoparticles. Higher mechanical strength of particleboard because of the nano-$SiO_2$ incorporation to UF adhesive was also noted by Yadav [229]. In the case of this study, the amount of 1.5% was considered as optimal. Moreover, mechanical properties of OSB boards made from paulownia wood were also improved due to the addition of small quantities (3%) of nano-silica [184]. Interesting results were obtained in terms of the use of nano-$SiO_2$-reinforced UF adhesive in the production of boards composed of wood particles partially substituted with rape straw and the boards made entirely from rape straw. It was concluded that the modification of resin had a positive effect on the strength features (MOR, MOE, and IB) of the resultant panels and furthermore, it allowed to extend the possibilities of their use through a reclassification in accordance with the requirements of EN 312 (2010) standard [56, 59]. Incorporation of nano-$SiO_2$ was also effective in increasing the strength of plywood glued with PF or melamine-urea-phenol-formaldehyde (MUPF) adhesives and PF resin-bonded LVL [58, 61, 176]. Authors agreed that the observed reinforcement probably resulted from the accelerated cross-linking process during the adhesive curing [57].

In addition, metal-oxide nanoparticles and nano-silver were also extensively studied for the use as reinforcing agents in wood-containing composites. Nano-sized titanium oxide was found to effectively improve the strength of boards made from the apple pruning and sunflower stalks. Studies have shown that the incorporation of nano-$TiO_2$ in the quantity of 8% led to the significant increase in the flexural modulus, flexural strength, and internal bond of the panels [74]. Moreover, both nano-$TiO_2$ and nano-ZnO added to PF and MUF adhesives contributed to the increase in bonding strength of glue lines between the layers of fir and aspen wood which indicates the great potential for implementation into the plywood production [167]. Small amount (1%) of nano-ZnO improved also the mechanical characteristics of particleboard such as MOR, MOE, and IB [193]. Interestingly, nano-silver showed highly enhancing effect on the mechanical properties of particleboard and fiberboard as well. As reported in literature, it positively influenced the outcomes of such qualities as MOR, MOE, IB, hardness, and screw withdrawal capacity [197, 203, 205]. The reinforcement connected with the introduction of MONPs and nano-silver can be attributed to their high thermal conductivity which consequently led to the acceleration of heat transfer. Therefore, the resultant panels were characterized by better integrity due to the more complete resin curing [107].

Despite the wide range of applications of carbon-based nanoparticles, there were only a few scientific attempts to incorporate them into production of wood-based panels and they mainly concerned with the aspect of fire resistance. Łukawski et al. [130] manufactured single-layer particleboard using CNT-coated particles. The application of 0.2% aqueous dispersion for wood impregnation led to a very slight increase in MOR, MOE, and internal bond determined after pre-treatment with boiling water (V-100). Authors stated that taking into account the observations of other authors which showed a similar effect [33, 53, 161], the concentration of CNT in a dispersion was too low to cause spectacular changes. Furthermore, the load of CNT up to 2% introduced to UF adhesive for MDF production resulted in a major improvement of its MOR, MOE, and IB. According to authors, the presence of nanoparticles increased bonding between wood fibers and adhesive [142]. In addition, Gul et al. [80] found that the use of CNT/UF adhesive positively influenced thermal conductivity during pressing and therefore, also a resin curing. Consequently, MOR, MOE, and IB of MDF panels increased by 28%, 30%, and 21%, respectively.

What is important, these favorable effects are in agreement with the outcomes of other research available in literature [52, 116]. GNP in the amount of 1.5% were also applied as a filler for melamine-formaldehyde (MF) adhesive in the production of plywood intended for flooring. Conducted investigations showed some promising results in terms of a major increase in both thermal conductivity and plywood shear strength [190]. It shows that concept of GNP incorporation into the wood-based materials manufacturing process has a great potential.

In recent years, much attention has been focused on the modification of formaldehyde-containing adhesives with the use of nanocellulose (NC). In the vast majority of cases, the reinforcement of resins with the small quantities of nano-sized cellulose led to the increase in the strength properties of manufactured wood-based panels [219]. Studies have shown that the use of nanocellulose-enhanced UF and MUF adhesives can contribute to the improvement in both particleboard and OSB board properties such as MOR, IB, fracture energy, and fracture toughness [15, 136, 174, 218]. Moreover, the nanocellulose was recently studied as a modifier of UF adhesive applied for rape strawboards production. Based on the outcomes, it was found that the improvement in mechanical properties (MOR, MOE, and IB) due to the modification was significant enough for strawboards to reach the same strength as standard particleboard [106]. The positive effect of nanocellulose introduction to UF adhesive was also noted in case of MDF production. Each of the studies performed by Moslemi et al. [146], Zhang et al. [245] and Khanjanzadeh et al. [109] showed the improvement of MOR, MOE, and IB of the boards. Furthermore, strength properties of layered wood-based composites made from veneers were also reinforced by the implementation of cellulosic nanoparticles. Shear strength of: UF-bonded LVL [19], UF-bonded plywood [104, 246], MUF-bonded plywood [100] and PF-bonded plywood [101, 122] was considerably improved. In addition, the observed increase in bonding strength and resin reactivity was so significant that it also allowed to reduce the adhesive consumption by up to 30% and to lower the plywood pressing parameters [102, 103, 105]. Such positive effects on the mechanical properties of wood-based materials were probably related to (i) enhanced ductility, fracture toughness, and fracture energy of adhesive bonds [219]; (ii) great interfacial interactions between the adhesive matrix and nanocellulose resulting from the high aspect ratio and large specific area of nanoparticles [145]; and (iii) improved transfer of stresses within the glue line [154].

While nanocellulose has been extensively studied as a modifier of formaldehyde-containing resins, in the case of nano-lignin no work can be found on the modification of commonly applied amino resins and the effect on the properties of wood-based materials. However, a recent study conducted by Yang et al. [233] indicates a potential of nano-lignin for application as an additive for phenolic resin. The results have shown that the incorporation of LNPs resulted in homogeneous dispersion, favored adhesive reactivity, and increased the shear strength of wood joints. Therefore, more studies are needed to determine the effect on the properties of produced wood-based panels.

Taking into account the increasing demands on the performance of wood-based materials, the improvement in their strength is a very desirable effect which is a subject of many studies appearing each year. Due to the remarkable features of nanomaterials, such as high thermal conductivity or tremendous specific surface, in most cases they positively affect the strength of the wood-based panels. The outcomes presented in Table 1 show that the addition of nanoparticles has a positive effect on the strength parameters (MOR, MOE, IB) of boards such as particleboard or MDF and the bonding quality of layered composites such as plywood or LVL.

## 3.2 Improvement of Water-Related Properties

The influence of water on the properties of wood-based materials is very important from the perspective of potential applications, especially in the case of composites used in the outdoor conditions. The swelling of the boards resulting from the swelling of the wood and its decompression is irreversible. It may depend on many factors related to both the wood used for the production of boards (species, extractives content) and technological aspects (geometry of particles, board density, type of resin, pressing parameters) [82, 143]. Taking into account that the strength of adhesive bonds in the board structure also has a significant impact on the water-related properties, it can be assumed that the modification of resins may considerably improve them.

As mentioned before, mineral nanoparticles have great potential in improving the properties of wood-based materials, for example, particleboards. Hosseyni et al. [87] investigated the effect of MMT addition to UF resin and discovered that 3% of nanoclay contributed to a major decrease in water absorption (WA) and thickness swelling (TS) by up to approx. 35%. Moreover, authors stated that further increase in the load of nanoparticles to 6% showed no additional effects. The positive effect was confirmed by the outcomes of experiments performed by Laosman et al. [119] who also observed a decrease in both WA and TS of particleboard made from wood wastes. In the case of OSB boards, it was found that the use of MMT/UF adhesive positively influenced dimensional stability of the panels soaked in water for 2 and 24 h. Authors stated that the amount of 5% was best suited for reaching the optimum effect and furthermore, both smaller and larger additions led to worst results [184]. Ashori and Nourbakhsh [17] confirmed that the favorable effect in terms of water-related properties can also be noted in the case of MDF. Studies have shown that as the amount of introduced nanoclay increased, the thickness swelling of boards proportionally decreased. Obtained value of TS was 20% lower than in the case of using non-modified adhesive. Positive effect of MMT on swelling-related properties of MDF was also observed by Zahed and Gholamian [241], Hosseini et al. [86], and Omrani et al. [164]. Interestingly, on the other hand, regardless of the amount the addition of MMT did not significantly affect the TS and WA of plywood glued with UF and MUF adhesive [54].

While none of the studies examined the effect of nano-sepiolite addition on water-related properties of wood-based materials, there are several reports on the effect of using nano-wollastonite. Based on conducted research, Yadav [229] found out that the modification of UF resin with NW can cause the improvement in TS and WA

**Table 1** The effect of applied nanoparticles on mechanical properties of wood-based materials bonded with formaldehyde-based adhesives

| Type of nanomaterial | Way of introduction | Effect on mechanical properties |
|---|---|---|
| MMT | UF adhesive modification | Improvement in MOR, MOE, and IB of particleboard |
| | UF adhesive modification | Improvement in MOR, MOE, and IB of OSB |
| | UF adhesive modification | Improvement in MOR, MOE, and IB of MDF |
| | Fibers modification | Improvement in MOR, MOE, and IB of MDF |
| | UF and PF adhesive modification | Improvement in bonding quality of plywood |
| SEP | UF adhesive modification | Improvement in hardness of OSL |
| NW | UF adhesive modification | Improvement in MOR, MOE, and IB of particleboard |
| | Surface treatment | Improvement in MOR, MOE, and IB of particleboard |
| | UF adhesive modification | Improvement in MOR, MOE, and IB of MDF |
| | Surface treatment | Improvement in MOR, MOE, and IB of MDF |
| | UF adhesive modification | Improvement in bonding quality of plywood |
| Nano-$SiO_2$ | UF adhesive modification | Improvement in MOR, MOE, and IB of particleboard |
| | UF adhesive modification | Improvement in MOR, MOE, and IB of OSB |
| | UF adhesive modification | Improvement in MOR, MOE, and IB of strawboard |
| | PF and MUPF adhesive modification | Improvement in bonding quality of plywood |
| | PF adhesive modification | Improvement in bonding quality of LVL |
| Nano-$TiO_2$ | UF adhesive modification | Improvement in MOR, MOE, and IB of particleboard |
| Nano-ZnO | UF adhesive modification | Improvement in MOR, MOE, and IB of particleboard |
| Nano-silver | UF adhesive modification | Improvement in MOR, MOE, and IB of MDF |
| CNT | Particle modification | No effect on MOR, MOE, and IB of particleboard |
| | UF adhesive modification | Improvement in MOR, MOE, and IB of MDF |

(continued)

**Table 1** (continued)

| Type of nanomaterial | Way of introduction | Effect on mechanical properties |
|---|---|---|
| GNP | MF adhesive modification | Improvement in bonding quality of plywood |
| NC | UF adhesive modification | Improvement in MOR, MOE, and IB of particleboard |
| | MUF adhesive modification | Improvement in MOR, MOE, and IB of OSB |
| | UF adhesive modification | Improvement in MOR, MOE, and IB of strawboard |
| | UF adhesive modification | Improvement in MOR, MOE, and IB of MDF |
| | UF adhesive modification | Improvement in bonding quality of plywood |
| | MUF adhesive modification | Improvement in bonding quality of plywood |
| | PF adhesive modification | Improvement in bonding quality of plywood |
| | UF adhesive modification | Improvement in bonding quality of LVL |

of the particleboard. The most satisfactory results were obtained using the additive in the amount of 2%. In this case, values of TS and WA after 2 h of soaking in water were reduced from 6.2% to 4.8% and from 33 to 26%, respectively. On the contrary, Taghiyari et al. [200] concluded that NW applied internally as the modifier of UF adhesive did not result in any significant changes both in TS and WA of the particleboard. However, it was found that surface application of NW together with water-based paint effectively improved dimensional stability of the resultant panels. In the case of MDF, applying the UF adhesive formulation involving NW caused a slight but statistically significant decrease in TS and WA, however, it was noticed again that the surface treatment of boards with NW-containing paint gives better results and led to more noticeable improvement [198, 204]. Nano-sized wollastonite positively influenced water-related properties also when it was incorporated to the production of composites obtained both by mixing wood fibers with camel thorn and wood fibers with rice straw [201, 209]. Moreover, what is promising, Taghiyari et al. [199] concluded that the TS of plywood can be effectively reduced by introducing 5% of wollastonite to the UF adhesive. The positive effect of implemented modifications probably resulted from the previously discussed increase in the strength of the bonds and moreover, according to Tichi et al. [209] it can be attributed also to the hydrophobic nature of nano-wollastonite itself.

In addition to mineral nanoparticles, nano-silica has also been studied for improving the water resistance of wood-based materials. According to Roumeli et al. [181] thickness swelling of the particleboard soaked in water for 24 h was decreased, especially when the low loading of nanoparticles (1–3%) was used which indicate that no high concentration of nano-silica is required to observe reinforcing effect. The favorable effect of nano-$SiO_2$ was confirmed by recent studies performed by Yadav [229] which showed a decrease in both TS and WA after 2 h of experiment. The results were particularly promising when the load of modifier ranged from 1 to 1.5%. Moreover, results obtained in the case of OSB have shown the improvement in TS and WA (both after 2 and 24 h) when the amount of introduced nanoparticles did not exceed 3%. However, authors stated that without the addition of any hydrophobic agent, the OSB made from paulownia wood did not reach the level required by EN standard [185]. Furthermore, the incorporation of 1% of nano-$SiO_2$ was found to have a positive effect on the TS and WA of rape strawboard [56].

The incorporation of UF adhesive containing 1% of ZnO nanoparticles also allowed for the improvement in water-related properties of particleboard. Studies have shown that the values of TS and WA after 24 h water immersion decreased from 22.2% to 14.9% and from 30.3% to 21.1%, respectively. According to the authors, the observed effect was a result of enhanced resin curing which led to the increase in internal adhesion between the particles and prevented penetration of water into the core layer of the board [193]. Reduced water uptake was also noted in the case of boards made from apple tree pruning and sunflower stalks glued with UF/$TiO_2$ adhesive. It was concluded that as the loading of the nanoparticles increased, both the TS and WA simultaneously decreased. In general, the water uptake of nano-modified boards was approx. 45% lower than in the case of boards bonded with neat UF adhesive. As explained by the authors, the application of nano-$TiO_2$ contributed to obtaining the mat characterized by greater compactness, which limited the access of water [74]. The reinforcing effect was also achieved by implementing the modification of MF and UF adhesives with nano-$TiO_2$ to the production of MDF. However, the reduction in TS and WA was observed only when 0.5% of modifier was added and the further increase in its loading worsened the performance of manufactured panels [45]. Additionally, the introduction of nano-silver to UF adhesive improved TS and WA of particleboard after 2 h of experiments as well. Interestingly, after 24 h of soaking in water the reinforcing effect was no longer visible. According to authors, the quality of bonding was increased due to the better transfer of heat and it slowed down the penetration of water, however, the 24-h soaking period was enough time to penetrate the particleboard structure despite its reinforcement.

Łukawski et al. [130] investigated the effect of coating wood particles with CNT on the properties of particleboard. In terms of water-related properties, it was found that the modification had no statistically significant effect on the TS and WA of the produced panels. Authors stated that the reason could be low concentration of CNT in a modifying dispersion and the application of hydrophilic sodium dodecylbenzene-sulfonate (SDBS) as a dispersant. Contrary to particleboard, research conducted in order to increase the dimensional stability of MDF showed more explicit and promising results. Mazaheri et al. [142] stated that TS of manufactured boards was lowered with increasing content of CNT and the best results were obtained by using UF adhesive formulation containing 3% of the nanoparticles. Moreover, investigations performed by Gul et al. [80] also showed that the introduction of CNT in the amount up to 5% led to the reduction of both TS and WA of the resultant MDF even by 45% and 29%, respectively. Therefore, it can be concluded that CNT have a great potential in improving water-related properties of MDF and moreover, more studies are needed to develop the effective protocol for improving the physical properties of particleboards by introduction of carbon-based nanoparticles.

Furthermore, as shown by the emerging research, the addition of nanocellulose also can contribute to the improvement in water-related properties of wood-based materials. Veigel et al. [218] found that the incorporation of UF/CNF adhesive led to decrease in TS of particleboard containing 1% of nano-additive. However, increasing the amount of modifier introduced to UF resin to 3% resulted in a considerable deterioration of dimensional stability of boards. Moreover, the addition of CNF in the amount of 1% to MUF adhesive was also effective in terms of improving physical properties of OSB such as TS and WA. However, on the other hand, research performed by Kawalerczyk et al. [106] showed that in the case of both particleboards and rape strawboards the enhancement of UF adhesive with NCC did not cause any statistically significant changes in their TS and WA. Therefore, the effect of nanocellulose-modified adhesive application on the water-related properties of particleboard is still inconclusive. The reasons for the observed differences may result from many factors related to the characteristic of the adhesive, features of introduced nanoparticles, or the technological aspects of boards production. However, the high potential of nanocellulose in terms of improving the water-related properties of wood-based materials can be indicated by the results obtained in the case of MDF. Khanjanzadeh et al. [109] introduced 0.5, 1, 1.5, and 2% of NCC to UF adhesive and investigated the properties of manufactured MDF panels. It was found that the highest load of cellulosic nanoparticles delivered the best results of TS and WA which decreased by 13.5% and 6.5%, respectively. According to authors, the reasons behind the observed changes could be, for example, high crystallinity of NCC, slowed diffusion process by the presence of NCC, and the improvement in mechanical strength of the bonds.

In summary, nanoparticles have a great potential in reducing the water uptake of wood-based materials. In most cases, this effect can be explained by the increase in the bonding strength in the structure of the boards, which limits the access of water to penetrate into them. This is particularly important since the improvement of water-related properties can extend the use of wood-based materials to structural and outdoor applications. The results are summarized in Table 2.

## 4 Improvement of the Properties of Wood-Based Materials Bonded with Bio-Based Adhesives

Considering the increasing environmental awareness, the reduction of fossil energy sources, and the requirements of current industrial production, there is a strong demand to develop green, sustainable, and low-cost bio-based adhesives [11, 242]. Figure 14 presents the most important of them according to Borges et al. [30].

**Table 2** The effect of applied nanoparticles on water-related properties of wood-based materials bonded with formaldehyde-based adhesives

| Type of nanomaterial | Way of introduction | Effect on water-related properties |
|---|---|---|
| MMT | UF adhesive modification | Improvement in TS and WA of particleboard |
| | UF adhesive modification | Improvement in TS and WA of MDF |
| | UF adhesive modification | No effect on TS and WA of plywood |
| | MUF adhesive modification | No effect on TS and WA of plywood |
| NW | UF adhesive modification | Improvement in TS and WA of particleboard |
| | Surface treatment | Improvement in TS and WA of particleboard |
| | UF adhesive modification | Improvement in TS and WA of MDF |
| | Surface treatment | Improvement in TS and WA of MDF |
| | UF adhesive modification | Improvement in TS of plywood |
| Nano-$SiO_2$ | UF adhesive modification | Improvement in TS and WA of particleboard |
| | UF adhesive modification | Improvement in TS and WA of OSB |
| | UF adhesive modification | Improvement in TS and WA of strawboard |
| Nano-$TiO_2$ | UF adhesive modification | Improvement in TS and WA of particleboard |
| | UF adhesive modification | Improvement in TS and WA of MDF |
| | MF adhesive modification | Improvement in TS and WA of MDF |
| Nano-ZnO | UF adhesive modification | Improvement in TS and WA of particleboard |
| Nan-silver | UF adhesive modification | Improvement in TS and WA of particleboard |
| CNT | Particles modification | No effect on TS and WA of particleboard |
| | UF adhesive modification | Improvement in TS and WA of MDF |
| NC | UF adhesive modification | Improvement in TS and WA of particleboard |
| | MUF adhesive modification | Improvement in TS and WA of OSB |

(continued)

**Table 2** (continued)

| Type of nanomaterial | Way of introduction | Effect on water-related properties |
|---|---|---|
| | UF adhesive modification | No effect on TS and WA of strawboard |
| | UF adhesive modification | Improvement in TS and WA of MDF |

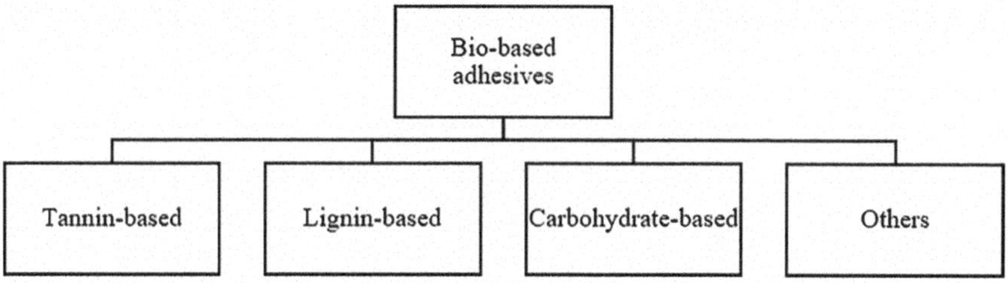


**Fig. 14** The most important groups of bio-based adhesives according to Borges et al. [30]

The great progress which can be observed in recent years in the field of bio-adhesives caused that a lot of attention in research is focused on their use in the production of wood-based materials as a replacement for the synthetic adhesives. However, there are several serious limitations which currently hinder their industrialization and the commercial use that should be considered. These limitations include: (i) poor availability of tannins; (ii) weak adhesion of starches; (iii) low water resistance of lignin and protein; and (iv) low strength properties limiting their potential use for non-structural applications [117]. For this reason, bio-based adhesives based on natural polymers such as starch, lignin, or tannins often require various modifications to achieve properties comparable to synthetic adhesives [202]. However, despite the promising solutions offered by nanotechnology, the amount of research focused on the incorporation of nanoparticles into bio-adhesives is very limited, especially when compared to formaldehyde-based adhesives.

## *4.1 Starch-Based Adhesives*

One of the natural polymers widely occurring in nature is starch. Due to its availability, renewable nature, high biodegradability, and low cost, it arouses great interest as a potential raw material in the production of adhesives, in most cases intended specifically for gluing paper and wooden surfaces. It is a polysaccharide composed of amylose and amylopectin in varying proportions depending on the botanical origin. These proportions can affect the characteristics of the starch itself and therefore, also the glue produced from it [141, 159, 223]. The disadvantage of starch as a raw material is its hydrophilicity which significantly hinders the use for the synthesis of water-resistant adhesives. The reason for this is the presence of numerous hydroxyl groups which easily form hydrogen bonds with water molecules [84]. Moreover, despite the advanced research, starch-based wood adhesives still do not exhibit the high enough bonding strength which is required from wood joints [222, 249]. One of the main challenges in the field of starch-based adhesives is also their poor stability during storage [159]. There were a few attempts which can be found in literature, to modify these types of adhesives in order to eliminate above-mentioned disadvantages, i.e., to increase their water resistance and to improve bonding quality. As might be expected, due to many exceptional properties, some of the nanomaterials were investigated as a modifiers used to develop high-performance ecological composite materials [223]. For example, the introduction of MMT, nano-$SiO_2$, and nano-$TiO_2$ as crosslinking or grafting agents to the adhesive formulation based on starch allowed to strengthen its molecular structure and to improve the interaction between the glue and wood, which consequently improves the bonding quality and water resistance of the resultant glue lines. Based on SEM photographs showed by Wang et al. [222] it was found that the nano-$SiO_2$-reinforced adhesive had much smoother and more compact structure in comparison with unmodified one and moreover, no clear microphase separation or agglomeration of applied nanoparticles was observed. What is interesting and promising, the incorporation of nano-sized particles to the starch-based adhesive contributed to the improvement in its thermal stability, rheological behavior, water resistance, and bonding strength [37, 125, 222, 249]. Additionally, the addition of MMT extends their shelf life up to 82 days. Moreover, the modification improved also the barrier properties, shear strength, pealing strength and led to lowering the water and oxygen vapor transmission rate of starch-based bonds [220].

The favorable effect of nanoparticles on the bonding properties of starch-based adhesives was addressed not only in the studies concerning wood bonding in general. Some efforts to incorporate them into the production of wood-based materials can also be found in research. Outcomes of the investigations performed by Abd Karim et al. [1] showed that particleboard bonded with starch/PVA adhesive reinforced with nano-$SiO_2$ was characterized by the improved both physical (TS, WA) and mechanical (IB) properties when compared to panels glued with a native starch. In contrast, research conducted by Chotikhun et al. [42] showed that the introduction of nano-$SiO_2$ in the amount exceeding 1% adversely affected mechanical properties of the boards. However, the implementation of formulation of nano-$SiO_2$-modified starch including 2% of urea-formaldehyde resin showed the improvement in bending strength and internal bond of the boards. In the case of starch-bonded plywood, Lubis et al. [128] achieved the improvement in bonding quality by the modification of adhesive with pristine-bentonite (P-BNT) and transition metal ion modified-bentonite (TMI-P-BNT) nanoclays. The research assumed the use of various contents of modifier, i.e., 1, 3 and 5%. Increasing amount of P-BNT and TMI-P-BNT caused the increase in adhesive viscosity and acceleration of its gel time. Moreover, the addition of 5% of nanoclay into the oxidized starch showed the lowest weight loss with higher degradation temperature which means that the presence of nanoparticles improved thermal stability by intercalation within the starch polymer. According to authors, the reason for the observed improvement could be the great barrier properties of the layered silicates. Furthermore, this method of starch modification allowed to increase the tensile shear strength of manufactured plywood by up to 15%. Therefore, it was concluded that using the nanoclay-reinforced starch adhesive made it possible to produce eco-friendly plywood with a bonding quality comparable or even higher than in the case of UF adhesive-bonded panels.

Moreover, the application of carboxylated NCC suspension (C-NCC) to starch-based adhesive seems to be a novel concept. Interestingly, the presence of carboxyl groups on the cellulose surface leads to increase in compatibility and dispersibility and therefore, it has a strong potential in limiting the tendency of nanoparticles to form the agglomerates [39]. The results presented by Benhamou et al. [28] showed a decrease in the starch adhesive viscosity caused by the presence of C-NCC. This effect was beneficial and it led to more even distribution of the glue on the wood particles. Therefore, it was found that the particleboard bonded with corn starch-mimosa tannin bio-adhesives containing up to 1% of C-CNC exhibited remarkable improvement in IB by 22%, in MOE by 24%, and in MOR by 58% when compared to neat adhesives without nanocellulose. Water-related properties of the boards were also enhanced, TS and WA were decreased by approx. 26% and 10%, respectively. The observed reinforcement probably resulted from the formation of hydrogen bonds between the hydroxyl groups of starch and carboxylate groups of C-NCC. However, as discovered by Amini et al. [10] nanocellulose itself does not have to be used only as an additive to bio-based or synthetic adhesives. Studies have shown that it can be also applied as the binding agent in the production of MDF and particleboard [207]. It was found that after gluing of wood particles, CNF was distributed on their surfaces in the form of separate fibrils or platelet-shaped agglomerates (Fig. 15). Moreover, part of CNF was able to penetrate into a wood structure and create a three-dimensional network during hot pressing which resulted in great strength and rigidity of the boards [10, 72, 207]. The outcomes presented by Amini et al. [10] indicate that particleboard glued with 3% of CNF suspension met the industrial requirements in terms of mechanical properties.

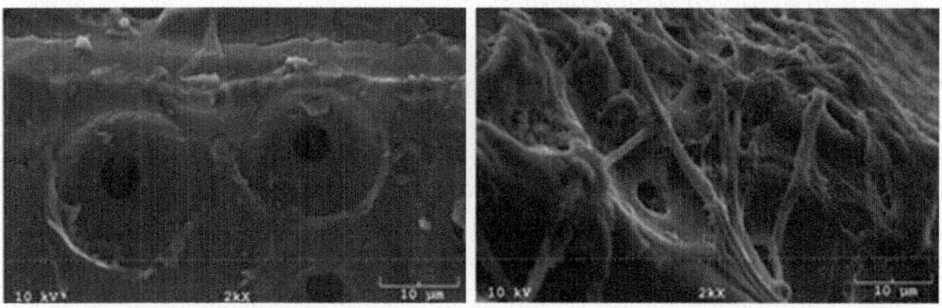

(a) (b)

**Fig. 15** SEM photographs of: **a** surface of wood particles before gluing; **b** surface of wood particles covered with CNF suspension [10] (Open access)

## 4.2 Tannin-Based Adhesives

Tannins are another interesting example of natural polymers showing a great potential, both in chemical and economic terms, for the production of bio-based glues for both wood and wood-based materials bonding. For example, they can be successfully used as a substitute for phenol in the adhesives [9, 13, 50, 97]. Tannins are obtained mainly from the bark of various tree species such as, e.g., pine, oak, willow, chestnut, or birch [12, 44]. They are natural polyphenols which can be sub-chemically divided into hydrolysable tannins and condensed tannins [12, 114]. Moreover, tannins show higher reactivity with formaldehyde than phenol, which is why the adhesives synthesized with their participation show a high curing degree and almost zero formaldehyde emission [95]. However, tannin-based adhesives, similarly to most of the bio-polymer-based adhesives, have limited ways of application due to their low strength especially in humid conditions [255]. These adverse characteristics can be improved, for example, by the modification with 5% of CNF. Studies have shown that the performance parameters of adhesive based on tannins obtained from Angico tree such as strength of wood-adhesive bonds and increased resistance to moisture were improved. This observation was confirmed by the research of Cui et al. [44] who modified tannin-based adhesive with different concentrations of CNF ranging from 1 to 3%. Based on the results it was found that introduction of 2% of CNF significantly improved mechanical properties of particleboards such as MOR, MOE, and IB. Additionally, it was shown that CNF, regardless of the amount, had no effect on the polycondensation reaction and gel time of the adhesive. The applied modification, however, led to a significant increase in the viscosity which according to authors can be attributed to the formation of strong hydrogen bond network involving flavanol units and CNF. The introduction of CNF into the tannin glue did not significantly affect the swelling properties of the particleboard after the long-term soaking in water.

## 4.3 Lignin-Based Adhesives

Another biopolymer demonstrating a great potential for the use in wood industry, including wood-based materials industry, is lignin. It can be applied for both the production of lignin-based adhesives and the production of synthetic resins as an additive or partial replacement [62]. However, its features such as high molecular weight, structural heterogenity, and low activity together with the difficulties in isolating lignin from wood can be listed as exemplary drawbacks for its implementation into wood bonding [78, 94]. According to Ferdosian et al. [66], lignin-based bio-adhesives can be divided into two groups: (i) lignin-phenol-formaldehyde adhesive where lignin is introduced to partially replace phenol and (ii) formaldehyde-free lignin-based adhesives in which case it reacts with chemical substances other different than formaldehyde to reduce a toxicity. A great example of formaldehyde-free lignin-based adhesive is the lignin-urea-glyoxal (LUG) adhesive developed by Younesi-Kordkheili [236]. Moreover, LUG adhesive was also modified with nanoclay and the research have shown shortening of gel time due to the presence of nano-modifier. Moreover, manufactured plywood was characterized by the higher values of shear strength, MOE, and lower WA in comparison with panels bonded with neat LUG adhesive. Simiarly promising results were obtained in the case of nanoclay-modified urea-glyoxalated lignin-formaldehyde (GLUF) adhesive used for particleboard manufacturing. The outcomes showed that the application of nanoclay contributed to the improvement in mechanical properties (MOR, MOE, IB) and a decrease in water absoption of the resultant panels [237].

Partial substitution of phenol with lignin to prepare lignin-phenol-formaldehyde (LPF) adhesive has been extensively investigated [64, 252]. What is important, the steric hindrance in the molecule of lignin makes its reaction sites covered comparing to phenol which is a restricting factor for the lignin application [38]. Fortunately, the improvement of the properties of LPF adhesives can be achieved by applying the modification with nanomaterials. For example, as reported by Liu et al. [126], the introduction of NCC to LPF adhesive improved the shear strength of plywood. Moreover, it was determined that optimal content of introduced nano-sized cellulose should be in the range from 0.25 to 0.5%. Another interesting concept which can be found in literature is the substitution of phenol with nano-lignin to obtain the nano-lignin-phenol-formaldehyde adhesive (NLPF). Chen et al. [38] developed a method for the production of NLPF using lignin nanoparticles characterized by the particles size of 300 nm. The obtained NLPF adhesive-bonded plywood was characterized by the bonding quality at the level meeting the requirements of the Chinese standard and lowered formaldehyde emission. Moreover, curing behavior was significantly improved by the substitution as shown by the reduced curing temperature from 186.2 to 145.4 °C.

Recently, a new concept in the field of nano-modified bio-based adhesives was presented by Younesi-Kordkheili and Pizzi [238], who used nano-lignin in the amount ranging between 10 and 40% instead of lignin bulk counterpart for the synthesis of phenol-lignin-glyoxal (PLG) adhesive. The investigations have shown that the use of LNPs had a favorable effect on the physicochemical characteristics of the resin such as, for example, gel time, viscosity, and curing temperature which was reduced from 170 to 159 °C. Furthermore, mechanical properties (MOR, MOE, IB) and water-related properties (WA) of particleboard improved with the increase in the share of nano-lignin in the PLG resin which is a very promising and perspective result.

In recent years, there has been an increased demand for bio-based adhesives all over the world which causes this market to rapidly grow. Table 3 presents the latest achievements in the field of improving properties of bio-based adhesive-bonded wood-based materials by implementation of various nanoparticles. However, the relatively small number of papers published on this subject and the promising results obtained so far indicate that it is necessary to carry out further research in this area.

**Table 3** The effect of applied nanoparticles on the properties of wood-based materials bonded with bio-based adhesives

| Type of nanomaterial | Type of adhesive | Type of wood-based material | Effect on the properties of wood-based materials |
|---|---|---|---|
| Nanoclay | Lignin-urea-glyoxal | Plywood | Improved MOR and bonding quality<br>Decreased water uptake |
| | Urea-glyoxalated lignin-formaldehyde | Particleboard | Improved MOR, MOE, and IB<br>Decreased water uptake |
| Nano-$SiO_2$ | Starch/PVA | Particleboard | Improved IB<br>Decreased water uptake |
| Nanocellulose | Lignin-phenol-formaldehyde | Plywood | Improved bonding quality |
| | Starch-tannin | Particleboard | Improved MOR, MOE, and IB |
| | Tannin | Particleboard | Improved MOR, MOE, and IB |
| Nano-lignin | Lignin-phenol-formaldehyde | Plywood | Improved bonding quality |
| | Phenol-lignin-glyoxal | Particleboard | Improved MOR, MOE, and IB |

## 5 Summary and Further Perspectives

Nanoscience is definitely one of the most exciting frontiers which plays a vital role in a general progress. It poses both scientific and technological challenge, which requires an interdisciplinary approach and different perspectives to succeed. It is a very challenging field because understanding and control of phenomena occurring at the nanoscale brings us closer to the atomic or molecular scale. Such large attention of the scientific world directed toward nanoparticles results from their extraordinary properties. Moreover, a constant progress opens up new possibilities and enables scientists to understand and consequently also use the unique chemical, mechanical, and physical properties of nanomaterials, for example, to improve the properties of commonly used wood-based materials. As shown on the basis of the literature review, nanoparticles have great potential to improve both strength and water-related properties of wood-based materials bonded with formaldehyde-based adhesives. In addition, despite that there are relatively few studies on the nanomodification of bio-adhesives, it can also be concluded that the outcomes are very promising. In most cases, the improvement in properties resulted from, i.e., tremendous surface area, high aspect ratio, high reactivity, and thermal conductivity of nanomaterials. Moreover, it can be expected that the discoveries in this field emerging every day will create new opportunities for the implementations into the wood-based materials industry.

In the case of further developments of nano-modification implemented in order to improve the characteristics of wood-based materials, the following directions can be distinguished: (i) research on the toxicity of nanomaterials and their impact on human health; (ii) research on bio-based adhesives reinforcement with nanoparticles; and (iii) research on functionalization of nanoparticles. It is known that people can be exposed to nanomaterials through dermal contact, inhalation, or ingestion, however, the most adverse aspect of nanotechnology is that the potential hazards are still not fully understood especially in the case of long-term exposure [162]. The task of defining potential hazards is particularly difficult due to the fact that the properties of nanoparticles and their impact on humans depend on many factors such as crystal structure, surface area, agglomeration level, shape, and state. However, taking into account the emerging reports indicating their cytotoxicity, carcinogenicity, and the ability to accumulate in organs, it can be concluded that research in this area is very important both from the perspective of scientists dealing with nanotechnology and people using implemented nano-modified materials [31, 70, 75, 189]. Therefore, the development of protocols describing precautionary measures is a very important direction of research in the field of nanotechnology. Another direction concerning nanotechnology applied in the wood industry should be the modification of bio-adhesives. As shown on the basis of the literature review, the number of scientific works on this subject is very scarce. Taking into account the progress observed in this field, statistics on the market of bio-adhesives, and technological challenges related to them, it can be assumed that this direction will be widely studied in the years to come. Furthermore, functionalization of nanoparticles poses new challenges and opens up new perspectives for their application, also in wood-based materials industry. It enhances the initial properties of nanoparticles through the surface modification and therefore, enables them to play a crucial role in the global technological progress.

## References

1. Abd Karim N, Lamaming J, Yusof M et al (2020) Properties of native and blended oil palm starch with nano-silicon dioxide as binder for particleboard. J Build Eng 29:101151

2. Abdo J, AL-Sharji H, Hassan E (2016) Effects of nano-sepiolite on rheological properties and filtration loss of water-based drilling fluids. Surf Interface Anal 48:522–526
3. Abitbol T, Rivkin A, Cao Y et al (2016) Nanocellulose, a tiny fiber with huge applications. Curr Opin Biotechnol 39:76–88
4. Ahari H, Anvar AA, Ataee M, Naeimabadi M (2021) Employing nanosilver, nanocopper, and nanoclays in food packaging production: a systematic review. Coatings 11:509
5. Ahmed L, Zhang B, Shen R et al (2018) Fire reaction properties of polystyrene-based nanocomposites using nanosilica and nanoclay as additives in cone calorimeter test. J Therm Anal Calorim 132:1853–1865
6. Akpan EI, Wetzel B, Friedrich K (2021) Eco-friendly and sustainable processing of wood-based materials. Green Chem 23:2198–2232
7. Alias AH, Norizan MN, Sabaruddin FA et al (2021) Hybridization of MMT/lignocellulosic fiber reinforced polymer nanocomposites for structural applications: a review. Coatings 11:1355
8. Alvarado JF, Rozo DF, Chaparro LM et al (2022) Synthesis and characterization of reproducible linseed oil-loaded silica nanoparticles with potential use as oxygen scavengers in active packaging. Nanomaterials 12:3257
9. Amari M, Khimeche K, Hima A et al (2021) Synthesis of green adhesive with tannin extracted from eucalyptus bark for potential use in wood composites. J Renew Mater 9:463–475
10. Amini E, Tajvidi M, Gardner DJ, Bousfield DW (2017) Utilization of cellulose nanofibrils as a binder for particleboard manufacture. BioResources 12:4093–4110
11. Antov P, Lee SH, Lubis MAR, Yadav SM (2022) Potential of nanomaterials in bio-based wood adhesives: an overview. In: Emerging nanomaterials: opportunities and challenges in forestry sectors. Springer, pp 25–63
12. Antov P, Savov V, Neykov N (2020) Sustainable bio-based adhesives for eco-friendly wood composites: a review. Wood Res 65:51–62
13. Arias A, González-García S, Feijoo G, Moreira MT (2022) Tannin-based bio-adhesives for the wood panel industry as sustainable alternatives to petrochemical resins. J Ind Ecol 26:627–642
14. Artioli G, Angelini I, Polla A (2008) Crystals and phase transitions in protohistoric glass materials. Ph Transit 81:233–252
15. Artner MA, de Cademartori PHG, Avelino F et al (2021) A novel design for nanocellulose reinforced urea–formaldehyde resin: a breakthrough in amino resin synthesis and biocomposite manufacturing. Cellulose 28:3435–3450
16. Aryapratama R, Pauliuk S (2019) Estimating in-use wood-based materials carbon stocks in Indonesia: towards a contribution to the national climate mitigation effort. Resour Conserv Recycl 149:301–311
17. Ashori A, Nourbakhsh A (2009) Effects of nanoclay as a reinforcement filler on the physical and mechanical properties of wood-based composite. J Compos Mater 43:1869–1875
18. Aubert M, Brumm A, Ramli M et al (2014) Pleistocene cave art from Sulawesi, Indonesia. Nature 514:223–227
19. Ayrilmis N, Kwon J-H, Lee S-H et al (2016) Microfibrillated-cellulose-modified urea-formaldehyde adhesives with different F/U molar ratios for wood-based composites. J Adhes Sci Technol 30:2032–2043
20. Azzaoui K, Barboucha B, Hammouti B, Touzani R (2022) Nanotechnology: history and various applications, a mini review. EHEI-J Sci Technol 2:22–33
21. Babicka M, Woźniak M, Bartkowiak M et al (2022) Miscanthus and Sorghum as sustainable biomass sources for nanocellulose production. Ind Crops Prod 186:115177
22. Babicka M, Woźniak M, Dwiecki K et al (2020) Preparation of nanocellulose using ionic liquids: 1-propyl-3-methylimidazolium chloride and 1-ethyl-3-methylimidazolium chloride. Molecules 25:1544
23. Baldwin RF (1995) Plywood and veneer-based products: manufacturing practices. Miller Freeman Books, pp 31–42
24. Barhoum A, García-Betancourt ML, Jeevanandam J et al (2022) Review on natural, incidental, bioinspired, and engineered nanomaterials: history, definitions, classifications, synthesis, properties, market, toxicities, risks, and regulations. Nanomaterials 12:177

25. Bayda S, Adeel M, Tuccinardi T et al (2019) The history of nanoscience and nanotechnology: from chemical–physical applications to nanomedicine. Molecules 25:112
26. Beisl S, Miltner A, Friedl A (2017) Lignin from micro-to nanosize: production methods. Int J Mol Sci 18:1244
27. Bekhta P, Müller M, Hunko I (2020) Properties of thermoplastic-bonded plywood: effects of the wood species and types of the thermoplastic films. Polymers 12:2582
28. Benhamou AA, Boussetta A, Kassab Z et al (2022) Elaboration of carboxylated cellulose nanocrystals filled starch-based adhesives for the manufacturing of eco-friendly particleboards. Constr Build Mater 348:128683
29. Bilotti E, Fischer HR, Peijs T (2008) Polymer nanocomposites based on needle-like sepiolite clays: effect of functionalized polymers on the dispersion of nanofiller, crystallinity, and mechanical properties. J Appl Polym Sci 107:1116–1123
30. Borges CS, Jalali S, Tsokanas P et al (2022) Sustainable development approaches through wooden adhesive joints design. Polymers 15:89
31. Brandelli A (2018) Toxicity and safety evaluation of nanoclays. In: Nanomaterials: ecotoxicity, safety, and public perception. Springer, pp 57–76
32. Brun N, Mazerolles L, Pernot M (1991) Microstructure of opaque red glass containing copper. J Mater Sci Lett 10:1418–1420
33. Bueno AF, Bañón MN, De Morentín LM, García JM (2014) Treatment of natural wood veneers with nano-oxides to improve their fire behaviour. In: IOP conference series: materials science and engineering. IOP Publishing, p 012021
34. Calvo-Flores FG, Dobado JA (2010) Lignin as renewable raw material. Chem Sus Chem 3:1227–1235
35. Can MF, Çınar M, Benli B et al (2010) Determining the fiber size of nano structured sepiolite using Atomic Force Microscopy (AFM). Appl Clay Sci 47:217–222
36. Castillo ADR, Pellegrini V, Ansaldo A et al (2018) High-yield production of 2D crystals by wet-jet milling. Mater Horiz 5:890–904
37. Chen L, Xiong Z, Xiong H et al (2018) Effects of nano-$TiO_2$ on bonding performance, structure stability and film-forming properties of starch-g-VAc based wood adhesive. Carbohydr Polym 200:477–486
38. Chen Y, Gong X, Yang G et al (2019) Preparation and characterization of a nanolignin phenol formaldehyde resin by replacing phenol partially with lignin nanoparticles. RSC Adv 9:29255–29262
39. Cheng M, Qin Z, Hu J et al (2020) Facile and rapid one–step extraction of carboxylated cellulose nanocrystals by $H_2SO_4$/$HNO_3$ mixed acid hydrolysis. Carbohydr Polym 231:115701
40. Chivrac F, Pollet E, Schmutz M, Avérous L (2010) Starch nano-biocomposites based on needle-like sepiolite clays. Carbohydr Polym 80:145–153
41. Choi W, Lahiri I, Seelaboyina R, Kang YS (2010) Synthesis of graphene and its applications: a review. Crit Rev Solid State Mater Sci 35:52–71
42. Chotikhun A, Hiziroglu S, Buser M et al (2018) Characterization of nano particle added composite panels manufactured from Eastern redcedar. J Compos Mater 52:1605–1615
43. Choudhary N, Hwang S, Choi W (2014) Carbon nanomaterials: a review. In: Handbook of nanomaterials properties. Springer, pp 709–769
44. Cui J, Lu X, Zhou X et al (2015) Enhancement of mechanical strength of particleboard using environmentally friendly pine (*Pinus pinaster* L.) tannin adhesives with cellulose nanofibers. Ann For Sci 72:27–32
45. da Silva APS, Ferreira BS, Favarim HR et al (2019) Physical properties of medium density fiberboard produced with the addition of ZnO nanoparticles. BioResources 14:1618–1625
46. Dai ZR, Bradley JP, Joswiak DJ et al (2002) Possible in situ formation of meteoritic nanodiamonds in the early Solar System. Nature 418:157–159
47. Da'na E (2022) Nano-silica modified with diamine for capturing azo dye from aqueous solutions. Molecules 27:3366
48. Del Buono D, Luzi F, Puglia D (2021) Lignin nanoparticles: a promising tool to improve maize physiological, biochemical, and chemical traits. Nanomaterials 11:846

49. Dhali K, Ghasemlou M, Daver F et al (2021) A review of nanocellulose as a new material towards environmental sustainability. Sci Total Environ 775:145871
50. Dhawale PV, Vineeth SK, Gadhave RV et al (2022) Tannin as a renewable raw material for adhesive applications: a review. Mater Adv 3:3365–3388
51. Di Giuseppe D, Scognamiglio V, Malferrari D et al (2021) Characterization of fibrous wollastonite NYAD G in view of its use as negative standard for in vitro toxicity tests. Minerals 11:1378
52. Dineshkumar S (2014) Use of multi-walled carbon nanotubes (MWCNTs) in medium density fiberboard to increase the fire retardancy. PhD Thesis. Universiti Malaysia Pahang
53. Dong Y, Yan Y, Zhang S et al (2015) Flammability and physical–mechanical properties assessment of wood treated with furfuryl alcohol and nano-$SiO_2$. Eur J Wood Wood Prod 73:457–464
54. Doosthoseini K, Zarea-Hosseinabadi H (2010) Using Na+ MMT nanoclay as a secondary filler in plywood manufacturing. J Indian Acad Wood Sci 7:58–64
55. Dorieh A, Ayrilmis N, Pour MF et al (2022) Phenol formaldehyde resin modified by cellulose and lignin nanomaterials: Review and recent progress. Int J Biol Macromol 222:1888–1907
56. Dukarska D (2013) The effect of an addition of nano-$SiO_2$ to urea resin on the properties of boards manufactured from rape straw. Ann Wars Univ Life Sci-SGGW. For Wood Technol 82:242–245
57. Dukarska D, Bartkowiak M (2016) The effect of organofunctional nanosilica on the cross-linking process and thermal resistance of UF resin. J Polym Res 23:1–8
58. Dukarska D, Czarnecki R (2016) Fumed silica as a filler for MUPF resin in the process of manufacturing water-resistant plywood. Eur J Wood Wood Prod 74:5–14
59. Dukarska D, Derkowski A (2014) Rape straw-wood particleboards resinated with UF resin and supplemented with nano-$SiO_2$. Ann Wars Univ Life Sci-SGGW. For Wood Technol 85:49–52
60. Dukarska D, Kawalerczyk J, Włodarczyk J (2022) Properties of the water-resistant plywood glued with pf resin with the addition of tannin filler and nanoclay. Wood Res 67(2):241–253
61. Dukarska D, Łęcka J (2010) Synthetic silica as a filler of phenolic resin in the manufacture of exterior plywood. Ann Wars Univ Life Sci-SGGW. For Wood Technol 71:152–156
62. Dunky M (2021) Wood adhesives based on natural resources: a critical review: Part III. Tannin-and lignin-based adhesives. Prog Adhes Adhes 6:383–529
63. Eglash R (2018) Nanotechnology and traditional knowledge systems. In: Nanotechnology and global sustainability. CRC Press, pp 74–97
64. Faris AH, Nasir M, Ibrahim M, Iqbal A (2017) Preparation of environmental friendly phenol-formaldehyde wood adhesive modified with kenaf lignin. Beni-Suef University Int J Basic Appl Sci 6:409–418
65. Farré M, Barceló D (2012) Introduction to the analysis and risk of nanomaterials in environmental and food samples. In: Comprehensive analytical chemistry. Elsevier, pp 1–32
66. Ferdosian F, Pan Z, Gao G, Zhao B (2017) Bio-based adhesives and evaluation for wood composites application. Polymers 9:70
67. Fitriani F, Aprilia S, Arahman N et al (2021) Isolation and characterization of nanocrystalline cellulose isolated from pineapple crown leaf fiber agricultural wastes using acid hydrolysis. Polymers 13:4188
68. Foldvari M, Bagonluri M (2008) Carbon nanotubes as functional excipients for nanomedicines: I. Pharmaceutical properties. Nanomed: Nanotechnol Biol Med 4:173–182
69. Fozia F, Ahmad N, Buoharee ZA et al (2022) Characterization and evaluation of antimicrobial potential of trigonella incise (Linn) mediated biosynthesized silver nanoparticles. Molecules 27:4618
70. Francis AP, Devasena T (2018) Toxicity of carbon nanotubes: a review. Toxicol Ind Health 34:200–210
71. Fridley KJ (2002) Wood and wood-based materials: current status and future of a structural material. J Mater Civ Eng 14:91–96
72. Gardner D, Tajvidi M (2016) Hydrogen bonding in wood-based materials: an update. Wood Fiber Sci 48:234–244

73. Gayarre FL, Boadella IL, Gonzalez JS et al (2021) Waste for aggregates in ultrahigh performance concrete (UHPC). In: Waste and byproducts in cement-based materials. Elsevier, pp 29–51
74. Ghofrani M, Haghdan S, NicKhah V, Ahmadi K (2015) Improvement of physical and mechanical properties of particleboard made of apple tree pruning and sunflower stalk using titanium oxide nanoparticles. Eur J Wood Wood Prod 73:661–666
75. Girigoswami K (2018) Toxicity of metal oxide nanoparticles. In: Cellular and molecular toxicology of nanoparticles. Springer, pp 99–122
76. Gogotsi Y, Presser V (2006) Carbon nanomaterials. CRC Press
77. Gonçalves C, Paiva NT, Ferra JM et al (2018) Utilization and characterization of amino resins for the production of wood-based panels with emphasis on particleboards (PB) and medium density fibreboards (MDF). A review. Holzforschung 72:653–671
78. Gong X, Meng Y, Lu J et al (2022) A review on lignin-based phenolic resin adhesive. Macromol Chem Phys 223:2100434
79. Gorgolis G, Galiotis C (2017) Graphene aerogels: a review. 2D Mater 4:032001
80. Gul W, Alrobei H, Shah SRA et al (2020) Effect of embedment of MWCNTs for enhancement of physical and mechanical performance of medium density fiberboard. Nanomaterials 11:29
81. Guo F, Aryana S, Han Y, Jiao Y (2018) A review of the synthesis and applications of polymer–nanoclay composites. Appl Sci 8:1696
82. Halligan AF (1970) A review of thickness swelling in particleboard. Wood Sci Technol 4:301–312
83. Hassegawa M, Van Brusselen J, Cramm M, Verkerk PJ (2022) Wood-based products in the circular bioeconomy: status and opportunities towards environmental sustainability. Land 11:2131
84. Heinrich LA (2019) Future opportunities for bio-based adhesives–advantages beyond renewability. Green Chem 21:1866–1888
85. Holban AM, Grumezescu AM, Andronescu E (2016) Inorganic nanoarchitectonics designed for drug delivery and anti-infective surfaces. In: Surface chemistry of nanobiomaterials. Elsevier, pp 301–327
86. Hosseini MJM, Faezipour M, Yousefi H (2012) Effect of nanoclay on the properties of medium density fiberboard (MDF). Iran J Wood Pap Sci Res 26:699–707
87. Hosseyni MJ, Rahimi S, Faezipour M, Faezipour MM (2014) Effect of nanoclay particles on the properties of particleboards. Int J Basic Appl Sci 4:280–287
88. Hu Y-C, Dai C-L, Hsu C-C (2014) Titanium dioxide nanoparticle humidity microsensors integrated with circuitry on-a-chip. Sensors 14:4177–4188
89. Hulla JE, Sahu SC, Hayes AW (2015) Nanotechnology: history and future. Hum Exp Toxicol 34:1318–1321
90. Hussin MH, Abd Latif NH, Hamidon TS et al (2022) Latest advancements in high-performance bio-based wood adhesives: a critical review. J Mater Res Technol 21:3909–3946
91. Hussin MH, Appaturi JN, Poh NE et al (2022) A recent advancement on preparation, characterization and application of nanolignin. Int J Biol Macromol 200:303–326
92. Iijima S (1991) Helical microtubules of graphitic carbon. Nature 354:56–58
93. Ijaz I, Gilani E, Nazir A, Bukhari A (2020) Detail review on chemical, physical and green synthesis, classification, characterizations and applications of nanoparticles. Green Chem Lett Rev 13:223–245
94. Islam MN, Rahman F, Das AK, Hiziroglu S (2022) An overview of different types and potential of bio-based adhesives used for wood products. Int J Adhes Adhes 112:102992
95. Jahanshaei S, Tabarsa T, Asghari J (2012) Eco-friendly tannin-phenol formaldehyde resin for producing wood composites. Pigment Resin Technol 41:296–301
96. Jiang F, Li T, Li Y et al (2018) Wood-based nanotechnologies toward sustainability. Adv Mater 30:1703453
97. Jorda J, Cesprini E, Barbu M-C et al (2022) Quebracho tannin bio-based adhesives for plywood. Polymers 14:2257

98. Jozefaciuk G, Skic K, Adamczuk A et al (2021) Structure and strength of artificial soils containing monomineral clay fractions. Materials 14:4688
99. Jung S-H, Oh S, Kim S-W, Moon J-H (2019) Effects of CNT dosages in cement composites on the mechanical properties and hydration reaction with low water-to-binder ratio. Appl Sci 9:4630
100. Kawalerczyk J, Dziurka D, Mirski R et al (2021) Possibility of use of NCC-reinforced melamine-urea-formaldehyde adhesive in plywood manufacturing. Drv Ind 72:279–289
101. Kawalerczyk J, Dziurka D, Mirski R et al (2020) The effect of nanocellulose addition to phenol-formaldehyde adhesive in water-resistant plywood manufacturing. BioResources 15:5388–5401
102. Kawalerczyk J, Dziurka D, Mirski R (2020) The possible reduction of phenol-formaldehyde resin spread rate by its nanocellulose-reinforcement in plywood manufacturing process. Ann Wars Univ Life Sci-SGGW. For Wood Technol 111:21–26
103. Kawalerczyk J, Dziurka D, Mirski R, Siuda J (2021) The reduction of adhesive application in plywood manufacturing by using nanocellulose-reinforced urea-formaldehyde resin. J Appl Polym Sci 138:49834
104. Kawalerczyk J, Dziurka D, Mirski R, Szentner K (2020) Properties of plywood produced with urea-formaldehyde adhesive modified with nanocellulose and microcellulose. Drv Ind 71:61–67
105. Kawalerczyk J, Dziurka D, Mirski R (2022) The effect of a phenol-formaldehyde adhesive reinforcement with nanocellulose on the pressing parameters of plywood. Wood Res 67:796–808
106. Kawalerczyk J, Walkiewicz J, Dziurka D et al (2022) APTES-modified nanocellulose as the formaldehyde scavenger for UF adhesive-bonded particleboard and strawboard. Polymers 14:5037
107. Kawalerczyk J, Walkiewicz J, Dziurka D, Mirski R (2022) Nanomaterials to improve fire properties in wood and wood-based composite panels. In: Emerging nanomaterials: opportunities and challenges in forestry sectors. Springer, pp 65–96
108. Khan ZI, Habib U, Mohamad ZB et al (2022) Mechanical and thermal properties of sepiolite strengthened thermoplastic polymer nanocomposites: a comprehensive review. Alex Eng J 61:975–990
109. Khanjanzadeh H, Behrooz R, Bahramifar N et al (2019) Application of surface chemical functionalized cellulose nanocrystals to improve the performance of UF adhesives used in wood based composites-MDF type. Carbohydr Polym 206:11–20
110. Klemm D, Heublein B, Fink H-P, Bohn A (2005) Cellulose: fascinating biopolymer and sustainable raw material. Angew Chem Int Ed 44:3358–3393
111. Kristak L, Antov P, Bekhta P et al (2022) Recent progress in ultra-low formaldehyde emitting adhesive systems and formaldehyde scavengers in wood-based panels: a review. Wood Mater Sci Eng, 1–20
112. Kuan C-F, Chiang C-L, Lin S-H et al (2018) Characterization and properties of graphene nanoplatelets/XNBR nanocomposites. Polym Polym Compos 26:59–68
113. Kučuk N, Primožič M, Knez Ž, Leitgeb M (2023) Sustainable biodegradable biopolymer-based nanoparticles for healthcare applications. Int J Mol Sci 24:3188
114. Kües U (2007) Wood production, wood technology, and biotechnological impacts. Universitätsverlag Göttingen, pp 15–30
115. Kumar A, Dixit CK (2017) Methods for characterization of nanoparticles. In: Advances in nanomedicine for the delivery of therapeutic nucleic acids. Elsevier, pp 43–58
116. Kumar A, Sharma KV, Gupta A et al (2017) Optimization of processing parameters of medium density fiberboard using response surface methodology for multiwalled carbon nanotubes as a nanofiller. Eur J Wood Wood Prod 75:203–213
117. Kumar C, Leggate W (2022) An overview of bio-adhesives for engineered wood products. Int J Adhes Adhes 118:103187
118. Kumar N, Kumbhat S (2016) Carbon-based nanomaterials. In: Essentials in nanoscience and nanotechnology. Wiley, pp 189–236

119. Laosman A, Norazman AN, Setepanus C et al (2020) Effect of nanoclay on the performance of particleboard made from mixed wood waste. In: Environmental development and sustainability beyond 2020. Faculty of Science and Natural Resources, pp 281–283
120. Le M-T, Huang S-C (2015) Thermal and mechanical behavior of hybrid polymer nanocomposite reinforced with graphene nanoplatelets. Materials 8:5526–5536
121. Lei H, Du G, Pizzi A, Celzard A (2008) Influence of nanoclay on urea-formaldehyde resins for wood adhesives and its model. J Appl Polym Sci 109:2442–2451
122. Lengowski EC, Bonfatti Júnior EA, Dallo R et al (2021) Nanocellulose-reinforced phenol-formaldehyde resin for plywood panel production. Maderas: Cienc Tecnol 23:1–10
123. Li X, Gao Q, Xia C et al (2019) Urea formaldehyde resin resultant plywood with rapid formaldehyde release modified by tunnel-structured sepiolite. Polymers 11:1286
124. Li X, Li J, Li J, Gao Q (2015) Effect of sepiolite filler in malamine-urea-formaldehyde resin on the properties of three-ply plywood. BioResources 10:6624–6634
125. Li Z, Wang J, Li C et al (2015) Effects of montmorillonite addition on the performance of starch-based wood adhesive. Carbohydr Polym 115:394–400
126. Liu Z, Zhang Y, Wang X, Rodrigue D (2015) Reinforcement of lignin-based phenol-formaldehyde adhesive with nano-crystalline cellulose (NCC): curing behavior and bonding property of plywood. Mater Sci Appl 6:567
127. Low LE, Teh KC, Siva SP et al (2021) Lignin nanoparticles: the next green nanoreinforcer with wide opportunity. Environ Nanotechnol Monit Manag 15:100398
128. Lubis MAR, Yadav SM, Park B-D (2021) Modification of oxidized starch polymer with nanoclay for enhanced adhesion and free formaldehyde emission of plywood. J Polym Environ 29:2993–3003
129. Lue JT (2007) Physical properties of nanomaterials. Encycl Nanosci Nanotechnol 10:1–46
130. Łukawski D, Grześkowiak W, Dukarska D et al (2019) The influence of surface modification of wood particles with carbon nanotubes on properties of particleboard glued with phenol-formaldehyde resin. Drewno: prace naukowe, doniesienia, komunikaty 62:93–105
131. Łukawski D, Hochmańska-Kaniewska P, Janiszewska D et al (2022) Enriching WPCs and NFPCs with carbon nanotubes and graphene. Polymers 14:745
132. Łukawski D, Lekawa-Raus A, Lisiecki F et al (2018) Towards the development of superhydrophobic carbon nanomaterial coatings on wood. Prog Org Coat 125:23–31
133. Luo T, Hao Y, Wang C et al (2022) Lignin nanoparticles and alginate gel beads: preparation, characterization and removal of methylene blue. Nanomaterials 12:176
134. Luyt AS, Dramićanin MD, Antić Ž, Djoković V (2009) Morphology, mechanical and thermal properties of composites of polypropylene and nanostructured wollastonite filler. Polym Test 28:348–356
135. Madhad HV, Mishra NS, Patel SB et al (2021) Graphene/graphene nanoplatelets reinforced polyamide nanocomposites: a review. High Perform Polym 33:981–997
136. Mahrdt E, Pinkl S, Schmidberger C et al (2016) Effect of addition of microfibrillated cellulose to urea-formaldehyde on selected adhesive characteristics and distribution in particle board. Cellulose 23:571–580
137. Maier C, Calafut T (1998) Fillers and reinforcements. In: Polypropylene. William Andrew Publishing Norwich, pp 49–56
138. Malanit P, Barbu MC, Frühwald A (2011) Physical and mechanical properties of oriented strand lumber made from an Asian bamboo (Dendrocalamus asper Backer). Eur J Wood Wood Prod 69:27–36
139. Mamatha BS, Jagadish RL, Aparna K, Anand N (2015) Durability and performance of particle board made using nanoclay reinforced resin. Indian J Adv Chem Sci 3:155–159
140. Masroor S (2022) Basics of metal oxides: properties and applications. In: Inorganic anticorrosive materials, Elsevier, pp 85–94
141. Maulana MI, Lubis MAR, Febrianto F, Hua LS, Iswanto AH, Antov P, Kristak L, Mardawati E, Sari RK, Zaini LH, Hidayat W, Lo Giudice V, Todaro L (2022) Environmentally friendly starch-based adhesives for bonding high-performance wood composites: a review. Forests 13:1614

142. Mazaheri M, Moghimi H, Taheri RA (2022) Urea impregnated multiwalled carbon nanotubes; a formaldehyde scavenger for urea formaldehyde adhesives and medium density fiberboards bonded with them. J Appl Polym Sci 139:51445
143. Medved S, Antonović A, Jambreković V (2011) Impact of resin content on swelling pressure of three-layer particleboard bonded with urea-formaldehyde adhesive. Drv Ind 62:37–42
144. Milewski JV, Katz HS (1987) Handbook of reinforcements for plastics. Van Nostrand Reinhold
145. Mondal S (2018) Review on nanocellulose polymer nanocomposites. Polym Plast Technol Eng 57:1377–1391
146. Moslemi A, Behzad T, Pizzi A (2020) Addition of cellulose nanofibers extracted from rice straw to urea formaldehyde resin; effect on the adhesive characteristics and medium density fiberboard properties. Int J Adhes Adhes 99:102582
147. Murty BS, Shankar P, Raj B et al (2013) Unique properties of nanomaterials. In: Textbook of nanoscience and nanotechnology. Springer, pp 29–65
148. Nageswara Rao T, Naidu TM, Kim MS et al (2019) Influence of zinc oxide nanoparticles and char forming agent polymer on flame retardancy of intumescent flame retardant coatings. Nanomaterials 10:42
149. Nasrollahzadeh M, Sajjadi M, Iravani S, Varma RS (2021) Starch, cellulose, pectin, gum, alginate, chitin and chitosan derived (nano) materials for sustainable water treatment: a review. Carbohydr Polym 251:116986
150. Natsuki J, Natsuki T, Hashimoto Y (2015) A review of silver nanoparticles: synthesis methods, properties and applications. Int J Mater Sci 4:325–332
151. Nazir MS, Mohamad Kassim MH, Mohapatra L et al (2016) Characteristic properties of nanoclays and characterization of nanoparticulates and nanocomposites. In: Nanoclay reinforced polymer composites: nanocomposites and bionanocomposites. Springer, pp 35–55
152. Negro F, Cremonini C, Zanuttini R et al (2011) A new wood-based lightweight composite for boatbuilding. Wood Res 56:257–266
153. Nelson KIM, Retsina T (2014) Innovative nanocellulose process breaks the cost barrier. TAPPI J 13:19–23
154. Ng H-M, Sin LT, Bee S-T et al (2017) Review of nanocellulose polymer composite characteristics and challenges. Polym Plast Technol Eng 56:687–731
155. Ng H-M, Sin LT, Tee T-T et al (2015) Extraction of cellulose nanocrystals from plant sources for application as reinforcing agent in polymers. Compos B Eng 75:176–200
156. Niemz P, Mannes D (2012) Non-destructive testing of wood and wood-based materials. J Cult Herit 13:S26–S34
157. Nine MJ, Cole MA, Tran DN, Losic D (2015) Graphene: a multipurpose material for protective coatings. J Mater Chem A 3:12580–12602
158. Noonan C, Tajvidi M, Tayeb AH et al (2019) Structure-property relationships in hybrid cellulose nanofibrils/Nafion-based ionic polymer-metal composites. Materials 12:1269
159. Norström E, Demircan D, Fogelström L et al (2018) Green binders for wood adhesives. Appl Adhes Bond Sci Technol 1:13–70
160. Novoselov KS, Geim AK, Morozov SV et al (2004) Electric field effect in atomically thin carbon films. Science 306:666–669
161. Nowaczyk-Organista M (2009) Protection of birch and walnut wood colour from the effect of exposure to light using ultra-fine zinc white. Drewno: prace naukowe, doniesienia, komunikaty 52:19–41
162. Olawoyin R (2018) Nanotechnology: the future of fire safety. Saf Sci 110:214–221
163. Olivato JB, Marini J, Pollet E et al (2015) Elaboration, morphology and properties of starch/polyester nano-biocomposites based on sepiolite clay. Carbohydr Polym 118:250–256
164. Omrani P, Taghiyari HR, Zolghadr M (2018) Effects of nano-clay on physical and mechanical properties of medium-density fiberboards made from wood and chicken-feather fibers and two types of resins. Drv Ind 69:329–337
165. Oskam G (2006) Metal oxide nanoparticles: synthesis, characterization and application. J Solgel Sci Techno 37:161–164

166. Owonubi SJ, Agwuncha SC, Malima NM et al (2021) Non-woody biomass as sources of nanocellulose particles: a review of extraction procedures. Front Energy Res 9:608825
167. Ozcifci A, Kara ME, Kaymakci A (2018) Impact of PF and MUF adhesives modified with $TiO_2$ and $SiO_2$ on the adhesion strength. Wood Res 63:75–84
168. Panin SV, Huang Q, Alexenko VO et al (2020) Design of wear-resistant UHMWPE-based composites loaded with wollastonite microfibers treated with various silane coupling agents. Appl Sci 10:4511
169. Papadopoulos AN (2019) Advances in wood composites. Polymers 12:1552
170. Parashar M, Shukla VK, Singh R (2020) Metal oxides nanoparticles via sol–gel method: a review on synthesis, characterization and applications. J Mater Sci: Mater Electron 31:3729–3749
171. Pędzik M, Janiszewska D, Rogoziński T (2021) Alternative lignocellulosic raw materials in particleboard production: a review. Ind Crops Prod 174:114162
172. Pędzik M, Kwidziński Z, Rogoziński T (2022) Particles from residue wood-based materials from door production as an alternative raw material for production of particleboard. Drv Ind 73:351–357
173. Pérez-Arantegui J, Molera J, Larrea A et al (2001) Luster pottery from the thirteenth century to the sixteenth century: a nanostructured thin metallic film. J Am Ceram Soc 84:442–446
174. Pinkl S, Veigel S, Gindl-Altmutter W et al (2022) Effects of fibrillar cellulosic additives on particleboard production and properties. Wood Mater Sci Eng 17:106–112
175. Poland S (2019) Statistical Yearbook of Forestry. Statistics Poland: Warsaw, Poland
176. Pour MF, Edalat H, Dorieh A et al (2022) Durability-related performance of reinforced bond-line by phenol formaldehyde/nano $SiO_2$ composite in Laminated Veneer Lumber (LVL). J Build Eng 60:105191
177. Radkau J (2012) Wood: a history. Polity, pp 13–34
178. Rafiq A, Tahir MA, Zia R et al (2022) Virus detection using nanobiosensors. In: Nanosensors for smart agriculture. Elsevier, pp 547–572
179. Rahman IA, Padavettan V (2012) Synthesis of silica nanoparticles by sol-gel: size-dependent properties, surface modification, and applications in silica-polymer nanocomposites—a review. J Nanomater 2012:8–8
180. Reh R, Kristak L, Antov P (2022) Advanced eco-friendly wood-based composites. Materials 15:8651
181. Roumeli E, Papadopoulou E, Pavlidou E et al (2012) Synthesis, characterization and thermal analysis of urea–formaldehyde/nano$SiO_2$ resins. Thermochim Acta 527:33–39
182. Sajid M (2022) Nanomaterials: types, properties, recent advances, and toxicity concerns. Curr Opin Environ Sci Health 25:100319
183. Sala CM, Robles E, Gumowska A et al (2020) Influence of moisture content on the mechanical properties of selected wood-based composites. BioResources 15:5503–5513
184. Salari A, Tabarsa T, Khazaeian A, Saraeian A (2012) Effect of nanoclay on some applied properties of oriented strand board (OSB) made from underutilized low quality paulownia (Paulownia fortunei) wood. J Wood Sci 58:513–524
185. Salari A, Tabarsa T, Khazaeian A, Saraeian A (2013) Improving some of applied properties of oriented strand board (OSB) made from underutilized low quality paulownia (Paulownia Fortunie) wood employing nano-$SiO_2$. Ind Crops Prod 42:1–9
186. Savun-Hekimoğlu B, Eren Z, Ince NH (2020) Photocatalytic destruction of caffeine on sepiolite-supported $TiO_2$ nanocomposite. Sustainability 12:10314
187. Schulz H (1993) The development of wood utilization in the 19th, 20th and 21st centuries. For Chron 69:413–418
188. Sedliacik J, Bekhta P, Potapova O (2010) Technology of low-temperature production of plywood bonded with modified phenol-formaldehyde resin. Wood Res 55:123–130
189. Sengul AB, Asmatulu E (2020) Toxicity of metal and metal oxide nanoparticles: a review. Environ Chem Lett 18:1659–1683
190. Seo J, Cha J, Kim S et al (2014) Development of the thermal performance of wood-flooring by improving the thermal conductivity of plywood. J Biobased Mater Bioen 8:170–174

191. Sharma V, Goyat MS, Hooda A et al (2020) Recent progress in nano-oxides and CNTs based corrosion resistant superhydrophobic coatings: a critical review. Prog Org Coat 140:105512
192. Sharon M (2019) History of nanotechnology: from prehistoric to modern times. Wiley
193. Silva LCL, Lima FO, Chahud E et al (2019) Heat transfer and physical-mechanical properties analysis of particleboard produced with ZnO nanoparticles addition. BioResources 14:9904–9915
194. Srivastava S, Bhargava A, Srivastava S, Bhargava A (2022) Green nanotechnology: an overview. In: Green nanoparticles: the future of nanobiotechnology. Springer, pp 1–13
195. Stefanidis SD, Kalogiannis KG, Iliopoulou EF et al (2014) A study of lignocellulosic biomass pyrolysis via the pyrolysis of cellulose, hemicellulose and lignin. J Anal Appl Pyrolysis 105:143–150
196. Suárez M, García-Romero E (2012) Variability of the surface properties of sepiolite. Appl Clay Sci 67:72–82
197. Taghiyari HR, Esmailpour A, Majidi R et al (2022) The effect of silver and copper nanoparticles as resin fillers on less-studied properties of UF-based particleboards. Wood Mater Sci Eng 17:317–327
198. Taghiyari HR, Ghorbanali M, Tahir PM (2014) Effects of the improvement in thermal conductivity coefficient by nano-wollastonite on physical and mechanical properties in medium-density fiberboard (MDF). BioResources 9:4138–4149
199. Taghiyari HR, Hosseini SB, Ghahri S et al (2020) Formaldehyde emission in micron-sized wollastonite-treated plywood bonded with soy flour and urea-formaldehyde resin. Appl Sci 10:6709
200. Taghiyari HR, Karimi A, Tahir PM (2013) Nano-wollastonite in particleboard: physical and mechanical properties. BioResources 8:5721–5732
201. Taghiyari HR, Mohammad-Panah B, Morrell JJ (2016) Effects of wollastonite on the properties of medium-density fiberboard (MDF) made from wood fibers and camel-thorn. Maderas: Cienc Tecnol 18:157–166
202. Taghiyari HR, Morrell JJ, Husen A (2022) Emerging nanomaterials for forestry and associated sectors: an overview. In: Emerging nanomaterials: opportunities and challenges in forestry sectors. Springer, pp 1–24
203. Taghiyari HR, Norton J (2014) Effect of silver nanoparticles on hardness in medium-density fiberboard (MDF). iForest 8:677
204. Taghiyari HR, Nouri P (2015) Effects of nano-wollastonite on physical and mechanical properties of medium-density fiberboard. Maderas: Cienc Tecnol 17:833–842
205. Taghiyari HR, Rangavar H, Bibalan OF (2011) Effect of nano-silver on reduction of hot-pressing time and improvement in physical and mechanical properties of particleboard. BioResources 6:4067–4075
206. Taghiyari HR, Soltani A, Esmailpour A et al (2020) Improving thermal conductivity coefficient in oriented strand lumber (OSL) using sepiolite. Nanomaterials 10:599
207. Tayeb AH, Amini E, Ghasemi S, Tajvidi M (2018) Cellulose nanomaterials—Binding properties and applications: a review. Molecules 23:2684
208. Thomas S, Shanks R, Joy J (2016) Micro-and nanostructured polymer systems: from synthesis to applications. CRC Press
209. Tichi AH, Bari E, Nicholas DD (2018) How nano-wollastonite can change the fundamental properties of a wood fibre and rice straw composites? IET Nanobiotechnol 12:1027–1030
210. Tichi AH, Bazyar B, Khademieslam H et al (2016) The effect of nano-wollastonite on biological, mechanical, physical, and microstructural properties of the composite made of wood-cement fiber. J Fundam Appl Sci 8:1466–1479
211. Tichi AH, Bazyar B, Khademieslam H et al (2019) Is wollastonite capable of improving the properties of wood fiber-cement composite? BioResources 14:6168–6178
212. Tolinski M (2009) Overview of fillers and fibers. In: Additives for polyolefins: getting the most out of polypropylene and TPO. Elsevier, pp 93–119
213. Tolochko NK (2009) History of nanotechnology. Encyclopedia of Life Support Systems (EOLSS)

214. Treusch O, Hofenauer A, Tröger F et al (2004) Basic properties of specific wood-based materials carbonised in a nitrogen atmosphere. Wood Sci Technol 38:323–333
215. ur Rahman O, Shi S, Ding J et al (2018) Lignin nanoparticles: synthesis, characterization and corrosion protection performance. New J Chem 42:3415–3425
216. Vahabi H, Brosse N, Abd Latif NH et al (2021) Nanolignin in materials science and technology—does flame retardancy matter? In: Biopolymeric nanomaterials. Elsevier, pp 515–559
217. Vansant EF, Van Der Voort P, Vrancken KC (1995) Characterization and chemical modification of the silica surface. Elsevier
218. Veigel S, Rathke J, Weigl M, Gindl-Altmutter W (2012) Particle board and oriented strand board prepared with nanocellulose-reinforced adhesive. J Nanomater 2012:1–8
219. Vineeth SK, Gadhave RV, Gadekar PT (2019) Nanocellulose applications in wood adhesives. Open J Polym Chem 9:63–75
220. Vishnuvarthanan M, Al Shahidah MF, Gobika NR et al (2021) Effect of alumina silicate (MMT K10) nanoclay on adhesion and barrier properties of cornstarch-based bioadhesive. SILICON 13:4315–4322
221. Wang H, Liu X, Zhang ZM (2013) Absorption coefficients of crystalline silicon at wavelengths from 500 nm to 1000 nm. Int J Thermophys 34:213–225
222. Wang Z, Gu Z, Hong Y et al (2011) Bonding strength and water resistance of starch-based wood adhesive improved by silica nanoparticles. Carbohydr Polym 86:72–76
223. Watcharakitti J, Win EE, Nimnuan J, Smith SM (2022) Modified starch-based adhesives: a review. Polymers 14:2023
224. Watkins D, Nuruddin M, Hosur M et al (2015) Extraction and characterization of lignin from different biomass resources. J Mater Res Technol 4:26–32
225. Winandy JE, Morrell JJ (2017) Improving the utility, performance, and durability of wood-and bio-based composites. Ann For Sci 74:1–11
226. Wu L, Liao L, Lv G (2015) Influence of interlayer cations on organic intercalation of montmorillonite. J Colloid Interface Sci 454:1–7
227. Wu Q, Miao W, Gao H, Hui D (2020) Mechanical properties of nanomaterials: a review. Nanotechnol Rev 9:259–273
228. Xian D (2012) Effect of nanoclay fillers on wood adhesives and particle board properties. University of British Columbia
229. Yadav R (2021) Development of low formaldehyde emitting particle board by nano particle reinforcement. J Appl Nat Sci 13:1187–1197
230. Yadav SM, Lubis MAR, Park B-D (2022) Modification of nanoclay with different methods and its application in urea-formaldehyde bonded plywood panels. Wood Mater Sci Eng 17:734–743
231. Yan Q-L, Gozin M, Zhao F-Q et al (2016) Highly energetic compositions based on functionalized carbon nanomaterials. Nanoscale 8:4799–4851
232. Yang M, Rosentrater KA (2020) Life cycle assessment of urea-formaldehyde adhesive and phenol-formaldehyde adhesives. Environ Process 7:553–561
233. Yang W, Rallini M, Natali M et al (2019) Preparation and properties of adhesives based on phenolic resin containing lignin micro and nanoparticles: a comparative study. Mater Des 161:55–63
234. Yao Q-Z, Yu S-H, Zhao T-L et al (2019) Enhanced potential toxic metal removal using a novel hierarchical $SiO_2$–$Mg(OH)_2$ nanocomposite derived from sepiolite. Minerals 9:298
235. Yetisen AK, Coskun AF, England G et al (2016) Art on the nanoscale and beyond. Adv Mater 28:1724–1742
236. Younesi-Kordkheili H (2017) Improving physical and mechanical properties of new lignin-urea-glyoxal resin by nanoclay. Eur J Wood Wood Prod 75:885–891
237. Younesi-Kordkheili H, Kazemi Najafi S, Behrooz R (2017) Influence of nanoclay on urea–glyoxalated lignin–formaldehyde resins for wood adhesive. J Adhes 93:431–443
238. Younesi-Kordkheili H, Pizzi A (2022) Improving the properties of phenol-lignin-glyoxal as a wood adhesive by lignin nanoparticles. Eur J Wood Wood Prod 1–6

239. Young RJ, Kinloch IA, Gong L, Novoselov KS (2012) The mechanics of graphene nanocomposites: a review. Compos Sci Technol 72:1459–1476
240. Yu Y, Qi S, Zhan J et al (2011) Polyimide/sepiolite nanocomposite films: preparation, morphology and properties. Mater Res Bull 46:1593–1599
241. Zahed R, Gholamian H (2011) The potential use of nanozycosil and sodium montmorillonite (NaMMT) nanoclay to decrease water absorption in MDF. Iran J Wood Pap Ind 1:69–81
242. Zeng G, Zhou Y, Wang T et al (2023) A bio-based adhesive reinforced with functionalized nanomaterials to build multiple strong and weak cross-linked networks with high strength and excellent mold resistance. Chem Eng J 453:139761
243. Zhang C, Chen M, Coasne B et al (2022) Hygromechanics of softwood cellulosic nanocomposite with intermolecular interactions at fiber-matrix interface investigated with molecular dynamics. Compos B Eng 228:109449
244. Zhang H, Mu Q, Yu X et al (2022) Preparation of copper ion adsorbed modified montmorillonite/cellulose acetate porous composite fiber membrane by centrifugal spinning. Polymers 14:5458
245. Zhang H, She Y, Song SP, Pu JW (2013) Modified nanocrystalline cellulose used for improving formaldehyde emission and bonding strength of urea formaldehyde resin adhesive. In: Key engineering materials. Trans Tech Publications, pp 846–851
246. Zhang H, Zhang J, Song S et al (2011) Modified nanocrystalline cellulose from two kinds of modifiers used for improving formaldehyde emission and bonding strength of urea-formaldehyde resin adhesive. BioResources 6:4430–4438
247. Zhang L, Xu J, Shen H et al (2021) Montmorillonite-catalyzed furfurylated wood for flame retardancy. Fire Saf J 121:103297
248. Zhang X-F, Liu Z-G, Shen W, Gurunathan S (2016) Silver nanoparticles: synthesis, characterization, properties, applications, and therapeutic approaches. Int J Mol Sci 17:1534
249. Zhang Y, Ding L, Gu J et al (2015) Preparation and properties of a starch-based wood adhesive with high bonding strength and water resistance. Carbohydr Polym 115:32–37
250. Zhang Z, Terrasson V, Guénin E (2021) Lignin nanoparticles and their nanocomposites. Nanomaterials 11:1336
251. Zhao J, Shao Q, Ge S et al (2020) Advances in template prepared nano-oxides and their applications: polluted water treatment, energy, sensing and biomedical drug delivery. Chem Rec 20:710–729
252. Zhao M, Jing J, Zhu Y et al (2016) Preparation and performance of lignin–phenol–formaldehyde adhesives. Int J Adhes Adhes 64:163–167
253. Zheng Y, Zheng Y (2006) Study on sepiolite-reinforced polymeric nanocomposites. J Appl Polym Sci 99:2163–2166
254. Zhuang C, Chen Y (2019) The effect of nano-$SiO_2$ on concrete properties: a review. Nanotechnol Rev 8:562–572
255. Zidanes UL, Dias MC, Lorenço MS et al (2021) Preparation and characterization of tannin-based adhesives reinforced with cellulose nanofibrils for wood bonding. Holzforschung 75:159–167
256. Zimmermann T, Pöhler E, Geiger T (2004) Cellulose fibrils for polymer reinforcement. Adv Eng Mater 6:754–761

# Nanocellulose Applications in Bio-based Wood Adhesives

**Mohd Ezwan Bin Selamat and Lum Wei Chen**

**Abstract** Nanocellulose is an example of natural-based material with a promising characteristic that makes it suitable for a wide range of applications. The cutting-edge properties of this bio-based material have gained attention toward the development of eco-friendly adhesives using nanocellulose. The primary composition of nanocellulose which is cellulose renders this material renewable since it can be extracted from various types of lignocellulosic materials and yet biodegradable. High mechanical strength, lightweight, and high surface area are the key points that distinguish nanocellulose compared to the polymeric or petroleum-based adhesives. This chapter discusses the previous and ongoing research efforts in optimizing nanocellulose-based wood adhesives with the aim to minimize the dependency to the formaldehyde-based adhesive in a wood panel industry.

**Keywords** Nanocellulose · Physical · Mechanical · High surface area · Biodegradability · Eco-friendly adhesive

## 1 Cellulose

Cellulose is a natural derived organic compound and considered as a major contribution in the structural component found in the cell walls of plants and trees. This material can be found abundantly in various types of vegetations [40, 53]. It is a complex component consisting of carbohydrate and a polysaccharide composed of linear chains of glucose molecules linked together by β-1,4-glycosidic bonds (Fig. 1). These long chains of glucose form microfibrils are bundled together to provide strength and rigidity to plant cells.

In general, cellulose can be divided into two main types which is crystalline and amorphous [45, 51] (Fig. 2). The crystalline cellulose is composed of well-organized long cellulose chains next to each other and connected by a strong hydrogen bond. This characteristic of crystalline cellulose contributes to the rigidity, resistance to

M. E. B. Selamat (✉) · L. W. Chen
Faculty of Bioengineering and Technology, Universiti Malaysia Kelantan, Jeli, Kelantan, Malaysia
e-mail: ezwan.s@umk.edu.my



P. Antov et al. (eds.), *Nanomaterials Additives in Bioadhesives for Wood Composites*, Smart Nanomaterials Technology, https://doi.org/10.1007/978-981-96-3400-2_3

**Fig. 1** The chemical structure of cellulose

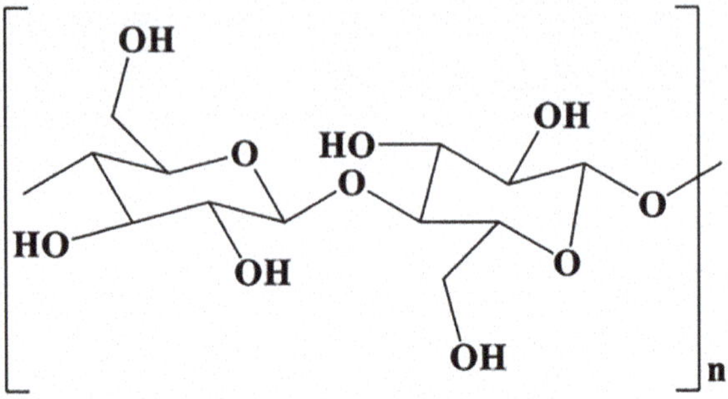


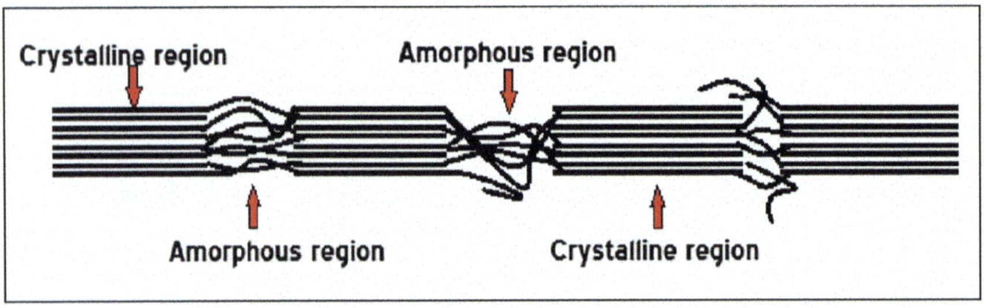


**Fig. 2** The crystalline and amorphous region of cellulose [7]

degradation, and makes it insoluble in water [51]. In addition, crystalline cellulose is less attractive to interact with other types of molecules, and the solvent makes it less favorable in the synthesis of new materials. For the amorphous cellulose, it consists of less organize cellulose chain and makes it less rigid and more flexible [45]. This type of cellulose possesses a better solubility and is prone to the enzymatic degradation compared to crystalline form. The advantage of amorphous cellulose is that it is susceptible to form a new linkage with other functional groups especially during the synthesis of a new chemical structure [28].

Cellulose is essential for the structural integrity of plants and plays a critical role in providing support and protection. It is one of the most abundant organic compounds on Earth, making up a significant portion of the biomass of plant materials. Cellulose possesses several important properties and makes it suitable to be used not only for human consumption. As presented in Table 1, cellulose which can be derived from various sources are suitable for.

## 2 What is Nanocellulose?

Nanocellulose is an example of promising nanomaterial derived from a cellulose fiber. In recent years, nanocellulose has emerged as a prominent environmentally friendly material, characterized by its myriad advantageous attributes. These include

**Table 1** The function of cellulose in a daily life

| Function | Benefit | References |
|---|---|---|
| Structural support | Provides rigidity and mechanical strength to plant cell walls, hence allowing plants to stand upright and resist the gravitational forces | Gupta et al. [14] |
| Dietary fiber | Generally derived from plant-based foods including fruits, grain, and vegetables. It significantly aids in digestion and suitable for a bulk diet | Li et al. [27] |
| Paper industry | Cellulose is the main component in the wood pulp industry and makes it suitable in the production of cardboard and paper | Provin et al. [41] |
| Textile industry | Has shown an excellent property to be further processed to produce rayon, viscose, lyocell as an alternative to the conventional cotton fibers and synthetic fibers | Provin et al. [41]<br>Rajnish et al. [42] |
| Biofuel production | A potential source in the production of biofuel due to its characteristic that can be break down into simpler form of sugars and fermented producing a bioethanol | Rahman et al. [43] |
| Biodegradable materials | Cellulose-based materials exhibit a remarkable potential to be further evaluated for the production of biodegradable and yet sustainable products | Liu et al. [28]<br>Li et al. [27] |
| Pharmaceuticals | Suitable to be used as a binder, disintegrant, and coating in the production of tablet | Tudoroiu et al. [49] |
| Biomedical | In the production of hydrogel for wound dressing | Tudoroiu et al. [49]<br>Piao et al. [38] |
| Food additives | Carboxymethyl cellulose is the example of common cellulose derivatives that is used to thicken or stabilize the ingredients mixture in food industry | Mishra, Kharkar and Pethe [31] |

its non-toxicity, biodegradability, high aspect ratio, exceptional mechanical properties, impressive optical characteristics, anisotropic shape, robust mechanical strength, excellent biocompatibility, and customizable surface chemistry [35]. It is composed of long chains of glucose molecules linked together by β-1,4-glycosidic bonds, forming microfibrils that give plants their structural strength. When cellulose is broken down to the nanoscale, it becomes nanocellulose, which exhibits unique properties due to its nanometer-sized dimensions. The term nano is referring to the size of cellulosic fibers that is in the range of 5–20 nm [5]. A nanocellulose includes nanocrystalline cellulose (NCC), nanofibrillated cellulose (NFC), and bacterial nanocellulose (BNC) [5, 48]. It has gained significant attention in recent years due to its unique properties and potential applications (Fig. 3) in various industries, including the development of bio-based wood adhesives [46].

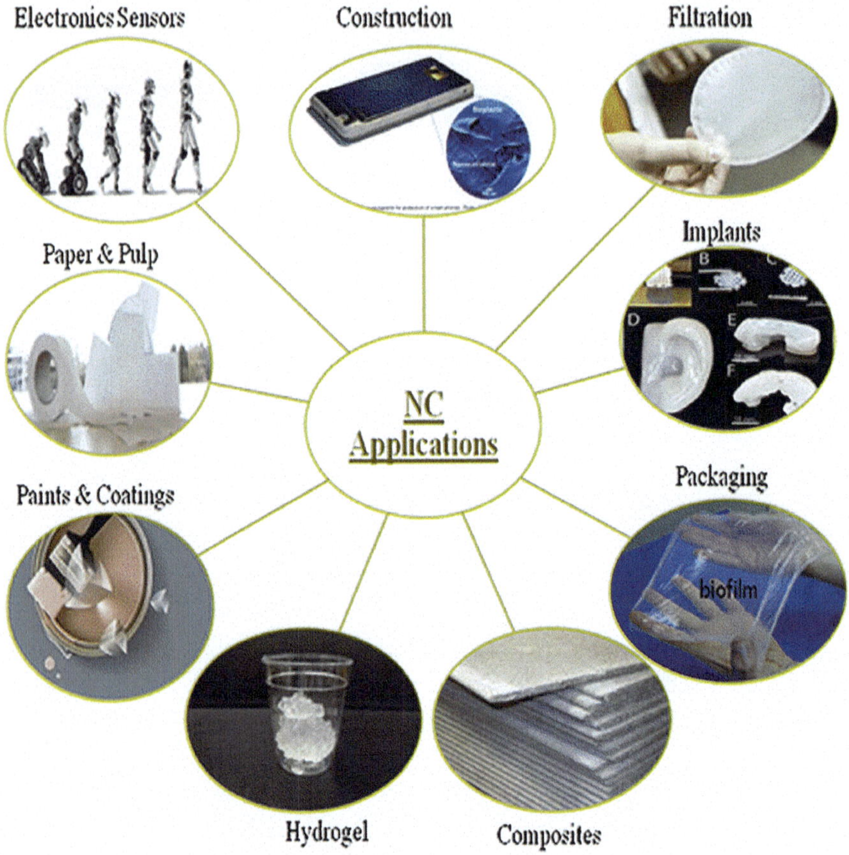


**Fig. 3** An example of nanocellulose promising applications [10]

## Types of Nanocellulose

**Nanocrystalline cellulose (NCC)** comprises highly crystalline, rod-shaped nanoparticles with dimensions on the order of a few nanometers in width and tens of nanometers in length [10, 19]. NCC is produced by a breakdown of cellulose fibers under controlled conditions producing a nanoscale crystalline particle. An example of NCC production is typically isolated from cellulose using acid hydrolysis or other chemical methods. The concentration of the acid is the crucial element to ensure the desired size of NCC fibers [12]. NCC exhibits exceptional mechanical strength, high surface area, and optical transparency, making it suitable for various applications. For wood-based composite applications, the potential of NCC as a natural-based adhesive was evaluated in several studies including the combination with the conventional resin [9, 22, 24]. Although the performance of the panels bonded with NCC are not comparable to the formaldehyde-based resins, it is shown that its suitability for the application does not require heavy load yet sufficient to comply with the standard requirements as stated in JIS A5908 for the [6, 21]. It has been shown the potential of this material derived from biomass indicates that the sole dependability to the hazardous formaldehyde resins slowly reduces, and it has shown a promising potential to be used as a wood-based composite binder.

**Nanofibrillated cellulose (NFC)** comprises long and flexible nanofibers with diameters in the nanometer range and lengths on the micrometer scale. NFC is often produced by mechanical fibrillation techniques, which physically disintegrate the cellulose fibers [10]. NFC has a high aspect ratio, good mechanical properties, and a large surface area, making it suitable for various applications as well.

**Bacterial nanocellulose (BNC)** is a newly developed cellulose-based material which exhibits a promising application not only limited to pharmaceutical and biomedical fields but also widely used in the food and engineering applications (Sharma et al., 16; [3]). The main characteristics that distinguish BNC from other types of nanocellulose is the production process which involves bacterial. The example of bacterial uses in the synthesis of BNC is the *Acetobacter* by a simple culturing method producing a homopolymer of β-1,4-glycosidic bonds [15].

## 3 Factors Influencing the Selection of Nanocellulose

In the production of wood-based composite, the selection of binder or adhesive is an integral part of manufacturing process. A significant effect on the selection of right adhesive will influence the quality of the product that will comply with the standard requirements. In general, the petroleum-based binders such as formaldehyde-based resins are the common types of adhesives used due to their characteristics such as cheap, availability, and excellent performance [23]. However, in recent years, the demand for bio-based adhesives as the substitution to the petroleum-based adhesives in a wood-based composite arises due to the consumer awareness toward the health and the indoor air pollution. One of the best options to align with this customer trend is by utilizing a nanocellulose as a bio-based adhesive in the production of wood-based composite. A unique characteristic of nanocellulose has gained considerable attention especially among the researchers to explore more potential of nanocellulose. For example, in the incorporation of nanocellulose shown an excellent property as a binder in the production of particleboard panels [1, 2] and in the preparation of fire-retardant composites [55]. The properties exhibited by nanocellulose can be tailored due to the characteristics of these versatile materials. The characteristics that make these nano-sized materials receive huge attention as well as consideration are displayed in Fig. 4.

High mechanical strength of nanocellulose exhibits remarkable mechanical properties, making it stronger than many other natural and synthetic materials. This is mainly due to the high aspect ratio which refers to the large length-to-width ratio [55]. A high aspect ratio will significantly affect the mechanical properties especially

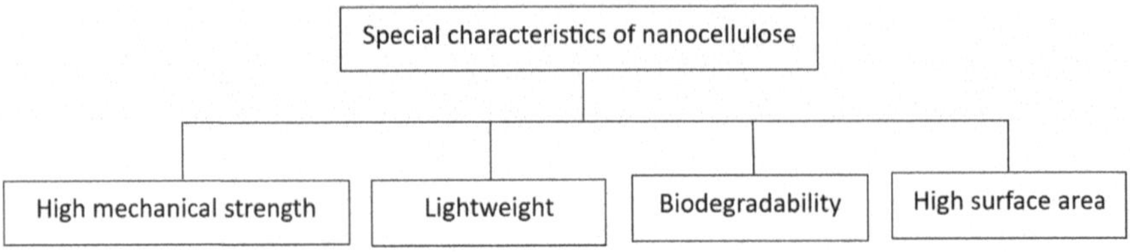


**Fig. 4** Factors influencing the selection of nanocellulose

on the stiffness and resilience to the materials. Previous studies that explained the effect of nanocellulose high aspect ratio are tabulated in Table 2.

Nanocellulose is exceptionally lightweight natural-based materials. This property becomes the advantage and makes it suitable for applications that produce a lightweight product. The example of applications that take the benefit from this characteristic is the production of nanocellulose-based composite as a medium to clean up the wastewater [52]. According to Yuan et al. [52] for this application, the main characteristics of the end product should be light and able to float on the water surface. The main reason it needs to have these properties is because in some cases the polluted industrial wastewater will cover the surface of the water forming a barrier that prevents aquatic life from getting sunlight and air [44]. In a wood-based panel production for example, the weight of each component that made up the panel is very important. This is because to ensure the material uses not just meet the standard requirement and reasonable price but also ease the transportation process [26].

The biodegradability properties are one of the key factors that influence the selection of nanocellulose compared to the man-made artificial materials. Nowadays, the world is focusing on green and environmentally friendly products so the utilizations of nanocellulose derived from renewable resources are the best approaches carried

**Table 2** A previous work showing the effect of nanocellulose in the composite material

| Title | Focus of the study | References |
|---|---|---|
| Biocomposites reinforced with cellulose nanocrystals derived from potato peel waste | A new approach that using potato peel waste as a source of cellulose nanocrystals for biocomposites | Chen et al. [7] |
| Effect of refining and homogenization on nanocellulose fiber development, sheet strength, and energy consumption | Investigates the relationship between energy consumption in the production of high aspect ratio and low diameter nanocellulose | Ang et al. [2] |
| Preparation and characterization of *Enteromorpha prolifera* nanocellulose/polyvinyl alcohol composite films | Evaluations on the potential of nanocellulose derived from *Enteromorpha prolifera* incorporated with PVA in the production of film | Zhang et al. [55] |
| Nanocellulose-based composite materials for wastewater treatment and waste-oil remediation | New formulation of composite for wastewater treatment | Yuan et al. [52] |

out by researchers [36, 47, 57]. Nanocellulose can be derived from the abundance of plant fibers, wood pulp, and even the agricultural waste such as oil palm trunk (OPT) or the empty fruit bunch (EFB). According to Trache et al. [48], the degradation of nanocellulose adhesives can be achieved by the microbial degradation using an enzyme to break down the cellulose molecules into a simpler compound. These compounds will be metabolized and utilized as a source of energy for microorganisms. In general, the biodegradability of nanocellulose adhesives for wood composites gives a minimal effect on the environment compared to conventional adhesives. The conventional petroleum-based resins such as urea formaldehyde (UF), phenol-formaldehyde (PF), and melamine formaldehyde (MF) are the common resins used in the production of wood-based composite. Even though it provides excellent physical and mechanical properties to the products, the main disadvantage of these adhesives is one of the sources of formaldehyde emission that results in indoor air pollution that is harmful to human health. By utilizing nanocellulose as a binder, it not only helps to dispose the abundance of agricultural waste but also explore the new potential of biomass for a new application. It is shown that the versatility together with the biodegradability of nanocellulose become the value added to nanocellulose.

Finally, nanocellulose is an example of materials with a high surface area [37]. A high surface area nanocellulose can be produced by several methods due to the versatility of this natural origin material. Mechanical treatment, chemical treatment, enzymatic treatment, combination methods, and controlled processing parameters are the examples of different treatments that can be carried out to produce a high surface area nanocellulose [12]. In combination methods, one of the approaches is by combining the chemical and mechanical treatments, and it can be either by acid hydrolysis or oxidation of nanocellulose followed by mechanical treatment to produce a high surface area [33]. By employing these methods and optimizing the processing parameters such as temperature, pressure, and the treatment duration, the aim to produce high surface area nanocellulose that is suitable for various applications in industries such as packaging, composites, biomedical materials, and electronics are achievable.

In a wood-based panel product, a significant impact on high surface area is the potential of nanocellulose to be applied as the reinforcement materials will enhance the dimensional stability and mechanical properties of the product [13, 34]. According to Ilyas and Sapuan [20], a larger surface area will provide a larger surface for the interfacial interaction not only between the binder but also with the filler that significantly influences the performance of wood composite. Previous work by Veigel et al. [50] indicates the incorporation of conventional formaldehyde-based resins with nanocellulose producing a particleboard and oriented strand board with a comparable performance but at lower percentage of urea formaldehyde (UF) and melamine urea formaldehyde (MUF) resins. The conventional adhesive mixed with nanocellulose significantly enhances the fracture energy of the adhesives and toughness resulting in increased mechanical strength of the wood-based panels (see Fig. 5).

Similar to the other applications, the improvement on the surface area will enhance the reactivity of the materials. The higher surface area will offer a larger site for the

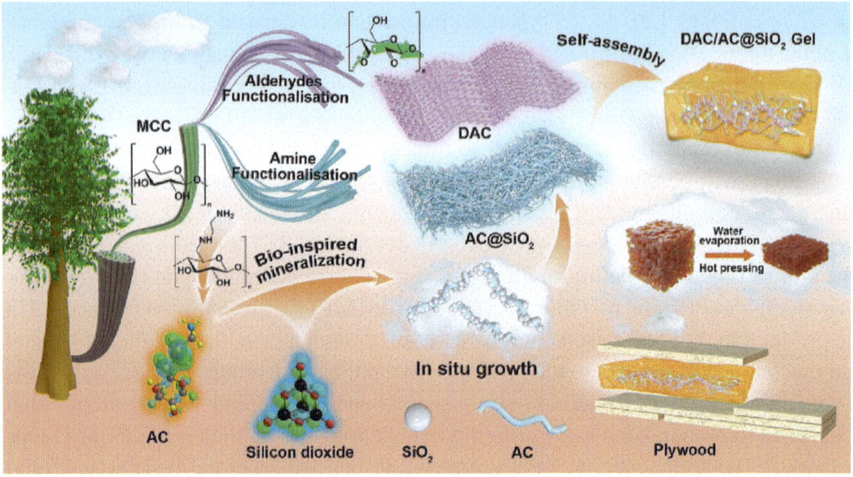


**Fig. 5** An example of chemical modification of nanocellulose reinforced with aldehyde and amine as bio-adhesive for plywood [34]

chemical interaction [20]. A material with a large or high surface area exhibits a high reactivity toward any chemical reaction due to the presence of highly reactive hydroxyl group (−OH). According to Ni et al. [34], high surface area of nanocellulose makes it suitable for various applications due to the availability of these materials to undergo a chemical modification. Abundance of hydroxyl group is the main factor that makes nanocellulose a versatile material that can suit various applications.

# 4 Types of Bonding Offers by Nanocellulose Based Adhesive

The bonding between nanocellulose-based bio-adhesives and wood involves a combination of hydrogen bonding, adhesion forces, interfacial interaction, and mechanical interlocking. Understanding these interactions or mechanisms is the key to creating a strong and durable bond between the nanocellulose-based adhesives and wood substrates. The bonding capabilities provided by nanocellulose can vary based on the method of chemical modification or its combination with other chemicals, ensuring alignment with specific purposes.

Nanocellulose and wood surfaces contain abundance of hydroxyl (−OH) groups. These hydroxyl groups will form hydrogen bonds with each other that will strengthen the mechanical properties of the adhesives [8, 13] (Fig. 6). This intermolecular bonding is a primary mechanism for adhesion between nanocellulose adhesives and wood resulting in the enhancement not only on the physical but also mechanical properties of the wood products. The hydroxyl groups on nanocellulose chains interact

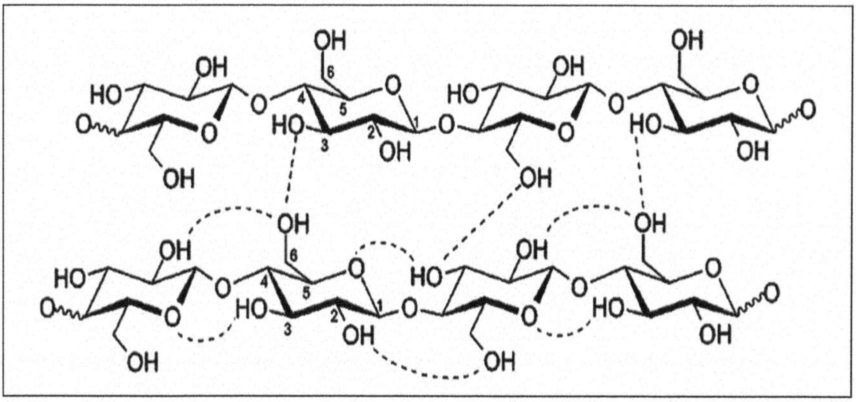


**Fig. 6** An illustration of the hydrogen bonds exists in nanocellulose [39]

with the hydroxyl groups present in the cellulose and lignin of wood, leading to strong hydrogen bonding between the two materials.

Van der Waals forces are weak, short-range forces that occur between all atoms and molecules [25]. In a wood-based panel product, it is related to the adhesion between nanocellulose adhesives and wood surfaces that will occur not only on the surface but also on pores present on wood. Although these forces are weaker compared to hydrogen bonds, Van der Waals forces still play a role in the overall adhesion especially when a load is applied to the sample.

Interfacial interaction of nanocellulose may exist between the binders and the wood or filler in composite [50]. This interaction is resultant of the high surface area of nanocellulose which is generally the size that can be controlled depending to the types of its manufacturing process. An average size of nanocellulose falls in the range of 5–20 nm [5]. A nano-sized material relatively has a higher surface area compared to the other materials with the same volume composed of bigger particle size [16, 19]. Nanocellulose with larger surface area will increase the contact of bio-adhesives with the other elements in the wood-based composite.

For mechanical interlocking, the adhesive penetrates the wood's surface irregularities, and the nanocellulose chains can intertwine with the wood's cellulose microfibrils, creating a mechanical interlock that enhances the adhesion [8, 25]. It is important to note that the quality and strength of the adhesive bond depend on factors such as the choice of nanocellulose material, its particle size, the wood species, the adhesive formulation, and the curing conditions. Researchers and engineers continue to explore various methods and formulations to optimize the chemical bonding between nanocellulose-based bio-adhesives and wood, with a focus on improving adhesion strength, water resistance, and durability for a wide range of applications in the wood industry.

## 5 Nanocellulose Application as Bio-based Adhesive for Wood-Based Product

Nanocellulose is increasingly being explored as a bio-based adhesive in wood-based products due to its unique properties and sustainability. Nanocellulose can be used as a bonding agent in the production of wood composites like particleboard, fiberboard, medium density fiberboard (MDF), and plywood. When incorporated into the adhesive formulation, nanocellulose can enhance the bonding strength between wood particles or veneers, improving the overall structural integrity of the composite (see Table 3).

Besides the application that involves a minor to medium load bearing structure, the wood-based panels bonded with nanocellulose have shown an excellent property in the production of non-load bearing structure. For example, in the production of soundproof and thermal insulation panels the introduction of nanocellulose-based bio-adhesives can enhance the bonding of materials used in soundproof panels for architectural and acoustic applications [30, 31].

**Table 3** The applications of nanocellulose in a wood-based panel product

| Product | Description |
|---|---|
| Particleboard and particleboard flooring | Nanocellulose adhesives can improve the bond strength between wood particles, enhancing the structural integrity of particleboard. This adhesive also will contribute to improved bonding and resistance to wear and tear |
| Medium Density Fiberboard (MDF) | The use of nanocellulose adhesives in MDF production can lead to stronger and more durable panels |
| High-Density Fiberboard (HDF) | Nanocellulose-based adhesives can enhance the bonding quality in high-density fiberboard, making it suitable for applications requiring increased strength |
| Oriented Strand Board (OSB) | Nanocellulose can be used to improve the adhesion of wood strands in the manufacturing of OSB, resulting in more robust panels for construction purposes |
| Plywood | The incorporation of nanocellulose in plywood adhesives can lead to superior bonding between veneer layers and improved resistance to delamination |
| Laminated Veneer Lumber (LVL) | Laminated veneer lumber is a common wood product used in construction. Nanocellulose-based adhesives can improve the bond quality between veneer layers, resulting in stronger and more durable LVL products |
| Cabinetry | In the construction of cabinets and cabinetry, nanocellulose-based adhesives can be used to bond wood components, ensuring a reliable and durable finish |
| Molded wood products | Nanocellulose adhesives can be applied in the production of molded wood products, such as decorative architectural elements and trim |
| Furniture manufacturing | In the production of furniture, nanocellulose adhesives can be used to bond wood components, such as joints and edges. The adhesive's strength and durability are essential for ensuring the longevity of furniture |

Other example of nanocellulose applications as bio-based adhesives is in the production of eco-friendly panels. In the manufacturing of eco-friendly panels, each raw material uses must possess 2 main properties which is it can be renewable and biodegradable [4, 6, 18]. The chosen of these natural-based materials mainly to minimize the dependency to the formaldehyde-based resins without compromising with the properties of the wood-based panels.

Previous findings have shown the incorporation of nanocellulose with conventional resins exhibits a promising performance and yet reduces the percentage of resin added [50]. Although their study involved the usage of conventional resins, the addition of nanocellulose as bio-adhesive successfully enhances the properties of low resin content samples and it indicates that lower resin content mixed with nanocellulose is sufficient to comply with the standard requirements. An example of previous findings involved the incorporation of nanocellulose with conventional composite resin with aim to diversify the utilization of nanocellulose as bio-based wood adhesives as tabulated in Table 4.

## 6 The Advantages of Nanocellulose as Bio-based Wood Adhesives

Nanocellulose-based adhesive derived from natural sources exhibits a comparable performance compared to the synthetic petroleum-based adhesives such as urea-formaldehyde (UF), phenol-formaldehyde (PF), and melamine-formaldehyde (MF). Although the formaldehyde-based resin gives an excellent physical and mechanical properties, the major disadvantages of these resins are they release a hazardous gas that may affect human health [11]. The main advantages of nanocellulose-based bio-adhesives are given in Fig. 7.

**Improved Adhesion Strength**: Nanocellulose, when incorporated into bio-based wood adhesives, can enhance the adhesion strength between wood surfaces. Its high surface area and strong hydrogen bonding capabilities contribute to better bonding performance, making it an effective reinforcement in wood adhesive formulations [4, 34].

**Reduced Formaldehyde Emissions**: One of the primary concerns with conventional wood adhesives, such as urea-formaldehyde or phenol-formaldehyde adhesives, is the emission of formaldehyde, a harmful volatile organic compound (VOC) [11]. By using nanocellulose in bio-based adhesives, it is possible to reduce or eliminate the need for formaldehyde-based resins, thereby lowering formaldehyde emissions.

**Rheological Control**: According to Cheng et al. [8], nanocellulose can modify the rheological properties of bio-based wood adhesives. It can act as a thickening agent, providing better control over the adhesive's viscosity, which is essential for application and bonding processes.

**Table 4** The examples of recent findings on the potential of nanocellulose incorporated with conventional wood-based composite resins

| Title | Application | References |
|---|---|---|
| Cross-linked entanglement of aldehyde and amine-functionalized nanocellulose reinforced with biomineralization to produce an all-bio-based adhesive | A nanocellulose reinforced with aldehyde and amine as a bio-based adhesive | Ni et al. [34] |
| Dialdehyde cellulose as a bio-based robust adhesive for wood bonding | Nanocellulose chemically modified with dialdehyde for wood adhesive | Zhang et al. [54] |
| Stability and rheological behavior of nanocellulose-modified UF resin compositions | A combination of urea formaldehyde resin and nanocellulose | Hochmanska and Janiszewska [17] |
| Preparation and characterization of tannin-based adhesives reinforced with cellulose nanofibrils for wood bonding | Nanocellulose incorporated with tannin for wood adhesives | Zidanes et al. [56] |
| Nano-modified adhesives for composite wood panel manufacturing | Modified nanocellulose as bio-adhesives for wood-based panels | Janiszewska, D., and Hochmańska, P. |
| Particle board and oriented strand board prepared with nanocellulose-reinforced adhesive | An incorporation of nanocellulose with formaldehyde-based resin for OSB and particleboard has shown promising physical and mechanical properties compared to the panels made from UF resin | Veigel, S., Rathke, J., Weigl, M., and Gindl-Altmutter, W. |

**Improved Water Resistance**: Nanocellulose can enhance the water resistance of bio-based wood adhesives. When added to the adhesive formulation, nanocellulose can create a barrier that reduces moisture infiltration, increasing the adhesive's durability and performance in damp or humid conditions.

**Enhanced Flexibility and Toughness**: Nanocellulose can improve the flexibility and toughness of bio-based wood adhesives. This is particularly beneficial in applications where the wood materials may experience stress, such as in engineered wood products like plywood, particleboard, and laminated veneer lumber [23, 29].

**Sustainable and Renewable Resource**: Nanocellulose is derived from cellulose, which is the most abundant organic polymer on Earth and can be sourced from various plant-based materials, including wood and agricultural residues [6]. This makes nanocellulose-based adhesives a more sustainable and environmentally friendly option compared to petroleum-based adhesives [29].

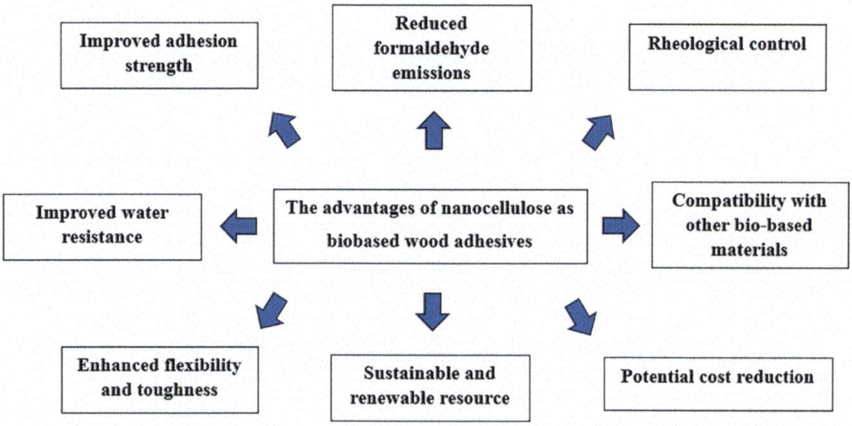


**Fig. 7** The advantages of nanocellulose bio-based adhesives

**Compatibility with Other Bio-Based Materials**: Nanocellulose can be easily integrated into formulations with other bio-based adhesives, such as soy-based or starch-based adhesives, to further enhance their properties and performance.

**Potential Cost Reduction**: While nanocellulose is relatively expensive compared to some other materials, the cost may decrease as production methods improve and economies of scale are realized. Its exceptional properties and benefits can justify the investment in certain high-performance wood adhesive applications [18, 32].

## 7 Conclusions

As the conclusions, nanocellulose has the potential to significantly improve the properties of wood-based composite. In addition, the substitution or reduced usage of formaldehyde-based resins gives a significant effect on the environment. Its use can lead to reduced formaldehyde emissions, better bonding strength, enhanced water resistance, and improved overall performance, making it a promising option for the wood industry. Researchers and manufacturers continue to explore and develop new formulations and applications for nanocellulose-based wood adhesives to minimize the dependence to the petroleum-based adhesives.

## References

1. Amini E, Tajvidi M, Gardner DJ, Bousfield DW (2017) Utilization of cellulose nanofibrils as a binder for particleboard manufacture. BioResources 12(2):4093–4110

2. Ang S, Haritos V, Batchelor W (2019) Effect of refining and homogenization on nanocellulose fiber development, sheet strength and energy consumption. Cellulose 26:4767–4786
3. Barja F (2021) Bacterial nanocellulose production and biomedical applications. J Biomed Res 35(4):310
4. Benhamou AA, Boussetta A, Kassab Z, Nadifiyine M, Sehaqui H, El Achaby M, Moubarik A (2022) Elaboration of carboxylated cellulose nanocrystals filled starch-based adhesives for the manufacturing of eco-friendly particleboards. Constr Build Mater 348:128683
5. Brand W, van Kesteren PC, Swart E, Oomen AG (2022) Overview of potential adverse health effects of oral exposure to nanocellulose. Nanotoxicology 16(2):217–246
6. Candan Z, Tozluoglu A, Gonultas O, Yildirim M, Fidan H, Alma MH, Salan T (2022) Nanocellulose: sustainable biomaterial for developing novel adhesives and composites. In Industrial applications of nanocellulose and its nanocomposites. Woodhead Publishing, pp 49–137
7. Chen D, Lawton D, Thompson MR, Liu Q (2012) Biocomposites reinforced with cellulose nanocrystals derived from potato peel waste. Carbohyd Polym 90(1):709–716
8. Cheng HN, Kilgore K, Ford C, Fortier C, Dowd MK, He Z (2019) Cottonseed protein-based wood adhesive reinforced with nanocellulose. J Adhes Sci Technol 33(12):1357–1368
9. Cheng L, Ren S, Lu X (2020) Application of eco-friendly waterborne polyurethane composite coating incorporated with nano cellulose crystalline and silver nano particles on wood antibacterial board. Polymers 12(2):407
10. de Amorim JDP, de Souza KC, Duarte CR, da Silva Duarte I, de Assis Sales Ribeiro F, Silva GS, de Farias PMA, Stingl A, Costa AFS, Vinhas GM, Sarubbo LA (2020) Plant and bacterial nanocellulose: production, properties and applications in medicine, food, cosmetics, electronics and engineering. A review. Environ Chem Letters 18:851–869
11. de Groot AC, Flyvholm MA (2020) Formaldehyde and formaldehyde-releasers. Kanerva's occupational dermatology, 521–542
12. Dhali K, Ghasemlou M, Daver F, Cass P, Adhikari B (2021) A review of nanocellulose as a new material towards environmental sustainability. Sci Total Environ 775:145871
13. Fujisawa S, Togawa E, Kimura S (2018) Large specific surface area and rigid network of nanocellulose govern the thermal stability of polymers: mechanisms of enhanced thermomechanical properties for nanocellulose/PMMA nanocomposite. Mater Today Commun 16:105–110
14. Gupta PK, Raghunath SS, Prasanna DV, Venkat P, Shree V, Chithananthan C, Choudry S, Surender K, Geetha K (2019) An update on overview of cellulose, its structure and applications. Cellulose 201(9):84727
15. Haldar D, Purkait MK (2020) Micro and nanocrystalline cellulose derivatives of lignocellulosic biomass: a review on synthesis, applications and advancements. Carbohyd Polym 250:116937
16. Hayat A, Sohail M, Hamdy MS, Taha TA, AlSalem HS, Alenad AM, Amin MA, Shah R, Palamanit A, Khan J, Nawawi WI, Mane SKB (2022) Fabrication, characteristics, and applications of boron nitride and their composite nanomaterials. Surf Interfaces 29:101725
17. Hochmanska P, Janiszewska D (2019) Stability and rheological behavior of nanocellulose-modified UF resin compositions. BioResources 14(1):1850–1866
18. Hu D, Ma W (2020) Nanocellulose as a sustainable building block to construct eco-friendly thermally conductive composites. Ind Eng Chem Res 59(44):19465–19484
19. Hubbe MA, Grigsby W (2020) From nanocellulose to wood particles: a review of particle size vs. the properties of plastic composites reinforced with cellulose-based entities. BioResources 15(1):2030–2081
20. Ilyas RA, Sapuan SM (2019) The preparation methods and processing of natural fibre biopolymer composites. Curr Org Synth 16(8):1068–1070
21. Janiszewska D, Hochmańska P (2018) Nano-modified adhesives for composite wood panels manufacturing. In: Living with modified wood, COST Action FP 1407, understanding wood modification through an integrated scientific and environmental impact approach (ModWoodLife), Final COST Action FP 1407 International Conference Belgrade, Serbia, 12–13 December 2018, p 138

22. Julkapli NM, Bagheri S (2017) Progress on nanocrystalline cellulose biocomposites. React Funct Polym 112:9–21
23. Kawalerczyk J, Dziurka D, Mirski R, Trociński A (2019) Flour fillers with urea-formaldehyde resin in plywood. BioResources 14(3):6727–6735
24. Khanjanzadeh H, Behrooz R, Bahramifar N, Pinkl S, Gindl-Altmutter W (2019) Application of surface chemical functionalized cellulose nanocrystals to improve the performance of UF adhesives used in wood based composites-MDF type. Carbohyd Polym 206:11–20
25. Lengowski EC, Bonfatti Júnior EA, Kumode MMN, Carneiro ME, Satyanarayana KG (2019) Nanocellulose-reinforced adhesives for wood-based panels. In Sustainable polymer composites and nanocomposites, pp 1001–1025
26. Liang D, Liu W, Zhong T, Liu H, Dhandapani R, Li H, Wang J, Wolcott M (2023) Nanocellulose reinforced lightweight composites produced from cotton waste via integrated nanofibrillation and compounding. Sci Rep 13(1):2144
27. Li T, Chen C, Brozena AH, Zhu JY, Xu L, Driemeier C, Dai J, Rojas OJ, Isogai A, Wågberg L, Hu L (2021) Developing fibrillated cellulose as a sustainable technological material. Nature 590(7844):47–56
28. Liu W, Liu K, Du H, Zheng T, Zhang N, Xu T, Pang B, Zhang X, Si C, Zhang K (2022) Cellulose nanopaper: fabrication, functionalization, and applications. Nano-Micro Lett 14(1):104
29. Mali P, Sherje AP (2022) Cellulose nanocrystals: Fundamentals and biomedical applications. Carbohyd Polym 275:118668
30. Midhun VC, Suresh S, Praveen B, Neethikumar R, Rajesh KS (2021) Preparation, characterisation and thermal property study of micro/nanocellulose crystals for vacuum insulation panel application. Therm Sci Eng Prog 25:101045
31. Mishra S, Kharkar PS, Pethe AM (2019) Biomass and waste materials as potential sources of nanocrystalline cellulose: comparative review of preparation methods (2016–till date). Carbohyd Polym 207:418–427
32. Mohamed AL, Hassabo AG (2015) Flame retardant of cellulosic materials and their composites. Flame retardants: polymer blends, composites and nanocomposites, pp 247–314
33. Nasir M, Hashim R, Sulaiman O, Asim M (2017). Nanocellulose: preparation methods and applications. In: Cellulose-reinforced nanofibre composites. Woodhead Publishing, pp 261–276
34. Ni K, Du G, Liu C, Wu Y, Yang H, Yin C, Su H, Ran X, Gao W, Yang L (2023) Cross-linked entanglement of aldehyde and amine functionalized nanocellulose reinforced with biomineralization to produce an all-bio-based adhesive. Chem Eng J 465:142888
35. Norizan MN, Shazleen SS, Alias AH, Sabaruddin FA, Asyraf MRM, Zainudin ES, Abdullah N, Samsudin MS, Kamarudin SH, Norrrahim MNF (2022) Nanocellulose-based nanocomposites for sustainable applications: a review. Nanomaterials 12(19):3483
36. Omran AAB, Mohammed AA, Sapuan SM, Ilyas RA, Asyraf MRM, Rahimian Koloor SS, Petrů M (2021) Micro-and nanocellulose in polymer composite materials: a review. Polymers 13(2):231
37. Phanthong P, Reubroycharoen P, Hao X, Xu G, Abudula A, Guan G (2018) Nanocellulose: extraction and application. Carbon Resour Convers 1(1):32–43
38. Piao X, Li J, Zhao Y, Guo L, Zheng B, Zhou R, Ostrikov KK (2022) Oxidized cellulose nanofibrils-based surimi gel enhancing additives: interactions, performance and mechanisms. Food Hydrocoll 133:107893
39. Pinkert A, Marsh KN, Pang S, Staiger MP (2009) Ionic liquids and their interaction with cellulose. Chem Rev 109(12):6712–6728
40. Polko JK, Kieber JJ (2019) The regulation of cellulose biosynthesis in plants. Plant Cell 31(2):282–296
41. Provin AP, dos Reis VO, Hilesheim SE, Bianchet RT, de Aguiar Dutra AR, Cubas ALV (2021) Use of bacterial cellulose in the textile industry and the wettability challenge—a review. Cellulose 28(13):8255–8274
42. Rajnish KN, Samuel MS, Datta S, Chandrasekar N, Balaji R, Jose S, Selvarajan E (2021) Immobilization of cellulase enzymes on nano and micro-materials for breakdown of cellulose for biofuel production-a narrative review. Int J Biol Macromol 182:1793–1802

43. Rahman MS, Hasan MS, Nitai AS, Nam S, Karmakar AK, Ahsan MS, Shiddiky MJA, Ahmed MB (2021) Recent developments of carboxymethyl cellulose. Polymers 13(8):1345
44. Raza A, Altaf S, Ali S, Ikram M, Li G (2022) Recent advances in carbonaceous sustainable nanomaterials for wastewater treatments. Sustain Mater Technol 32:e00406
45. Salem KS, Kasera NK, Rahman MA, Jameel H, Habibi Y, Eichhorn SJ, French AD, Pal L, Lucia LA (2023) Comparison and assessment of methods for cellulose crystallinity determination. Chem Soc Rev 52:6417–6446
46. Sharma A, Thakur M, Bhattacharya M, Mandal T, Goswami S (2019) Commercial application of cellulose nano-composites–a review. Biotechnol Rep 21:e00316
47. Sultana T, Sultana S, Nur HP, Khan MW (2020) Studies on mechanical, thermal and morphological properties of betel nut husk nano cellulose reinforced biodegradable polymer composites. J Compos Sci 4(3):83
48. Trache D, Tarchoun AF, Derradji M, Hamidon TS, Masruchin N, Brosse N, Hussin MH (2020) Nanocellulose: from fundamentals to advanced applications. Front Chem 8:392
49. Tudoroiu EE, Dinu-Pîrvu CE, Albu Kaya MG, Popa L, Anuța V, Prisada RM, Ghica MV (2021) An overview of cellulose derivatives-based dressings for wound-healing management. Pharmaceuticals 14(12):1215
50. Veigel S, Rathke J, Weigl M, Gindl-Altmutter W (2012) Particle board and oriented strand board prepared with nanocellulose-reinforced adhesive. J Nanomater, 1–8
51. Yao W, Weng Y, Catchmark JM (2020) Improved cellulose X-ray diffraction analysis using Fourier series modeling. Cellulose 27:5563–5579
52. Yuan B, Li L, Murugadoss V, Vupputuri S, Wang J, Alikhani N, Guo Z (2020) Nanocellulose-based composite materials for wastewater treatment and waste-oil remediation. ES Food Agrofor 1(7):41–52
53. Zainal SH, Mohd NH, Suhaili N, Anuar FH, Lazim AM, Othaman R (2021) Preparation of cellulose-based hydrogel: a review. J Mater Res Technol 10:935–952
54. Zhang H, Liu P, Musa SM, Mai C, Zhang K (2019) Dialdehyde cellulose as a bio-based robust adhesive for wood bonding. ACS Sustainable Chem Eng 7(12):10452–10459
55. Zhang Z, Liu X, Wang H, He H, Bai R (2021) Preparation and characterization of Enteromorpha prolifera nanocellulose/polyvinyl alcohol composite films. Polym Compos 42(4):1712–1726
56. Zidanes UL, Dias MC, Lorenço MS, da Silva Araujo E, Ferreira e Silva MJ, Sousa TB, Ferreira SR, Ugucioni JC, Denzin Tonoli GH, Bianchi ML, Mori FA (2021) Preparation and characterization of tannin-based adhesives reinforced with cellulose nanofibrils for wood bonding. Holzforschung 75(2):159–167
57. Zinge C, Kandasubramanian B (2020) Nanocellulose based biodegradable polymers. Eur Polym J 133:109758

# Nanolignin in Bioadhesives for Wood Composites

**Widya Fatriasari, Asma Sohail, Muhammad Rasyidur Ridho, Elvara Windra Madyaratri, Swetha Gillela, Sumit Manohar Yadav, Tuan Sherwyn Hamidon, and Mohd. Hazwan Hussin**

**Abstract** The synthesis of lignin into lignin nanoparticles (LNPs) is an approach for facilitating new excellent properties, including great specificity, rich functional groups, solubility, and compatibility, for further application in biocomposite systems. There are two types of LNP preparation techniques: top-down (grinding/mechanical milling, chemical etching, sputtering, laser ablation, and electroexplosion) and bottom-up (spinning, template support synthesis, plasma/flame spraying synthesis, laser pyrolysis, chemical vapor deposition, and atomic/molecular condensation) methods. Successful LNP synthesis results in a variety of morphologies, including quasi-spherical, hollow, smooth colloidal, and spherical. The incorporation of LNPs into adhesives has the potential to modify the physical, mechanical, and thermal characteristics of biocomposites. These qualities may include decreased water content, resistance to water absorption, greater tensile and flexural strength, enhanced thermal stability, and nonflammability. This chapter provides insight into the use of LNPs as adhesives for biocomposites, the methods used to synthesize LNPs, the properties of LNPs in combination with some advanced relevant equipment, and the improved characteristics of biocomposites achieved by introducing LNPs to bioadhesives. Knowledge of the characteristics of LNPs will influence the

W. Fatriasari (✉) · M. R. Ridho · E. W. Madyaratri
Research Center for Biomass and Bioproducts, National Research and Innovation Agency (BRIN), Cibinong-Bogor, Indonesia
e-mail: widy003@brin.go.id

A. Sohail
Department of Chemistry, Lahore College for Women University, Lahore, Pakistan

M. R. Ridho · E. W. Madyaratri
Faculty of Forestry and Environment, Department of Forest Products, IPB University, Bogor, Indonesia

S. Gillela · S. M. Yadav
Department of Forest Product and Utilization, Forest College and Research Institute, Telangana, Mulugu, India

T. S. Hamidon · Mohd. H. Hussin
Materials Technology Research Group (MaTReC), School of Chemical Sciences, Universiti Sains Malaysia, Minden, Penang, Malaysia

 
P. Antov et al. (eds.), *Nanomaterials Additives in Bioadhesives for Wood Composites*, Smart Nanomaterials Technology, https://doi.org/10.1007/978-981-96-3400-2_4

biocompatibility between LNPs and bioadhesives. Thus, by addressing these issues, the future outlook of LNPs can be optimized and widened for other fields.

**Keywords** Biocomposite · Bioadhesive · Biocompatibility · Lignin nanoparticle · Incorporation

## 1 Introduction

Nanotechnology is one of the emerging fields of the twenty-first century [1], providing engineered nanomaterials with high potential for producing products with significantly improved performance [43]. Because of their inherent characteristics, such as annual renewability, low-density biodegradability, high thermal and mechanical properties, dimensional stability, and excellent reinforcing capacity, biobased nanomaterials derived from abundant and lignocellulosic bioresources represent a viable and "green" option [4, 45, 61]. The addition of nanomaterials to composites has been shown to improve durability, flexural strength, and elasticity [83]. Because of material and technological advances, the wood-based panel industry has undergone significant changes in recent years and has become one of the world's fastest-growing manufacturing industries [44, 61], reaching ~380 million $m^3$ by 2021 [92]. Thus, it will be beneficial to use nanotechnology for creating advanced biocomposites by introducing nanomaterials into the adhesive system. Some nanosized biopolymers, such as nanolignin, nanocellulose, and nanoclay, have been reported to have high potency as binders and reinforcing agents in panel composites [4].

Lignin is the second most commonly used natural polymer in green biomaterials after cellulose. Technical lignins such as kraft, organosolv, soda, hydrolysis, and lignosulfonate are primarily produced by pulp and paper mills and emerging cellulosic biorefineries and are present as byproducts or waste [32, 75]. Chauhan [10] reported that 62 million tons of lignin were generated from bioethanol plants, and these data indicated that lignin is an abundant waste biopolymer for raw materials in value-added bioproducts. Lignin in biocomposites has many advantages, including antioxidant activity, improved thermal stability, a high carbon content, excellent mechanical strength, favorable stiffness [4, 16], enhanced bonding strength and decreased formaldehyde emission [16], and antimicrobial properties [82].

However, different kinds of lignin exhibit distinct variations in terms of elemental composition, functional groups, and molecular weight [4]. The investigation of the potential uses of lignin presents opportunities and challenges due to the variety of lignin sources and the intricacy of the structures. The amount of hydroxyl, carboxyl, phenolic, and sulfonate functional groups present, as well as how they affect lignin structure, is heavily influenced by the isolation procedure [51, 75]. For use in wood adhesives, these functional groups of lignin can facilitate its integration into biobased thermoplastic and thermosetting polymers [4]. The development of lignin-based nanomaterials will provide a new approach for extending the high-value use of lignin [106].

Nanolignin/lignin nanoparticles (LNPs) are one of the greatest strategies for improving lignin blending properties because they have novel characteristics, such as a large surface area and improved properties compared to those of the initial material [4]. Additionally, because of the abundance of chemically modifiable functional groups such as aliphatic hydroxyl, thiol, and phenolic groups, the surface of these materials can be freely altered, increasing their applicability [87].

Recently, similar to other nanomaterials, various kinds of LNPs have been synthesized from different resources using bottom-up or top-down methods [70, 74], resulting in various sizes on the nanoscale ranging 1–100 nm. These LNPs can be used as additives, stabilizers, and biobased adhesives when dissolved in water [53]. The origin, structure, chemical modification, and function of LNPs are all factors that affect reactivity [32]. Without undergoing any chemical alterations, LNPs can be integrated into the polymer matrix of various wood adhesives. Due to their tiny size compared to that of technical lignin, LNPs are able to improve the performance of composites by interacting with polymer matrices more effectively. Nevertheless, the esterification and etherification of LNPs increase the accessibility of the OH group [32].

Understanding the properties of LNPs can improve the biocompatibility of LNPs with resin in composites. Morphological observation of the LNPs was performed to assess the shape type and size of the resulting LNPs. However, the stability of LNPs during storage is needed to ensure stable properties with high charge density and prevent agglomeration. In composite systems, LNPs are also subjected to thermal treatment; thus, behaviors such as weight loss, transition glass temperature ($T_g$), and temperature maximum ($T_{max}$) are often observed. In addition, after synthesis, it is necessary to determine the structure of the LNPs in addition to the chemical content, such as the yield, as a tool to assess the effectiveness of the method.

Based on these outstanding properties and the ability to master synthesis methods, LNPs have the potential for use as bioadhesives. This chapter presents comprehensive and updated information on the roles of LNPs in bioadhesives for biocomposites based on existing published articles. The structure of this paper started with the synthesis method for LNP preparation, the determination of the properties of LNPs using advanced equipment techniques, and the evaluation of biocomposites loaded with LNPs. The physical, mechanical, thermal, and flammability effects after the incorporation of LNPs were evaluated in detail.

## 2 Recent Methods for Lignin Nanoparticle Production in Bioadhesives

In general, several methods for preparing nanomaterials can be divided into two categories: top-down and bottom-up methods [70, 74]. The top-down approach refers to physically reducing the larger molecular dimension of the original materials to a smaller size to convert them into nanoparticles [43]. Grinding/mechanical milling, chemical etching, sputtering, laser ablation, and electroexplosion are included in this technique. The bottom-up technique, known as the building approach, involves the formation of nanomaterials from simpler substances. Some examples of this technique include spinning, template support synthesis, plasma/flame spraying synthesis, laser pyrolysis, chemical vapor deposition (CVD), and atomic/molecular condensation. Biological synthesis via bacteria, fungi, algae, plants, and yeasts also includes bottom-up synthesis [43]. The nanoscale allows for the advancement of properties due to the large surface area and chemical and physical properties [57].

## 2.1 Bottom-Up Method

Lignin can only dissolve in certain solvents, which limits its large-scale or industrial-scale use. However, it has been reported that LNPs can be well dispersed in water [21, 72]. These LNPs dispersed in water may be used as crop additives, stabilizers, and biobased adhesives [53]. This is one of the first studies of an aqueous dispersion of LNPs in which ethylene glycol was used as a solvent during dialysis to manufacture LNPs [21]. Solvent exchange, in which a nonsolvent is added to a lignin solution or vice versa, is the most popular technique for creating LNPs. Solvent exchange causes the creation of spherical particles by minimizing surface energy [93]. Several studies have used solvents, such as dioxane [50], ethanol [48, 49], acetone [18, 84, 102], aqueous and nonaqueous tetrahydrofuran (THF) [53, 80], and dimethyl sulfoxide (DMSO) [84, 85], to prepare nanoparticles. However, water is the most frequently utilized nonsolvent to create spherical particles.

LNPs can be obtained using several methods, namely chemical, physical, and mechanical methods. Several types of lignin can be made into LNPs, such as kraft lignin, alkaline lignin, acetylation, and phosphorylation [19]. Colloidal spheres of acetylated alkali lignin (ACL) were created by gradually causing the hydrophobic aggregation of ACL molecules by continually adding water to the ACL-THF solution. Following rotational solvent evaporation, colloidal spheres produced a hydrodynamic radius <Rh> of 110 nm and a polydispersity ($\mu_2/\Gamma^2$) of 0.22 [72]. Prevention of LNP aggregation can be achieved using the acid precipitation technique, which can dissolve lignin with ethylene glycol and stir for 6 h. Hydrochloric acid (0.025 M) was added at a rate of 0.5 mL/min. The solution was dialyzed for 3 days and precipitated at pH 2 with 1 N HCl, after which the solid phase was washed and dried. The results showed the size of the LNPs to be 50–250 nm (i.e., narrow size distribution) [28]. Another acid precipitation method was reported in a previous study [73], which compared three solvents: castor oil, ethylene glycol, and water. Lignin was dissolved in each solvent (20 wt.% in a round bottom flask (neck 3 and stirred until homogeneous,20 mL of 1 M hydrochloric solution was added dropwise in a nitrogen atmosphere and incubated for 4 h at 50 °C; then, the LNPs were washed until they reached pH 7. The experiment produced LNPs measuring 15–20 nm using ethylene glycol and castor oil solvents.

Dioxane lignin nanoparticles (DLNPs) and alkali lignin nanoparticles (ALNPs) were produced using the nanoprecipitation approach from two distinct sources of lignin: hardwood dioxane lignin (DL) and softwood alkali lignin (AL). The obtained DLNPs and ALNPs have an average size of 80–104 nm and a spherical shape [102]. The mechanism of lignin nanoprecipitation and subsequent self-assembly was clarified by examining the production of LNPs from aqueous ethanol (Fig. 1). Large lignin molecules were discovered to produce the crucial initial nuclei of an LNP creation process that was kinetically controlled in terms of nucleation and growth. Lignin precipitation from high to low molecular weight starts when water is added to the lignin solution, which produces particles. At ethanol concentrations ranging from 62 to 36%, 40–70 nm LNPs were obtained [79]. Ju et al. [35] created a microchannel reactor that used polyvinylpyrrolidone (PVP)/sodium dodecyl sulfate (SDS) as a stabilizer for the continuous manufacture of LNPs. Using this technique, a sphere with an average size of 13 nm and a better stability of 60 days is obtained [39]. LNPs of 48.9 ± 16.4 nm were added to the phenol-formaldehyde adhesive at 5 wt.% of the total resin weight. The results show that adding LNPs increases the shear strength from 8.7 to 10.9 MPa, using nanoscale lignin in wood adhesives may cause crosslinking, which affects adhesion performance [96].

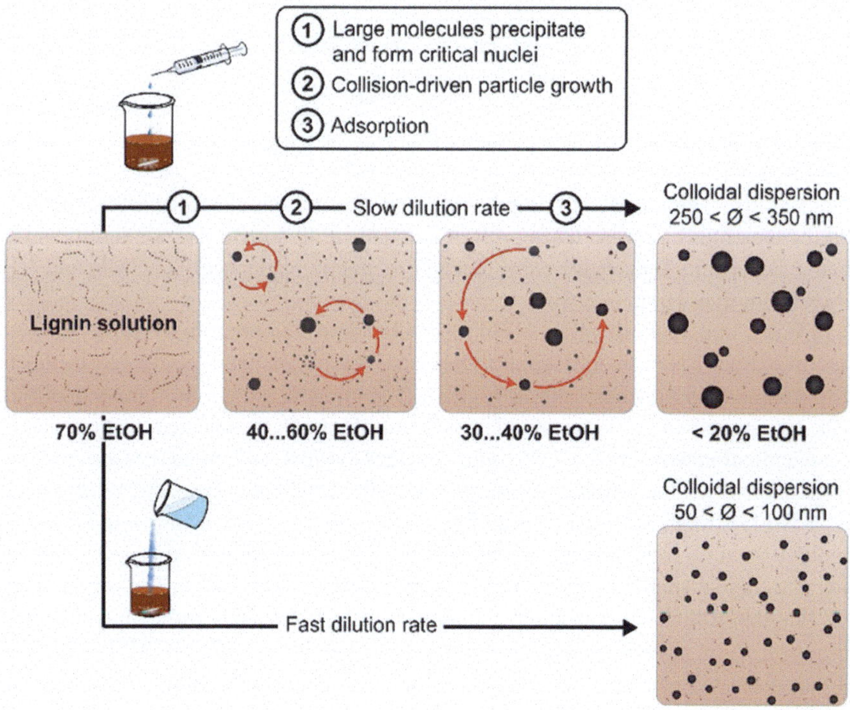


**Fig. 1** Lignin nanoparticle production in an ethanol and water system [79]. Copyright © 2018 American Chemical Society

## 2.2 Top-Down Method

In addition to chemical processes, mechanical or physical methods can also be used to create LNPs because they are faster and less complicated. The milling process in the mechanical technique of making LNPs has the disadvantage of nonuniform particle size and board particle size distributions [59]. The physical method of ultrasonication was carried out by Gilca et al. [24], lignin was suspended in water (0.7%) and sonicated for 1 h at a frequency of 20 kHz and a power of 600 W. The diameter of the produced LNPs was approximately 10–50 nm. The results of ultrasonication suggested lignin depolymerization. Gonzales et al. [25] reported that ultrasonic treatment at a frequency of 20 kHz, a 130 W power, and a 96% oscillation amplitude for 6 h produced LNPs with sizes ranging from 10 to 50 nm. It exhibited outstanding water dispersion stability because no precipitate had formed for over 5 months. Moreover, ultrasonic treatment for 1 h at a frequency of 25 kHz and a power of 800 W resulted in an average LNP size of 200 nm and a polydispersity index of approximately 0.25 [13]. Additionally, ultrasonication irradiation ensures uniform dispersion without chemical treatment [24, 53].

Lignin extracted from giant reeds was treated with ultrasonication using a 25 mm diameter probe (Sonics and Materials Inc., Model: VCX 500) with a frequency of 20 kHz, 40% oscillation amplitude, and a power of 500 W for 3 h. LNPs with sizes of approximately 10–50 nm were obtained. However, even with longer sonication times, the size of the lignin particles did not decrease. This might be because the LNPs had little to no impact on cavitation [33]. By using a high shear homogenizer (IKA Ultra-Turrax T-50 Homogenizer) for 4 h, kraft lignin from softwood kraft black liquor was successfully converted into LNPs, resulting in particle sizes less than 100 nm. The chemical composition between the initial kraft lignin particles and the nanolignin produced after 4 h of mechanical treatment did not significantly change according to analyses using nuclear magnetic resonance $^{13}$C NMR and $^{31}$P NMR techniques [64].

In Juikar and Vigneshwaran's [36] research, controlled microbial hydrolysis was used to extract LNPs from coconut fibers, and the results were compared to those of mechanical approaches such as high-shear homogenization and ultrasonication. Bulk lignin suspended in water (7% concentration) was subjected to high-shear homogenization with an IKA T 25 digital ULTRA-TURRAX® at 10,000 rpm for 1 h. At the same time, the ultrasonication process (ELMA®, India) used 30 W of power and a 37 kHz frequency for 1 h. Bulk lignin was used as a carbon source in the basal medium for the fungal isolate *Aspergillus* sp. during the microbial process, which took place under shaking conditions at 110 rpm at 31 °C for 15 days. The supernatant was filtered through a 1 μm filter to obtain LNPs. The results demonstrated that the homogenization process produced the greatest amount of LNPs, at 81.4%, with a particle size of 26–88 nm, while the ultrasonication process produced LNPs at a rate of 64.3% and a diameter of 35–45 nm. The microbial process produced the fewest LNPs, at 58.4%, and had the smallest particle size (22–28 nm).

The LNPs created have a negative charge on their surface, which makes them electrostatically stable. Surface crosslinking and other surface charge modifications can further increase the material stability [21]. The functionalization of the generated LNPs improved their dispersion in the polymer matrix due to their better compatibility. They are widely used in the field of polymeric nanocomposites with varied polarities. By combining LNPs with an adhesive that can offer superior interfacial adhesion through hydrogen bonding, LNPs can be used as a wood bioadhesive [9, 34]. Lignin has been utilized as a nanofiller in some recent studies to reinforce polysaccharides, proteins, natural rubber, and synthetic polymers such as polypropylene (PP), poly(methyl methacrylate) (PMMA), polystyrene (PS), polyaniline (PANi), poly(acrylonitrile) (PAN), poly(vinyl alcohol) (PVA), and epoxy polymer (Ep) [19].

## 3 Assessment of the Properties of Nanolignin in Bioadhesives

### 3.1 *Morphological Analysis*

When used for LNP synthesis, each approach has advantages and restrictions. To produce LNPs of the right sizes and forms, the synthesis technique should be carefully designed [55]. Successful LNP synthesis results in a variety of morphologies, including quasi-spherical, hollow, smooth colloidal, and spherical forms (Fig. 2). LNPs have exceptional thermal stability and have potential uses as antioxidants and reinforcement materials [106].

LNPs of various sizes and shapes were created using various ultrasonic intensities [89]. It can be assumed that applying high-intensity ultrasound may enhance the homogeneity and surface shape of LNPs while reducing their size (Fig. 3a). The

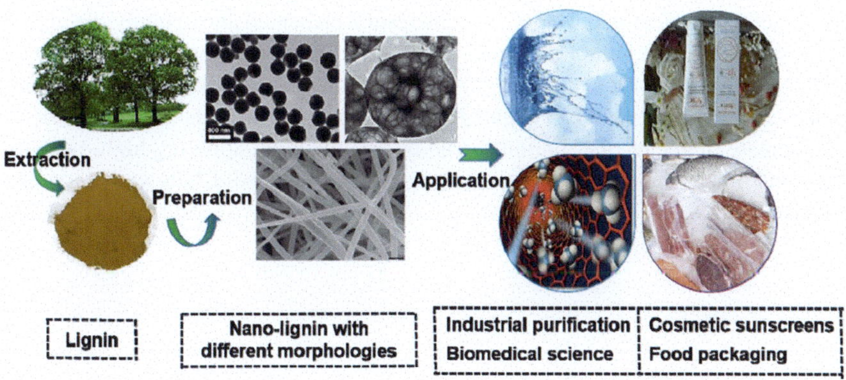


**Fig. 2** LNPs with various morphologies and their uses in various industries [106]. Copyright © 2018 MDPI, Basel, Switzerland

**Fig. 3** SEM images of LNPs from alkali lignin subjected to high-intensity ultrasonic treatment. **a** Wang et al. [89]. Copyright © 2018 American Chemical Society; enzymatic hydrolysis of lignin, **b** Pang et al. [66]. Copyright © 2020 American Chemical Society; LNPs were prepared using spray freezing, **c** Mishra and Ekielski [60]. Copyright © 2019 MDPI, Basel, Switzerland

LNPs obtained from the residual lignin fraction were insoluble in 9 and 80% ethanol solutions (F3) and had relatively consistent particle sizes of approximately 50 nm and regular spherical shapes (Fig. 3b). In addition, none of the large particles from F3 had hollow or incompletely spherical shapes, and all the LNPs were solid. According to the quality factor, lignin with a high molecular weight and low polydispersity index (PDI) is more likely to create tiny LNPs with a narrow distribution [66]. The variance in droplet size from a hand-held spray bottle can be predicted to contribute to the high PDI value [60]. Figure 3c shows that the resulting particles were in the nanometer range.

The morphology or shapes of the selected LNPs were investigated using TEM. Images of the LNPs from different sources created using various techniques are shown in Fig. 4a–i. LNPs created with the acid precipitation technique resemble aggregations or clusters of smaller molecules, whereas LNPs created using 70% ethanol and acetone/water (3:1) produced quasi-spherical and spherical LNPs, respectively (Fig. 4a–f) [20]. Based on Pang et al. [66], the residual lignin fraction, which yielded LNPs and was insoluble in both 95 and 80% ethanol solutions, was chosen because it had a relatively uniform particle size of approximately 50 nm and a regular spherical shape (Fig. 4h). The results presented above imply that lignin with a low PDI and a high molecular weight is susceptible to the creation of LNPs with tiny sizes and restricted size distributions. Figure 4i shows that ultrasonic intensities of 200 W were used to create LNPs with shapes with satisfactory sphericity and adequate diameter distribution. It can be assumed that applying high-intensity ultrasound may enhance the homogeneity and surface shape of LNPs while reducing their size [89].

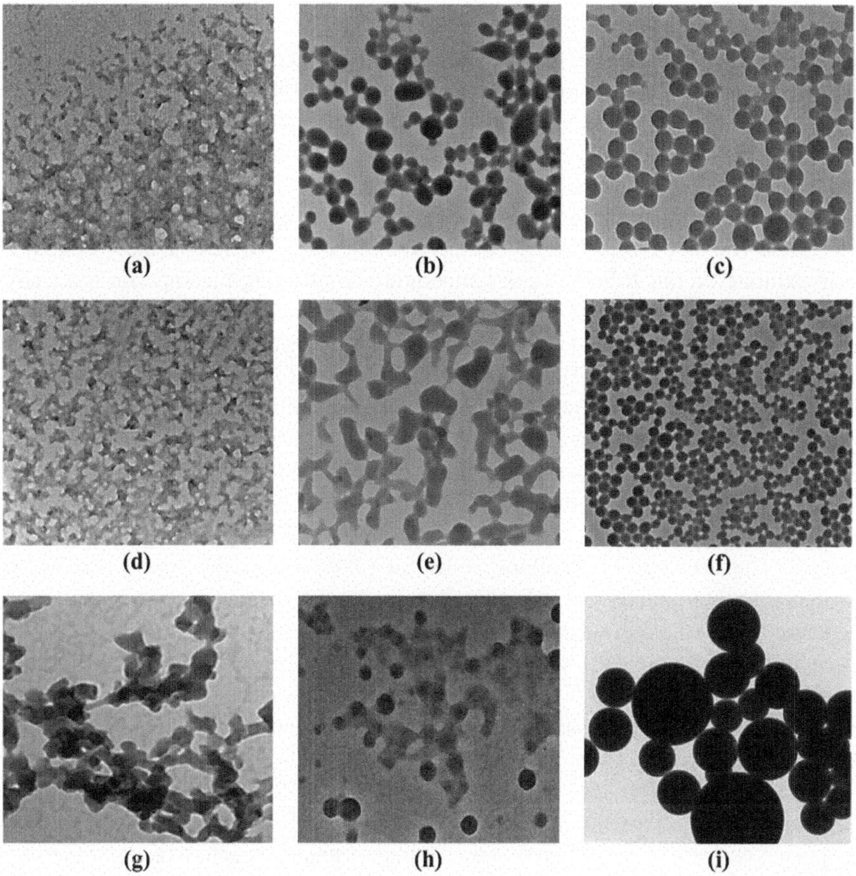


**Fig. 4** TEM images of the LNPs from softwood with acid preparation (**a**), 70% ethanol (**b**), and acetone/water (3:1) (**c**) and hardwood with acid preparation (**d**), 70% ethanol (**e**), and acetone/water (3:1) (**f**) Figueiredo et al. [20]. Copyright © 2021 by Wiley–VCH GmbH. Alkali lignin hydrochloric acid process (**g**) Lizundia et al. [54]. Copyright © 2020 American Chemical Society. Enzymatic hydrolysis lignin (**h**) Pang et al. [66]. Copyright © 2020 American Chemical Society. Alkali lignin with a high-intensity ultrasound treatment (**i**) Wang et al. [89]. Copyright © 2018 American Chemical Society

## *3.2 Thermal Properties*

The thermosetting properties (curing behavior) of LNP-based bioadhesives can be evaluated via differential scanning calorimetry (DSC). At the same time, thermogravimetric analysis (TGA) can infer the thermal stability of LNP-based bioadhesives. For instance, Chen et al. [11] researched a novel mild and green approach to formulate LNP phenol formaldehyde (LNPF) thermosetting resin for use as a wood adhesive for artificial wood boards to replace alkali lignin phenol formaldehyde (LPF) resin. Their study used LNPs with a particle size of approximately 300 nm and a specific surface area of 5.3765 $m^2\ g^{-1}$ as the raw material to partially substitute for phenol. The results revealed that substituting 40 wt.% phenol in LNPF resin with LNPs resulted in a dry bond strength of 1.30 ± 0.08 MPa, which is 1.85 times greater than the national grade 1 plywood standard. The results of the analysis of the curing behavior of the LNPFs, LPFs, and PFFs are shown in Fig. 5. The curing reaction yielded exothermic peaks for all three resins. The 40% LNPF, 40% LPF, and phenol-formaldehyde resins exhibited exothermic peak temperatures of 145.4, 186.1, and 180.3 °C, respectively. Compared to the other two resins, the 40% LNPF resin exhibited a much lower peak temperature, confirming the superior reactivity of the LNPs with formaldehyde in the curing reaction. The presence of a wider exothermic peak corresponds to the occurrence of the curing reaction throughout a broader temperature range. The researchers highlighted that the lower curing temperature of LNPF allows for lower temperature hot pressing, which also benefits from reducing the production cost of artificial wood boards. Younesi-Kordkheili and Pizzi [103] reported a similar observation for synthesized phenol-lignin-glyoxal (PLG) resin as a wood adhesive, where a DSC study established that incorporating LNPs decreased the curing temperature of PLG resin from 170 °C (neat PLG resin) to 159 °C. The excellent crosslinking ability of LNPs during the curing reaction is attributed to their smaller particle size, which causes hydroxyl groups in the lignin structure to become more exposed to the LNP monomer [40].

Figure 6 shows the thermal stability of the LNPF, LPF, and PF resins. At higher temperatures (800 °C), 40% LPF resin resulted in a greater solid residue mass (45 wt.%) than 40% LPF resin (40 wt.%). The higher content of solid residue is attributed to the better reactivity of LNPs with formaldehyde and their high surface area, which expedites a greater degree of polymerization of the LNPF resin.

Yang and coworkers [96] investigated the role of LNPs and microlignin (LMP) in the curing performance of phenol-formaldehyde resol as an adhesive. A DSC study (Fig. 7a) revealed the curing behavior of the neat resol and blended composites,

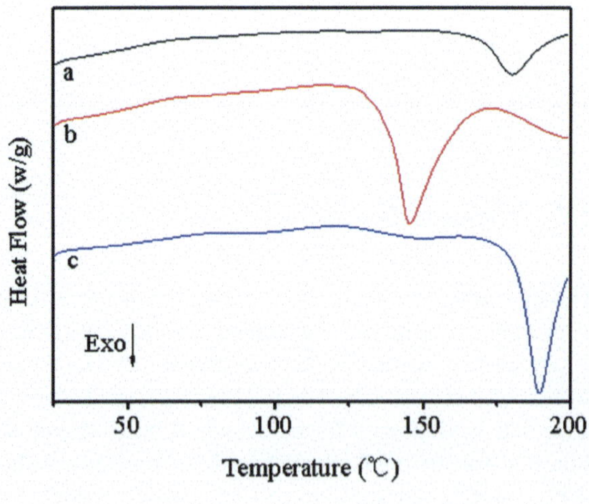


**Fig. 5** DSC diagrams of (a) phenol formaldehyde, (b) 40% NLPF, and (c) 40% LPF resins [11]. Copyright © 2019 Royal Society of Chemistry under CC BY-NC 3.0

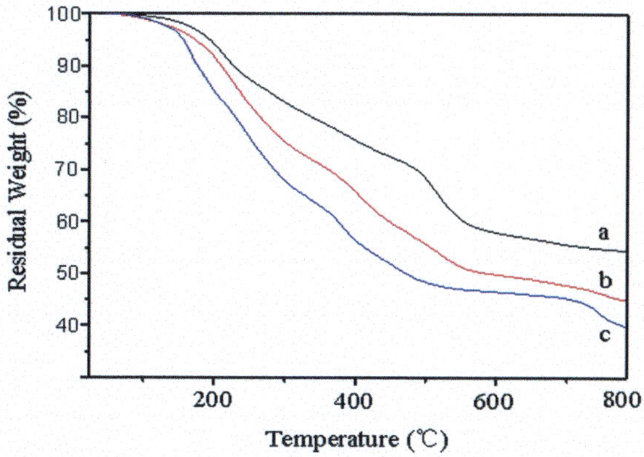


**Fig. 6** TGA curves of (a) phenol formaldehyde, (b) 40% LNPF, and (c) 40% LPF resins [11]. Copyright © 2019 Royal Society of Chemistry under CC BY-NC 3.0

whereas TG analysis (Fig. 7a) revealed their thermal stability. DSC results indicated that blending LNPs with formaldehyde led to a lower curing temperature and heat of the curing reaction compared to the addition of LMPs due to their smaller size and improved reactivity. They also observed that the amount of LNPs and LMP dispersed in the phenolic resol affects the curing temperature and heat of thermal curing. In general, the addition of LNPs to polymer matrices offers an improved thermal curing reaction because of the availability of abundant phenylpropane units. Phenylpropane units serve as active substituents on phenolic groups that react with formaldehyde to yield methylol groups. These substituents can be condensed into methylene bridges, which lowers the curing temperature and heat release during the curing reaction. TG analysis revealed that phenol resol blended with LNPs produced more char residue at 900 °C than did LMP-based resol, confirming that LNP-based resol composites enhanced thermal degradation. The enhanced thermal resistance could be related to the removal of more aliphatic structures when preparing LNPs than when preparing LMPs. Thus, LNPs retain a high proportion of aromatic structures that are more stable than aliphatic structures.

Zhao et al. prepared nanolignin polyol-based epoxy resins employing functionalized nanoalkali LNP obtained through the acid precipitation self-assembly process in an ethylene glycol solution of hydroxymethylated lignin [107]. TG analysis confirmed the superior thermal stability of the epoxy resins as a result of the introduction of modified nanoalkali lignin particles. TG curves revealed that nanoalkali lignin particles modified with a polyol liquefaction agent (nanoalkali lignin polyols) shifted the maximum thermal decomposition temperature of the epoxy resin to 332.3 °C from the 263.8 °C attained for unmodified nanoalkali lignin-based epoxy resin.

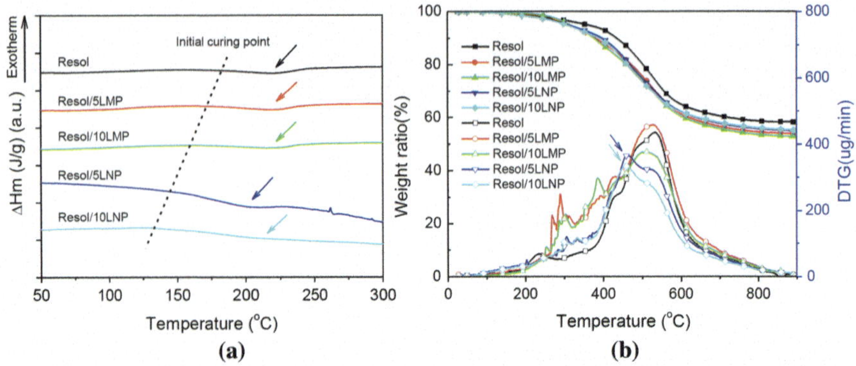


**Fig. 7** **a** DSC diagrams and **b** TG/DTG curves of the uncured neat resol and composites containing 5 and 10 wt.% LMP/LNP [96]. Copyright © 2019 Elsevier under CC BY-NC-ND 4.0

## 3.3 Structural Elucidation

Infrared absorption spectra of LNPs exhibit complicated bands due to the wide range of chemical bond vibration modes of the lignin structure. Kauldhar et al. extracted LNPs from the waste of sugarcane bagasse and observed that the primary structural aromatic vibrations of lignin corresponded to the peak at 1550 $cm^{-1}$. The C–H interactions in the vibrational deformation of the benzene rings caused the stretching at 1057 $cm^{-1}$. The presence of p-hydroxyphenyl groups caused a sharp peak at 825 $cm^{-1}$ [42]. The primary LNP spectrum showed peaks at 2938 $cm^{-1}$ (C–H stretching of methylene group), 1700 $cm^{-1}$ (C=O stretching), 1220 $cm^{-1}$ (aromatic skeletal vibration), and (C–O stretching in ether and phenol) [65]. The peaks at 1514 and 1426 $cm^{-1}$ indicated vibrations due to aromatic skeletons, suggesting that lignin was present in the spectra. An O–H group from phenol and alcohol was responsible for the stretching vibrational bands that produced the broad absorption band at 3421 $cm^{-1}$ [105]. For instance, the asymmetrical vibrations of $-CH_2$ are associated with the peak at 2900 $cm^{-1}$, and the stretching vibrations of C=C in the benzene ring are associated with the peak at 1600 $cm^{-1}$. Gupta et al. reported that throughout the formation of LNPs, the free hydroxyl groups in the mixture adhere to their interface, shifting the absorption peak of the material's infrared spectrum to a longer wavelength [27]. Formaldehyde's C=O group contains a partly positively charged carbon atom, which can readily react with LNPs. LNPs are porous, have a greater specific surface area, are small in size, and have a reduced charge compared with alkali lignin [12].

Additionally, it has been suggested that adding a –OH group may boost the interaction between LNPs and formaldehyde [98]. WeijunYang et al. synthesized phenolic resin-based bioadhesives using LNPs and analyzed them using FT-IR; Fig. 8 shows the FT-IR spectra of the pure substance and the nanocomposites containing 5 and 10 wt.% LNPs and LMP, respectively. The chemical compositions of lignin-containing nanocomposite materials and neat resol are comparable. The new peak at

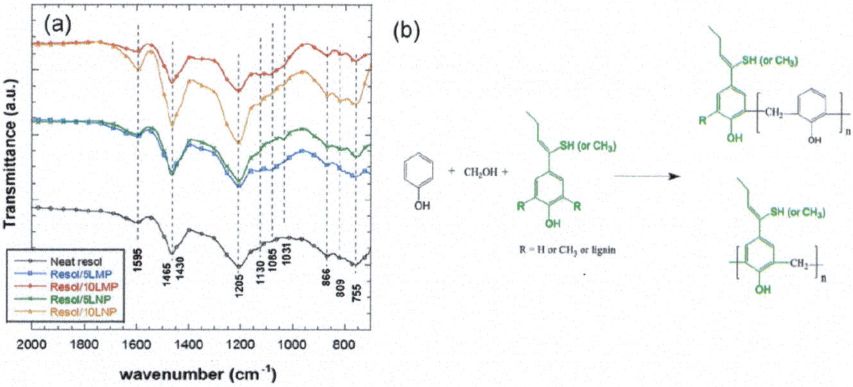


**Fig. 8** FTIR spectra of nanocomposites containing different concentrations of micro lignin and LNPs [96]. Copyright © 2019 with CC-BY license

1031 $cm^{-1}$ for the nanocomposites can be attributed to aromatic C–H deformation along with C=O stretching and guaiacyl-type aryl-ether bonds [96].

Nair et al. presented a novel high-shear homogenizer-based approach for producing LNPs. After 4 h of mechanical shear, the kraft LNP, which ranged in size from large micron-sized to nanoparticles, were homogenized into LNPs with radius smaller than 100 nm [64]. Nuclear magnetic resonance (NMR) tests revealed that the biochemical characteristics of the raw lignin and the LNPs produced after four hours of treatment were not significantly different [2]. Since LNPs lose their basic structure in most solvents, it has been common practice to explore the structure of lignin using NMR. However, NMR of LNPs in solution may also provide significant knowledge regarding the surface chemistry of LNPs, especially if a solvent is used that does not destroy the structure of lignin and assumes that the solvent readings are attenuated or do not interfere with the readings of the LNPs. The uppermost layers of numerous nanosized molecules have been well described using NMR. Conclusions about the structural components found on the surface of the LNPs were based on the signals in the $^{1}H$ NMR spectra of the LNPs and the characteristic features of the lignin structure in the raw form.

To be subjected to 1H NMR, nanoparticles must be dispersed in solvents without changing their fundamental structure. Due to the potential of the LNP structure for degradation, the use of organic solvents may not be appropriate. Although lignin is insoluble in water, the only option to address this restriction thus far has been to employ an aqueous suspension. The molar concentration is another factor to consider when using $^{1}H$ NMR effectively. The raw lignin content and the LNPs differed significantly according to $^{1}H$ NMR analysis. This difference in signals might be related to aromatic and aliphatic compounds. The aliphatic area was more prominent in the LNP data; only a small signal from aromatic protons was observed.

NMR was used to explain the process by which the shell of the LNPs comprised a relatively low-density, reduced-molecular-weight fraction composed of a relatively high-molecular-weight fraction of the LNP shell and its core. Accordingly, it is proposed that a shell made of a low-molecular-weight fraction of LNPs accounts for the unusual sharpening of NMR signals [71].

The weak reactivity of lignin, caused by the blockage of ortho- and para-active centers on the phenyl ring and a deficiency of the O–H group, must be resolved via chemical modification procedures [78]. Lignin is only used as a filler in bioadhesives and other composites without being altered, which could reduce its effectiveness and subsequent processing efficiency [3]. To increase the interaction of lignin molecules for the manufacture of bioadhesives, several chemical modification techniques have been developed [46]. The most popular chemical modification technique for enhancing the reactive region in lignin is phenolation. According to the $^1$H NMR study, phenolation preferentially occurs at the phenol ortho site rather than at the para site. After that, a three-step polymerization process was used to create a phenol-formaldehyde PF adhesive by substantially substituting phenol with rice husk lignin. This adhesive was more water- and ultraviolet-resistant than unmodified lignin. The number of phenolic hydroxyl groups in commercial alkali lignin increased approximately threefold after phenolation [97]. According to Lee et al., phenolated organosolv oak wood lignin-based phenol formaldehyde (POLPF) resin demonstrated a superior binding strength to traditional phenol-formaldehyde adhesives when 10% phenol was replaced with lignin [47].

## 3.4 Chemical Content

Different methods for producing LNPs can result in different yields. The highest yield of LNPs produced was obtained in the high-shear homogenization process at 81.4%; for the ultrasonication process, it was 64.3%; and the most negligible yield obtained using the microbial process was 58.4%. The low yield is due to the fungus *Aspergillus*, which can consume most of the lignin for its growth process [36, 37]. Wang et al. reported the manufacture of LNPs from acetylated lignin utilizing ultrasonication with THF, revealing that ultrasonic treatment resulted in more LNPs than without ultrasonication [89]. An increase in lignin concentration resulted in the production of LNPs of approximately 82.3%, which is lower than that reported in a previous study, which produced 88% LNPs [17], the yield of acetylated lignin was greater (50.8% than that of unmodified lignin 10.1%). According to these findings, the acetylation procedure increased the lignin concentration, and ultrasonic processing created LNPs with outstanding quality, including high yield, uniform size, good dispersion, and regular shape [28, 90]. The use of the prevalent dialysis method resulted in different yields of LNPs on three types of recalcitrant cellulolytic enzyme lignin steam-pretreated, namely lignin from hardwood poplar (90.9%), corn stover (81.8%), and the lowest amount of softwood lodgepole pine (41.0%). According to the findings presented here, the capacity to generate LNPs for substrates that have undergone steam pretreatment may not be restricted by such lignin condensation [85].

UV-vis spectroscopy was used to determine the compositional development of the LNPs after ultrasonic treatment. During the 6-h ultrasonication process, variations in the UV-vis absorption response can produce changes in the type and number of conjugated groups in LNPs. Figure 9 depicts noncondensed phenolic groups in guaiacyl units visible at 280 nm, which are characteristic of softwood lignin [23, 68], and the loss of certain conjugated groups during ultrasonication, such as α-carbonyl groups and esters of hydroxamic acids visible at 340 nm [25]. The specific surface area of LNPs is 35.74 $m^2/g$, but that of raw lignin is only 4.38 $m^2/g$; increasing the specific surface area and decreasing the particle size can increase the dissolution of lignin. The test results show that only 2.5% of the raw lignin is solubilized, compared to 31% of the LNPs. The solubility of LNPs is 12.4 times greater than that of raw lignin [56].

LNPs have lower potential antioxidant activity than commercial antioxidants and raw lignin, with an $IC_{50}$ value (half maximum inhibitory concentration) of 0.088 mg/mL. In comparison, raw lignin antioxidants and the commercial antioxidant *butylated hydroxytoluene* (BHT) were obtained at 0.027 mg/mL and 0.024 mg/mL, respectively [22]. According to Lu et al., the superoxide radical scavenging activity of LNPs was 1.41 ± 0.21 mg/mL, that of raw lignin was 15.49 ± 0.17 mg/mL, and that of the antioxidant butylated hydroxyanisole (BHA) was 0.36 ± 0.11 mg/mL, these findings indicate that LNPs have high superoxide radical scavenging activity. It can be concluded that increasing the solubility of LNPs can significantly increase their antioxidant activity [56]. Testing for purity and phenolic content was only carried out on lignin samples.

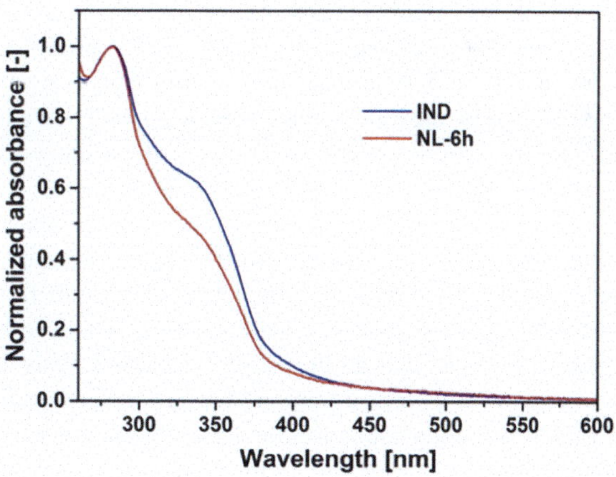


**Fig. 9** UV-Vis spectra of lignin and nanolignin after ultrasonication for 6 h [25]. Copyright © 2017 Wiley Periodicals, Inc.

### 3.5 Size and Stability Characteristics

Depending on the feedstock type, the core structural components of lignin (guiacyl, syringyl, and para hydroxyphenyl) and the number of connections may differ substantially [7]. The physicochemical characteristics of lignin can differ greatly between plantation regions, even for a single kind of biomass [41]. As a result, depending on the origin of the lignin, the physical and chemical characteristics, such as the solubility, size, and thermal stability of the produced lignin macromolecules, may differ significantly. The particle size of the kraft lignin ranged from large micron-sized to nanostructures, whereas after 4 h of mechanical shearing, the LNPs had sizes less than 100 nm. The lignin particles that were larger than 500 nm in size constituted more than 50% of the total particles. After being mechanically sheared for 4 h, the lignin fragments were completely homogenized, and the lignin nanoparticles had sizes smaller than 100 nm.

Additionally, there are significant differences in other physical characteristics, such as the glass transition temperature ($T_g$), hydrophilicity, and solubility. These variable elements have a significant influence on the stability of the lignin content, which has an impact on the biocompatibility of lignin-based bioadhesives. Compared with lignin/PVA mixtures, LNPs exhibit greater thermal stability when combined with polyvinyl alcohol (PVA) [5].

Commercially available lignin from wheat straw and Sarkanda grass was used in ultrasonication to produce LNPs in a colloidal lignin dispersion (0.7%) for 60 min. The mean diameter of LNPs after ultrasonic irradiation markedly decreased from 110 to 105 nm. A facilitated microbial hydrolysis procedure, high-shear homogenization, and ultrasonic radiation exposure procedures were used to obtain LNPs from coconut fibers [36, 37]. The mean sizes of the LNPs synthesized via the different techniques were 26.5, 7.5, 35.8, 7.8, and 22.3, 6.5 nm, respectively. The mean size of the LNPs was affected by the lignin concentration, antisolvent addition rate, and pH. When the concentration of lignin was 5 mg $mL^{-1}$, the average size of the LNPs was 52.6 nm, and when the concentration of lignin was 20 mg $mL^{-1}$, the maximum production of 87.0% was achieved, and the mean size of the LNPs was 210.8 nm [52].

According to one report, prior to resin synthesis, lignin purification might be an efficient way to augment the heat resistance of lignin-based PF adhesives [91]. Sphere-shaped lignin nanostructures were produced by dissolving softwood kraft lignin in THF and then adding water by dialysis. Water acts as a nonsolvent, reducing the range of motion of lignin and forcing the encapsulation of hydrophobic groups within the growing nanoparticles. The relationship between time, salt concentration, and pH affected the stability of the nanoparticle dispersion. The dispersion of lignin nanostructures in pure water was stable for more than two months at room temperature, and it caused agglomeration at low pH levels or high salt concentrations. The average diameter of the LNPs was 49.16 nm, whereas the particle size of the pure lignin ranged primarily from 50 to 100 nm [15]. LNPs were produced by precipitating them from an ethane 1,2 diol solution using a diluted acidic aqueous solution.

Lignin nanostructures were utilized to create wheat gluten (WG) nanocomposites with homogeneous particle sizes and excellent stability in aqueous media. Additionally, the LNPs could efficiently assimilate ultraviolet radiation and enhance the moisture responsiveness, mechanical behavior, and heat resistance of the bionanocomposites. The diameter of the pure LNPs ranged from 10 to 50 nm. For practical use at extremely low pH, lignin nanostructures from kraft and organosolv lignin with controllable physicochemical characteristics were produced [86]. The authors produced particles with a size range of 45–250 nm using customized growth reaction rates and interdependencies on mass balance. The stimulated pH drop flash precipitation method was used to create kraft LNPs, which were electrostatically stable due to their highly negative surface charge and colloid stability. Lignin spherical nanoparticles with diameters less than 100 nm were produced by flash precipitation. In the pH range from 3.2 to 8.5, stable nanostructures with a size of approximately 80 nm and a potential of approximately 45 mV were produced [86]. In ethane 1,2 diol and water, spheroid lignin nanostructures with sizes of 15–20 nm were found. The mean size of the LNPs was not more than 100 nm in the aqueous solution, 20–30 nm in the ethane 1,2 diol medium, and 15–20 nm in the castor oil medium. Epoxy coatings with dispersed lignin nanostructures may shield the core material and exhibit a superior shielding effect to that of epoxy adhesives [38].

# 4 Recent Development of Nanolignin in Bioadhesives for Biocomposites

## *4.1 Physical and Mechanical Properties of the Biocomposites*

LNPs are used as natural fillers in various polymer matrices to fabricate biocomposites to improve their properties. There are different types of LNPs, such as organosolv lignin particles, lignin polycomplexes, and lignin micromiscelles [31, 34, 104]. The addition of a small number of LNPs to both synthetic and biobased adhesives improved the overall performance of the composites [8, 19, 32]. The physical and mechanical properties of biocomposites with LNPs are dependent on many factors, including the fiber content, fiber length, fiber alignment, type of polymer, moisture absorption, fiber-matrix bonding, use of nanofillers, proportion of nanofillers, and proper distribution of nanofillers [4, 19].

### 4.1.1 Physical Properties of the Biocomposites

#### Moisture Content

The moisture content is one of the most important physical properties and is determined by considering the initial weight and oven-dried weight of the samples. Perera et al. analyzed the moisture content of nanocomposite films fabricated with thermoplastic starch (TPS)/LNP and nanocrystalline cellulose (NCC) based on weight measurements [69]. The samples were sized to 1 mm × 2 mm. The weights of the samples that were oven-dried at 50 °C (w1) and conditioned at 100% relative humidity (RH) for one day (w2) were measured. At 2 wt.%, the LNPs showed a decrease in moisture content of 35.1% compared to that of the neat films. However, 4 wt.% LNPs and NCC reduced the moisture content to 34.6% because of the hydrophobic nature of the LNPs and the high molecular order of NCC, which reduced the amount of hydrogen bonding in the starch matrix [26, 63, 104]. Yang et al. [63] incorporated 3 wt.% LNPs in a wheat gluten (WG) matrix and reported a decrease in the moisture content from 12.64 to 10.14% due to the presence of nonpolar hydrocarbon chains in the LNPs.

**Water Absorption/Swelling Ratio**

The water absorption (WA)/swelling ratio was determined by measuring the initial dry weight of the samples and the weight of the samples after they were submerged in distilled water under specific conditions. The water absorption behaviors of composites produced from biopoly(trimethylene terephthalate) (bio-PTT)/1.5 wt.% LNPs and bio-PTT/1.5 wt.% LNPs/7 wt.% vapor-grown carbon fibers (VGCFs) were examined by Gupta et al. according to ASTM-D 570 [29]. They used a sample size of 60 × 60 × 3 mm and dried it in an oven at 80 °C for 6 h. Thereafter, the initial weight of the samples was recorded. After that, the samples were submerged in distilled water and maintained at 23 ± 2 °C for 60 days. They were withdrawn from distilled water every 10 days, gently cleaned with a neat cloth or tissue paper, and reweighed.

The bio-PTT/1.5 wt.% LNP nanocomposites had a WA of 17% because of poor polymer matrix adhesion, whereas the bio-PTT/1.5 wt.% LNP/7 wt.% VGCF had a WA of 12% due to the presence of hydrophobic VGCF, which contributed to the reduced tendency of water absorption. Perera et al. calculated the swelling ratio of nanocomposite films manufactured from TPS/LNP and NCC. The initial weights of the samples were taken after drying at 50 °C for 48 h. Thereafter, the samples were soaked in distilled water for 1 day, and the weights of the samples were measured after the removal of the films from the water. Finally, they found that films with 2 wt.% LNPs showed lower water penetration, which reduced the swelling ratio up to 27%. On the other hand, compared with the neat film, the films with TPS/4 wt.% LNPs/4 wt.% NCC showed a significant decrease in the swelling ratio [69]. The water absorption of composites produced by WG/LNP was evaluated by Yang et al. [100]. The initial weight and final weight of the samples were measured before and after they were submerged in 30 mL of distilled water for 10 min. They discovered that the water uptake of composites with 1 wt.% LNPs/WG was 138%, and that with 3 wt.% LNPs/WG was 103.21%. These results suggested that the addition of 3 wt.% LNPs improved the strength of the bonding between WG and the LNPs. However, proper dispersion of LNPs also reduced the water absorption of the WG/LNP composites.

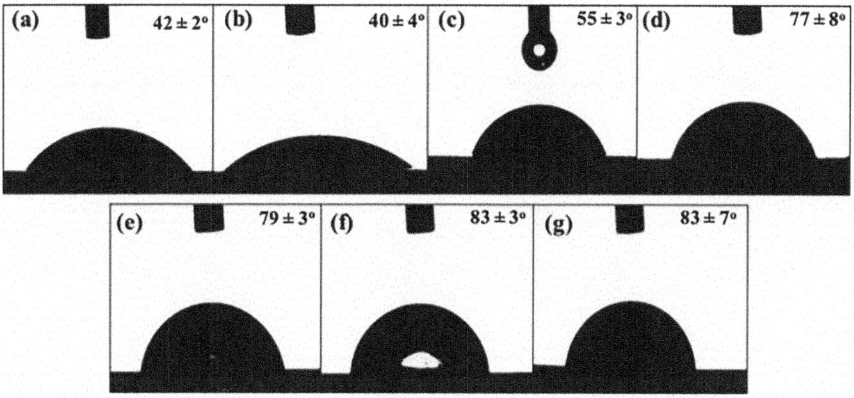


**Fig. 10** Contact angle of the thermoplastic starch composites **a** neat, **b** 2NCC, **c** 2 L, **d** 2NL, **e** 1NL1NCC, **f** 2NL2NCC, and **g** 4NL4NCC [69]. Copyright © 2023 Springer under exclusive license to Springer Nature Singapore Pte Ltd.

#### Contact Angle

The contact angle was studied to investigate the wetting behavior and surface hydrophobicity of biocomposites with LNPs. Perera et al. used a goniometer for contact angle measurements of TPS/LNP composite films containing 2 wt.% LNPs and 4 wt.% LNPs + 4 wt.% NCC [69], as shown in Fig. 10. Initially, the samples were fixed to a glass plate, and distilled water droplets (4 μL) were sprayed on the films. Finally, the images were captured. Thereafter, the contact angle was measured using DROPimage software. They discovered that the contact angle of nanocomposite films with 4 wt.% LNPs and 4 wt.% NCC was 83° due to the blending of NCC and LNPs, which reduced the hydrophilicity of the starch molecules, and that of films with 2 wt.% LNPs was 77°.

Makri et al. observed the contact angle of biobased polylactic acid (PLA)/LNP films with different LNP contents (0.5–5 wt.%) [58]. They noted that the contact angle was stable at 81–82°, which is independent of the LNP loading level. However, it increased with increasing lignin content from 75 to 83° due to the effectiveness of lignin as a filler for enhancing surface hydrophobicity. The physical properties, including moisture content, water absorption, swelling ratio, and contact angle, of the different biocomposites with varying proportions of LNPs are listed in Table 1.

### 4.1.2 Mechanical Properties of Biocomposites with Nanolignin

Mechanical property testing was conducted according to the American Standard Testing Method (ASTM) standards. The most common mechanical tests, such as impact, tensile, and flexural tests, were performed in accordance with the ASTM D638 and ASTM D790 standards. Table 2 summarizes the mechanical properties of the biocomposites with LNPs.

**Table 1** Physical properties of the biocomposites containing nanolignin

| Biocomposites | Nanolignin (NL) content (wt.%) | Moisture content (%) | Water absorption/ Swelling ratio | Contact angle (°) | References |
|---|---|---|---|---|---|
| TPS/LNP | 2 | 35.1 | 27.0 | 77 | Perera et al. [69] |
| TPS/LNP/NCC | NL-4<br>NCC-4 | 34.6 | 31.1 | 83 | Perera et al. [69] |
| WG/LNP | 3 | 10.14 | 103.21 | | Yang et al. [100] |
| Biobased PLA/ LNP | 2.5 | | | 81 | Makri et al. [58] |
| Bio-PTT/LNP | 2.5 | | 17 | | Gupta et al. [29] |
| Bio-PPT/LNP/ Vapor Grown Carbon Fiber (VGCF) | LNP-1.5<br>VGCF-7 | | 12 | | Gupta et al. [29] |

Gupta et al. examined the mechanical properties of Bio-PTT/LNP (LNP) nanocomposites according to ASTM D 638 standards [29]. They discovered that the addition of 1.5 wt.% of LNPs in bio-PTT enhanced the mechanical properties. The tensile strength, flexural strength, tensile modulus, flexural modulus, and impact strength of the nanocomposites with 1.5 wt.% LNPs were 59.16 MPa, 88.25 MPa, 2227.21 MPa, 2626.23 MPa, and 38.52 J/m, respectively. However, the incorporation of 1.5 wt.% LNPs + 7 wt.% VGCF in the bio-PTT matrix resulted in composites with the greatest increase in mechanical properties. These composites exhibited 61.74 MPa of tensile strength, 93.73 MPa of flexural strength, 2308.84 MPa of tensile modulus, 3086.18 MPa of flexural modulus, and 43.17 J/m of impact strength. Perera et al. evaluated the tensile strength and elongation at break of TPS/NL/NCC composite films using a TAXT2 texture analyzer. They reported that a higher tensile strength of 13.15 MPa and a lower elongation at break (9%) were obtained with the incorporation of 4 wt.% LNPs/4 wt.% NCC in a starch matrix compared to films with 2 wt.% LNPs alone, which exhibited a tensile strength of 6.88 MPa and an 18% elongation at break [69]. They concluded that at high loading levels of LNPs, up to 8 wt.% of both LNPs+NCC contributed to an increase in the tensile strength of the composite films. Yang et al. [100] discovered that adding 3 wt.% LNPs to a WG matrix enhanced the tensile properties of WG/LNP composites. However, it also reduced the elongation at break to 8%. Dynamic mechanical analysis of WG/ 3 wt.% LNP composites using an ARES N2 instrument showed increased storage modulus (E′) and tan δ values and exhibited strong reinforcement between the filler and matrix by Yang et al. [100].

Hosseinmardi et al. evaluated the mechanical characteristics of natural rubber latex (NR)/organosolv lignin (OSL) nanocomposites at various loading levels of OSL (1,

**Table 2** Mechanical properties of the biocomposites containing nanolignin

| Biocomposites | NL content (wt.%) | Tensile strength (MPa) | Tensile modulus (MPa) | Flexural strength (MPa) | Flexural modulus (MPa) | Impact strength (J/m) | Elongation at break (%) | Modulus at 300% (MPa) | Hardness (shore A) | References |
|---|---|---|---|---|---|---|---|---|---|---|
| Bio-poly (trimethylene terephthalate) (Bio-PTT)/LNP | 1.5 | 59.16 | 2227.21 | 88.25 | 2626.23 | 38.52 | – | – | – | Gupta et al. [29] |
| TPS/LNP | 2 | 6.88 | – | – | – | – | 18 | – | – | Perera et al. [69] |
| Thermoplastic starch/LNP/NCC | NL-4 NCC-4 | 13.15 | – | – | – | – | 9 | – | – | Perera et al. [69] |
| Bio-PPT/LNP/VGCF | LNP-1.5 VGCF-7 | 61.74 | 2308.89 | 93.73 | 3086.16 | 43.17 | – | – | – | Gupta et al. [29] |
| | 3 | 13.3 | 553.2 | – | – | – | 28 | – | – | Yang et al. [100] |
| Biobased PLA/Nanolignin | 2.5 | 23.86 | 1373 | – | – | – | 2.3 | – | – | Makri et al. [58] |
| Natural rubber latex/Organosolv Lignin (OSL) particles | 5 | 39.3 | – | – | – | – | 1798 | 2.1 | 59.8 | Hosseinmardi et al. [31] |
| Natural rubber/LigninPoly Complexes (LPCs) | 7 | 29.24 | – | – | – | – | 658 | 2.95 | 44 | Jiang et al. [34] |

2, 5, 10, and 20 wt.% of OSL) according to ASTM D638 standards. Composites prepared without dusting or leaching at the incorporation of 5 wt.% OSL in rubber latex exhibited a tensile strength of 39.3 MPa, a modulus of 2.1 MPa at 300%, an elongation of 1798% at break, and a hardness of 59.8 at shore A, as depicted in Fig. 11. Moreover, NR/5 wt.% OSL nanocomposites fabricated after dusting and leaching exhibited a decrease in tensile strength and an increase in other mechanical properties. Therefore, the results suggested that the dusting and leaching process affected the mechanical properties of the NR/OSL nanocomposites [31]. Jiang et al. [34] discovered that NR/lignin composites showed a decrease in tensile properties, whereas increasing lignin polycomplexes (LPCs) to 7 wt.% resulted in the highest tensile properties while retaining the elongation of break-in NR/LPC composites due to strong crosslinking between the LNPs and the matrix. The dynamic mechanical properties of NR/LPCs and NR/lignin composites were determined by Jiang et al. [34]. They suggested that the tangent delta (tan δ) peak height was lower and the storage modulus (E′) was greater than the glass transition temperature in NR/LPCs because of the strong interaction between the polymer and LNPs, which resulted in an increase in the stiffness compared to that of the NR/lignin composites.

Makri et al. investigated the tensile strength, tensile modulus, and elongation at break of biobased-polylactic acid (PLA)/LNP composites. The results showed that increasing the LNP content from 0 to 2.5 wt.% increased the tensile strength by 23.86 MPa, increased the tensile modulus by 1373 MPa, and reduced the elongation at break to 2.3% because of the proper dispersion of LNPs in the PLA matrix. However, the addition of LNPs up to 5 wt.% reduced the tensile properties due to the improper dispersion of LNPs in the biobased PLA matrix [58].

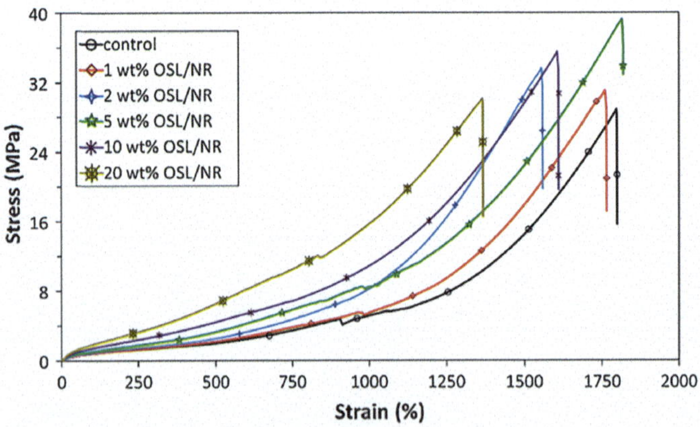


**Fig. 11** Tensile stress–strain curves of the control and OSL/NR nanocomposites without dusting and leaching [31]. Toughening of natural rubber nanocomposites by incorporating LNPs combined with an industrially relevant leaching process. *Industrial Crops and Products*, *159*, 113063

## 4.2 Thermal Properties of the Biocomposites

Thermogravimetric analysis (TGA) and differential scanning calorimetry (DSC) were used to evaluate the thermal properties of the biocomposites with LNPs. TGA can provide data related to thermal parameters, including peak temperatures and residue wt.%, and the derivative thermogravimetric (DTG) curve is used to determine the weight loss and highest thermal deterioration temperature ($T_{max}$) of samples [19, 95, 94]. DSC analysis mainly focuses on the $T_g$ and crystallization of samples. The addition of LNPs as fillers at lower loading levels can improve the thermal stability of biocomposites due to strong crosslinking [4, 49, 104]. Commonly used instruments for evaluating thermal properties are the thermogravimetric analyzer (TGA 550) and the TA Q200 series DSC instrument [69, 100]. During TGA, approximately 8–10 mg of sample was initially heated in a nitrogen atmosphere at a temperature ranging from 25 to 800 °C at a rate of 10 °C/min. Finally, TG/DTG curves were obtained to estimate the thermal parameters of the samples [81].

DSC analysis can be performed by taking 5–8 mg of each sample, placing it in a sealed pan, and heating it from 20 to 210 °C under a nitrogen environment at a ramping rate of 10 °C/min. During the first scan, the samples were cooled at an average rate of 100 °C/min and reheated to 210 °C in the second scan. The two main protocols employed in DSC analysis estimate the degree of biocomposite crystallization [58, 69, 100]. Perera et al. [69] investigated the thermal properties of TPS/LNP/NCC composite films by using a thermogravimetric analyzer (TGA 550) and a TA Instruments DSC Q200 instrument. As shown in Fig. 12, the TG curve revealed three major peaks, which were associated with moisture evaporation (60–100 °C), glycerol breakdown (160–210 °C), and starch decomposition (240–280 °C). As the LNPs and NCCs absorbed more space in the starch matrix and slowed the evaporation of water and glycerol, the temperature at which water evaporated ($T_{peak1}$) increased in the films composed of 1NL/1NCC and 2NL/2NCC [99, 101]. From the DTG curves, it was observed that the final residue quantities at 800 °C were lower for the 2 L (8%) and 2NL (9%) composites than for the 2NCC film (11%) due to their lower degree of crystallization. Moreover, the 4NL/4NCC film produced the most residue because it contained the greatest amount of nanomaterials.

DSC analysis revealed that the $T_g$ of the 2NL film was 52 °C higher than that of the neat films (40 °C) and 2 L films (42 °C). However, the $T_g$ increased with increasing nanofiller loading in the TPS/4NL/4NCC composites. The results suggested that the incorporation of 4 wt.% LNPs and 4 wt.% NCC into the starch matrix improved the thermal stability due to the high crystallization rate and strong bonding. Table 3 shows the different thermal properties of the biocomposites with LNPs through TGA and DSC analysis.

Yang et al. reported the thermal properties of WG/LNP composites with varying LNP contents (0–3 wt.%) [100]. They found that composites containing LNPs displayed a four-step decomposition pattern during thermal deterioration under nitrogen, consistent with the removal of ambient moisture and glycerol, the disintegration of LNPs, and the breakdown of WG protein. The DTG curve data revealed

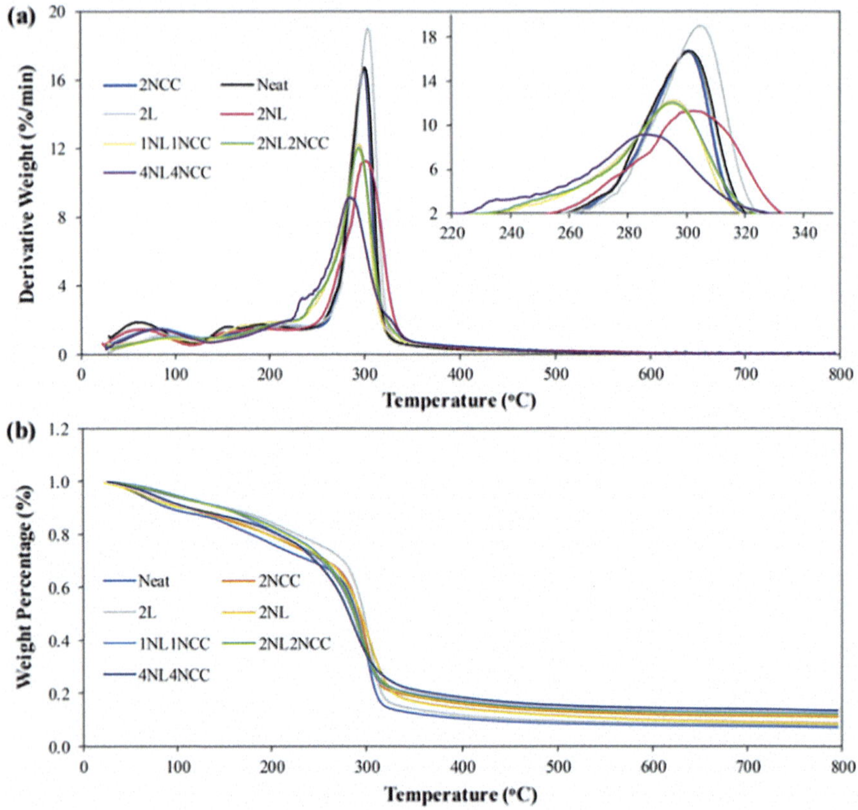


**Fig. 12** TGA analysis of the TPS/LNP/NCC nanocomposites: **a** derivative weight curves and **b** residual mass curves [69]. Copyright © 2023 Springer under exclusive license to Springer Nature Singapore Pte Ltd.

that the $T_{peak1}$ values (°C) below 100 °C ascribed to the elimination of water by the WG/0LNP, WG/1LNP, and WG/3LNP composites were 66.3, 78.5, and 82.4 °C, respectively. The second peak attributed to the evaporation of glycerol in peak 2 was attributed to the increase in the LNP concentration (0–3 wt.%), and peak 2 increased from 225.4 to 248.5 °C. The breakdown of WG occurred at 310 °C ($T_{max}$), which was attributed to the degradation of LNPs, as the LNPs have a larger temperature range for deterioration [62]. Moreover, the WG matrix with a greater initial onset of degradation is more thermally stable. The reversible heat flow of the WG/LNP bionanocomposites from the DSC thermal scan is shown in Fig. 13d. However, the addition of LNPs did not affect the residue weight but improved the thermal stability of the WG composites. DSC analysis suggested that the Tg increased from 90.2 to 106.8 °C when the LNP content increased from 0 to 3 wt.%. Makri et al. reported that the $T_g$ temperature was not significantly impacted by the presence of either filler (lignin or LNPs) in biobased PLA/Lignin (L)/LNP composites obtained through

**Table 3** Thermal properties of biocomposites with nanolignin

| Biocomposites | LNP content (wt.%) | $T_{peak1}$ (°C) | $T_{max}$ (°C) | $T_{50\%}$ (°C) | Residue (%) | $T_g$ (°C) | References |
|---|---|---|---|---|---|---|---|
| TPS/LNP | 2 | 67 | 304 | | 9 | 52 | Perera et al. [69] |
| TPS/LNP/NCC | NL-4 NCC-4 | 78 | 286 | | 14 | 61 | Perera et al. [69] |
| WG/LNP | 3 | | 310 | | 21.1(@600 °C) | 106.8 | Yang et al. [100] |
| Biobased PLA/LNP | 2.5 | | | | 46 (24 h) | 58 | Makri et al. [58] |
| Bio-poly (trimethylene terephthalate) (Bio-PTT)/LNP | 1.5 | | 447 | 386 | | | Gupta et al. [29] |
| Bio-PPT/LNP/VGCF | LNP-1.5 VGCF-7 | | 542 | 397 | | | Gupta et al. [29] |
| Natural rubber/LigninPoly Complexes (LPCs) | | | 386.5 | 390.7 | | | Jiang et al. [34] |
| Chitosan/LNP | | | | | 30.20@800 °C | | Sohni et al. [81] |

DSC analysis. They evaluated the temperatures of $T_g$, cold crystallization ($T_{cc}$), and melting ($T_m$) of biobased PLA/LNP/L composites. They concluded that cold crystallization resulted in incomplete crystallization due to more nucleation compared to melt crystallization under both cold and melt crystallization conditions recorded at low temperatures. However, the addition of LNPs prevented the proportional change in heat capacity from occurring in the PLA composites [58].

By TGA analysis, Jiang et al. demonstrated the thermal stability and thermo-oxidative decomposition of pure natural rubber (NR), NR/lignin, and NR/lignin poly-complex (LPC) composites. The DTG curves indicated that natural rubber oxidation was the key factor contributing to the first major degradation peak (358.8 °C). The results showed that the thermal decomposition at 5% (312 °C), 50% weight loss (390.7 °C), and $T_{max}$ (386.5 °C) were greatest for the NR/LPC composites with a loading of 3 wt.% LPC due to proper dispersion. All other specific temperatures during the thermal-oxidative decomposition of NR/3LPC composites, excluding the $T_{max,}$ were inferior to those of NR/lignin composites [34]. These detailed results achieved in thermal-oxidative decomposition suggested that the phenolic hydroxyl groups found in lignin are inhibited and can serve as stabilizing agents for reactions carried by oxygen and its radical species. Sohni et al. [81] investigated the impact of LNPs in a chitosan matrix by examining the thermal stabilities of chitosan (Chit)

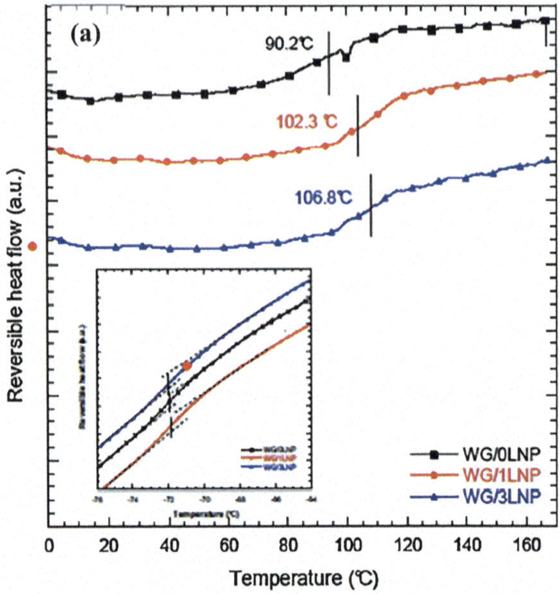


**Fig. 13** DSC analysis of WG/LNP nanocomposites filled with 0, 1, or 3 wt.% LNPs. (a) DSC thermal scan (reversible heat flow) [100]. Yang, W., Kenny, J. M., and Puglia, D. [100]. Structure and properties of biodegradable wheat gluten bionanocomposites containing lignin nanoparticles. *Industrial Crops and Products, 74,* 348–35

and Chit/LNP composites. The TG/DTG curves for the native chitosan and Chit/LNP composites were remarkably similar. However, Chit/LNP had a higher thermolytic onset temperature (268.76 °C) than did native chitosan (257.41 °C) due to the moderating effect of the lignin nanospheres. The residue weight at 800 °C in Chit/LNP composites (30.20%) was greater than that in chitosan (11%). They concluded that incorporating 0.2 g of LNPs into the chitosan matrix formed a strongly crosslinked network at high temperatures and enhanced the thermal stability of the Chit/LNP composites.

Gupta et al. analyzed the thermal characteristics of biopoly(trimethylene terephthalate) (Bio-PTT)/LNP/VGCF nanocomposites. By TGA, they reported that the addition of 1.5 wt.% LNPs to bio-PTT increased the temperature of initial degradation (337 °C), weight loss (at 50% [386 °C]), and final degradation (447 °C) compared to those of the native bio-PTT composites. The results showed that LNPs were impermeable to tiny molecules produced during heat degradation. Therefore, LNPs reduced the release rate of byproducts during decomposition and exhibited a thermal shielding effect that improved the thermal stability of the bio-PTT/LNP composites. However, 7 wt.% VGCF loaded in the bio-PTT/1.5 wt.% LNP composite had the highest initial degradation temperature (344 °C) and final degradation temperature (542 °C) due to strong crosslinking between the LNPs and VGCFs, which restricted the thermal action of the bio-PTT molecules in the bio-PTT/1.5 wt.% LNP/7 wt.% VGCF composites [29].

### 4.3 Flammability Behavior of the Biocomposites

One of the drawbacks of polymeric materials is that they can be prone to fire incidents, resulting in economic and human losses. Polymeric matrices are susceptible to burning because of their high calorific value [77]. To overcome this issue, additives or reinforcing agents such as flame retardants are added to biocomposites [6]. Villamil Watson and Schiraldi [88] reported that biomolecules such as lignin significantly influence the combustibility, flammability, and char formation rates of biocomposites. The char-forming ability of the lignin biopolymer makes it a viable candidate for flame retardancy-based applications, whereas nanometric lignin offers even more potential in this regard. The high porosity, specific surface area, and aromatic carbon structure (i.e., the availability of hydroxyl functional groups) of LNPs allow them to be further developed as efficient additives for enhancing the fire performance of biocomposites. Due to the large surface area of LNPs, their incorporation in a polymer matrix to achieve excellent flammability characteristics is much lower than that of lignin. Thus, in recent years, LNPs have gained increasing attention owing to their flame retardancy properties because the nanoscale dimension allows nanometric-level dispersion in the polymer [87]. Dorieh et al. highlighted that nanofiller-based polymers improve properties such as flame retardancy, ablation resistance, and thermal conductivity owing to the high specific surface area, surface-to-volume ratio, and nanometric size of nanofillers [16]. NPs can impart an outstanding flame-retardant effect by proper dispersion in polymer matrices due to reduced heat release during the burning process [14]. Although LNPs possess superior properties to lignin pertaining to fire retardancy applications, research studies on the flammability features of LNPs are scarce [87].

He and coworkers [30] developed highly efficient polyurethane (PUE) elastomers functionalized with LNPs, which exhibited enhanced flame-retardant, thermal, and mechanical properties. Initially, lignin was chemically modified (FL) to incorporate phosphorus and nitrogen functionalities via a mild grafting reaction. Afterward, the solvent exchange method was adopted to prepare LNPs containing nitrogen (N)-phosphorus (P) moieties via subsequent ultrasonication. Ultrasonic irradiation facilitates homogeneous dispersion of LNPs, resulting in improved interactions with the polymer matrix [67]. Chemically modified LNPs obtained through this effective strategy have the potential to partially substitute polyols for flame-retardant PUEs. Their combustion behavior was evaluated via a cone calorimetry test (CCT). This study demonstrated that the addition of functionalized nanolignin (Nano-FL) lowered the heat release rate along with the total heat release. N- and P-grafted LNPs also reduced the amount of total smoke generated during the course of combustion.

Figure 14 depicts the fire risk assessment of PUEs containing lignin, FL, and various compositions of nano-FL investigated through CCT analysis. As shown in Fig. 14, the smoke production rate and total smoke production rate of the PUEs decreased due to the addition of nano-FL. In addition, compared with the other tested samples, the PUEs with various compositions of nano-FL showed delayed ignition

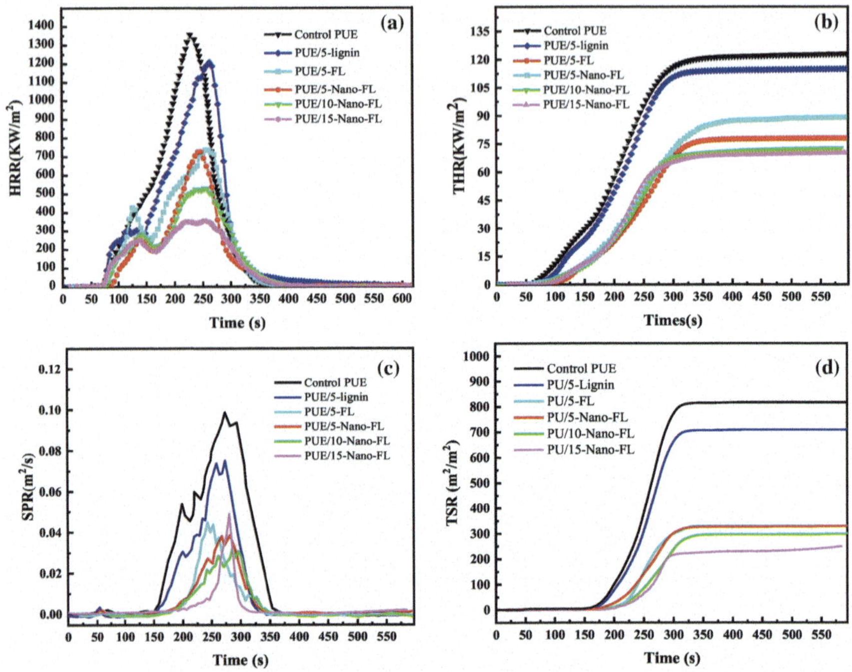


**Fig. 14** Cone calorimetry test results pertaining to PUEs. **a** Heat release rate (HRR) curves, **b** total heat release rate (THR) curves, **c** smoke production rate (SPR) curves, and **d** total smoke production rate (TSP) curves [30]. Reproduced with permission from Elsevier

times. The underlying flame prevention mechanism was confirmed through morphological observation of carbon char residues. SEM images showed that the PUE with modified LNPs contained a comparatively greater amount of carbon residue (9.08 wt.%) than did the PUE without any additives (2.46 wt.%), confirming the enhanced fire retardance performance. The incorporation of 15 wt.% nano-FL resulted in the highest limited oxygen index (LOI) of 30.8%, demonstrating the lowest flammability. The flame retardant effect of LNPs, i.e., fire prevention, was attributed to the formation of a dense char (carbon) layer and the release of ammonia gas (smoke) during the burning process. The study concluded that functionalized lignin at the nanoscale could be a potential green flame-retardant additive for a wide array of polymers, including elastomers, adhesives, coatings, and foams.

Chollet et al. investigated the ability of LNPs to enhance the flame-retarding effect of polylactide acid (PLA). LNPs were obtained from kraft lignin microparticles (LMPs) via a dissolution–precipitation process. Initially, LNPs were phosphorylated via diethyl chlorophosphate and diethyl (2-(triethoxysilyl)ethyl) phosphonate and incorporated into PLA by melt blending. The cone calorimeter test revealed that functional LNP blends obtained through diethyl (2-(triethoxysilyl)ethyl) phosphonate phosphorylation offered the highest resistance to ignition due to the grafting of a higher phosphorus content, which resulted in a longer combustion time and a

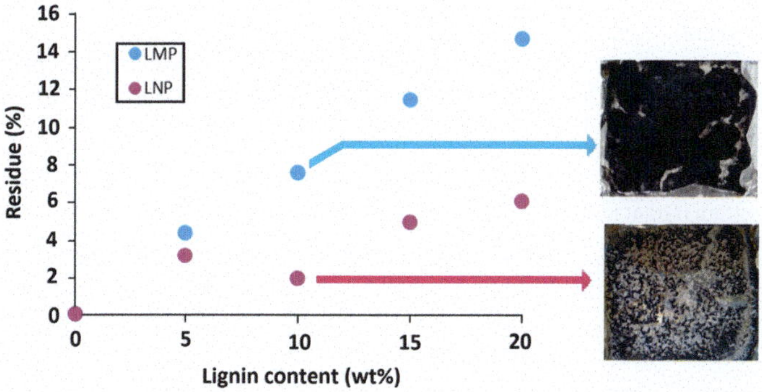


**Fig. 15** Influence of lignin particle size and content on the amount of carbon residue produced upon cone calorimeter testing [14]. Copyright © 2019 MDPI under CC BY 4.0

lower char residue. The amount of remaining char residue after being subjected to a cone calorimeter test also depends on the particle size of lignin [14]. Figure 15 reveals that LMP retained a greater amount of carbon residue than did LNP due to its larger particle size. The authors highlighted that smaller particles, in this case, LNPs, induced carbon residue degradation after the flame was removed. It was also noted that proper dispersion of LNPs in the polymer matrix results in a reduced rate of heat release during ignition.

## 5 Conclusion

Nanotechnology plays a significant role in facilitating the creation of biobased polymers with excellent properties, such as lignin. Lignin is a viable option for nonformaldehyde adhesives due to its functional groups from aliphatic and aromatic phenolic structures. Modification of lignin can enhance its reactivity. Lignin is employed solely as a filler in bioadhesives and other composites without being altered, which may decrease the effectiveness of these materials and the efficiency of subsequent processing. However, the conversion process for fully replacing existing commercial adhesives is still challenging. The synthesis of lignin into nanolignin (LNPs) is an approach for facilitating new excellent properties, including great specificity, abundance of functional groups, solubility, and compatibility, for further application in biocomposite systems. The mechanical milling approach is a popular method for preparing LNPs that includes a top-down technique, while self-assembly using a suitable solvent is an example of a bottom-up method. The active substances of LNPs support biocomposites with better thermal stability, high solubility, high fire resistance, and improved physical and mechanical properties. LNPs with controllable size and stability are still being developed to optimize bioadhesives through the introduction of LNPs.

**Acknowledgements** The authors are grateful for funding from the Join Collaboration Program of the Research Organization for Life Sciences and Environment National Research and Innovation Agency (BRIN) FY 2024 (No. 1/III.5/HK/2024). The authors are grateful for the scientific and technical support provided by the Integrated Laboratory of Bioproducts (iLaB) BRIN through the E-Layanan Sains BRIN, Indonesia. The authors are grateful for the research assistant scheme (Grant number: 60/II/HK/2023) of the National Research and Innovation Agency-BRIN and Degree by Research Program (Grant number: 174/H/2021) of Indonesian Institute of Sciences for Muhammad Rasyidur Ridho, S. Hut.

## References

1. Abid N, Khan A, Shujait S, Chaudhary K, Ikram DM, Imran M, Haider J, Khan M, Khan Q, Maqbool M (2021) Synthesis of nanomaterials using various top-down and bottom-up approaches, influencing factors, advantages, and disadvantages: a review. Adv Colloid Interface Sci 300(issue):102597
2. Agarwal M, Koelling KW, Chalmers JJ (1998) Characterization of the degradation of polylactic acid polymer in a solid substrate environment. Biotechnol Prog 14(issue):517–526
3. Ang AF, Ashaari Z, Lee SH, Tahir PM, Halis R (2019) Lignin-based copolymer adhesives for composite wood panels–a review. Int J Adhes Adhes 95(issue):102408
4. Antov P, Lee SH, Lubis MA, Yadav S (2022) Potential of nanomaterials in bio-based wood adhesives: an overview, pp 25–63
5. Avelino F, Almeida SL, Duarte EB, Sousa JR, Mazzetto SE, de Souza Filho MDSM (2018) Thermal and mechanical properties of coconut shell lignin-based polyurethanes synthesized by solvent-free polymerization. J Mater Sci 53(issue):1470–1486
6. Azman Mohammad Taib MN, Hamidon TS, Garba ZN, Trache D, Uyama H, Hussin MH (2022) Recent progress in cellulose-based composites toward flame retardancy applications. Polymer 244(issue):124677
7. Boerjan W, Ralph J, Baucher M (2003) Lignin biosynthesis
8. Cavallo E, He X, Luzi F, Dominici F, Cerrutti P, Bernal C, Foresti ML, Torre L, Puglia D (2020) UV protective, antioxidant, antibacterial and compostable polylactic acid composites containing pristine and chemically modified lignin nanoparticles. Molecules 26(issue)
9. Chauhan PS (2020) Lignin nanoparticles: eco-friendly and versatile tool for new era. Bioresour Technol Rep 9(issue)
10. Chauhan PS (2020) Lignin nanoparticles: eco-friendly and versatile tool for new era. Bioresour Technol Rep 9(issue):100374
11. Chen Y, Gong X, Yang G, Li Q, Zhou N (2019) Preparation and characterization of a nanolignin phenol formaldehyde resin by replacing phenol partially with lignin nanoparticles. RSC Adv 9(issue):29255–29262
12. Chen Y, Gong X, Yang G, Li Q, Zhou N (2019) Preparation and characterization of a nanolignin phenol formaldehyde resin by replacing phenol partially with lignin nanoparticles. RSC Adv 9(issue): 29255–29262
13. Chen Y, Zheng K, Niu L, Zhang Y, Liu Y, Wang C, Chu F (2019) Highly mechanical properties nanocomposite hydrogels with biorenewable lignin nanoparticles. Int J Biol Macromol 128(issue):414–420
14. Chollet B, Lopez-Cuesta J-M, Laoutid F, Ferry L (2019) Lignin nanoparticles as a promising way for enhancing lignin flame retardant effect in polylactide. Mater 12(13):2132. https://doi.org/10.3390/ma12132132
15. Deng J, Sun S-F, Zhu E-Q, Yang J, Yang H-Y, Wang D-W, Ma M-G, Shi Z-J (2021) Submicro and nanolignin materials: Small size and rapid progress. Ind Crop Prod 164(issue):113412

16. Dorieh A, Ayrilmis N, Farajollah Pour M, Ghafari Movahed S, Valizadeh Kiamahalleh M, Shahavi MH, Hatefnia H, Mehdinia M (2022) Phenol formaldehyde resin modified by cellulose and lignin nanomaterials: review and recent progress. Int J Biol Macromol 222(issue):1888–1907
17. Farooq M, Zou T, Riviere G, Sipponen MH, Österberg M (2018) Strong, ductile, and waterproof cellulose nanofibril composite films with colloidal lignin particles. Biomacromolecules 20(issue):693–704
18. Farooq M, Zou T, Riviere G, Sipponen MH, Österberg M (2019) Strong, ductile, and waterproof cellulose nanofibril composite films with colloidal lignin particles. Biomacromolecules 20(issue):693–704
19. Feldman D (2016) Lignin nanocomposites. J Macromol Sci, Part A: Pure Appl Chem 53(issue):382–387
20. Figueiredo P, Lahtinen MH, Agustin MB, de Carvalho DM, Hirvonen S-P, Penttilä PA, Mikkonen KS (2021) Green fabrication approaches of lignin nanoparticles from different technical lignins: a comparison study. ChemSusChem 14(issue):4718–4730
21. Frangville C, Rutkevičius M, Richter AP, Velev OD, Stoyanov SD, Paunov VN (2012) Fabrication of environmentally biodegradable lignin nanoparticles. ChemPhysChem 13(issue):4235–4243
22. Freitas FMC, Cerqueira MA, Gonçalves C, Azinheiro S, Garrido-Maestu A, Vicente AA, Pastrana LM, Teixeira JA, Michelin M (2020) Green synthesis of lignin nano- and microparticles: physicochemical characterization, bioactive properties and cytotoxicity assessment. Int J Biol Macromol 163(issue):1798–1809
23. Gellerstedt G (2015) Softwood kraft lignin: raw material for the future. Ind Crop Prod 77(issue):845–854
24. Gilca IA, Popa VI, Crestini C (2015) Obtaining lignin nanoparticles by sonication. Ultrason Sonochemistry 23(issue):369–375
25. Gonzalez MNG, Levi M, Turri S, Griffini G (2017) Lignin nanoparticles by ultrasonication and their incorporation in waterborne polymer nanocomposites. J Appl Polym Sci 134(issue):1–10
26. González K, Retegi A, González A, Eceiza A, Gabilondo N (2015) Starch and cellulose nanocrystals together into thermoplastic starch bionanocomposites. Carbohydr Polym 117(issue):83–90
27. Gupta AK, Mohanty S, Nayak SJ (2014) Synthesis, characterization and application of lignin nanoparticles (LNPs). Mater Focus 3(issue):444–454
28. Gupta AK, Mohanty S, Nayak SK (2014) Synthesis, characterization and application of lignin nanoparticles (LNPs). Mater Focus 3(issue):444–454
29. Gupta AK, Mohanty S, Nayak SK (2015) Influence of addition of vapor grown carbon fibers on mechanical, thermal and biodegradation properties of lignin nanoparticle filled biopoly(trimethylene terephthalate) hybrid nanocomposites. RSC Adv 5(issue):56028–56036
30. He T, Chen F, Zhu W, Yan N (2022) Functionalized lignin nanoparticles for producing mechanically strong and tough flame-retardant polyurethane elastomers. Int J Biol Macromol 209(issue):1339–1351
31. Hosseinmardi A, Amiralian N, Hayati AN, Martin DJ, Annamalai PK (2021) Toughening of natural rubber nanocomposites by the incorporation of nanoscale lignin combined with an industrially relevant leaching process. Ind Crop Prod 159(issue):113063
32. Hussin MH, Appaturi JN, Poh NE, Latif NHA, Brosse N, Ziegler-Devin I, Vahabi H, Syamani FA, Fatriasari W, Solihat NN, Karimah A, Iswanto AH, Sekeri SH, Ibrahim MNM (2022) A recent advancement on preparation, characterization and application of nanolignin. Int J Biol Macromol 200(issue):303–326
33. Ingtipi K, Moholkar VS (2019) Sonochemically synthesized lignin nanoparticles and its application in the development of nanocomposite hydrogel. Mater Today: Proc 17(issue):362–370
34. Jiang C, He H, Jiang H, Ma L, Jia DM (2013) Nanolignin filled natural rubber composites: Preparation and characterization. Express Polym Lett 7(issue):480–493
35. Ju T, Zhang Z, Li Y, Miao X, Ji J (2019) Continuous production of lignin nanoparticles using a microchannel reactor and its application in UV-shielding films. RSC Adv 9(issue):24915–24921

36. Juikar SJ, Vigneshwaran N (2017) Extraction of nanolignin from coconut fibers by controlled microbial hydrolysis. Ind Crop Prod 109(issue):420–425
37. Juikar SJ, Vigneshwaran N (2017) Microbial production of coconut fiber nanolignin for application onto cotton and linen fabrics to impart multifunctional properties. Surf Interfaces 9(issue):147–153
38. Jędrzejczak P, Collins MN, Jesionowski T, Klapiszewski Ł (2021) The role of lignin and lignin-based materials in sustainable construction–a comprehensive review. Int J Biol Macromol 187(issue):624–650
39. Kai D, Tan MJ, Chee PL, Chua YK, Yap YL, Loh XJ (2016) Toward lignin-based functional materials in a sustainable world. Green Chem 18(issue):1175–1200
40. Kalami S, Chen N, Borazjani H, Nejad M (2018) Comparative analysis of different lignins as phenol replacement in phenolic adhesive formulations. Ind Crop Prod 125(issue):520–528
41. Kaplan DL (1998) Introduction to biopolymers from renewable resources. Biopolymers from renewable resources. Springer, pp 1–29
42. Kauldhar BS, Sooch BS, Rai SK, Kumar V, Yadav SK (2021) Recovery of nanosized silica and lignin from sugarcane bagasse waste and their engineering in fabrication of composite membrane for water purification. Environ Sci Pollut Res 28(issue):7491–7502
43. Khan I, Saeed K, Khan I (2019) Nanoparticles: properties, applications and toxicities. Arab J Chem 12(issue):908–931
44. Krišťáková S, Neykov N, Antov P, Sedliačiková M, Reh R, Halalisan A-F, Hajdúchová I (2021) Efficiency of wood-processing enterprises—evaluation based on DEA and MPI: a comparison between Slovakia and Bulgaria for the period 2014–2018. Forests 12(issue):1026
45. Köse K, Mavlan M, Youngblood J (2020) Applications and impact of nanocellulose based adsorbents. Cellulose 27(issue)
46. Laurichesse S, Avérous L (2014) Chemical modification of lignins: toward biobased polymers. Prog Polym Sci 39(issue):1266–1290
47. Lee SJ, Kim HJ, Cho EJ, Song Y, Bae H-J (2015) Isolation and characterization of lignin from the oak wood bioethanol production residue for adhesives. Int J Biol Macromol 72(issue):1056–1062
48. Leskinen T, Smyth M, Xiao Y, Lintinen K, Mattinen ML, Kostiainen MA, Oinas P, Österberg M (2017) Scaling up production of colloidal lignin particles. Nord Pulp Pap Res J 32(issue):586–596
49. Li H, Deng Y, Liu B, Ren Y, Liang J, Qian Y, Qiu X, Li C, Zheng D (2016) Preparation of nanocapsules via the self-assembly of kraft lignin: a totally green process with renewable resources. ACS Sustain Chem Eng 4(issue):1946–1953
50. Li H, Deng Y, Wu H, Ren Y, Qiu X, Zheng D, Li C (2016) Self-assembly of Kraft lignin into nanospheres in dioxane-water mixtures. Holzforschung 70(issue):725–731
51. Liao JJ, Latif NHA, Trache D, Brosse N, Hussin MH (2020) Current advancement on the isolation, characterization and application of lignin. Int J Biol Macromol 162(issue):985–1024
52. Lievonen M, Valle-Delgado JJ, Mattinen M-L, Hult E-L, Lintinen K, Kostiainen MA, Paananen A, Szilvay GR, Setälä H, Österberg M (2016) A simple process for lignin nanoparticle preparation. Green Chem 18(issue):1416–1422
53. Lievonen M, Valle-Delgado JJ, Mattinen ML, Hult EL, Lintinen K, Kostiainen MA, Paananen A, Szilvay GR, Setälä H, Österberg M (2016) A simple process for lignin nanoparticle preparation. Green Chem 18(issue):1416–1422
54. Lizundia E, Armentano I, Luzi F, Bertoglio F, Restivo E, Visai L, Torre L, Puglia D (2020) Synergic effect of nanolignin and metal oxide nanoparticles into poly(l-lactide) bionanocomposites: material properties, antioxidant activity, and antibacterial performance. ACS Appl Bio Mater 3(issue):5263–5274
55. Low LE, Teh KC, Siva SP, Chew IML, Mwangi WW, Chew CL, Goh B-H, Chan ES, Tey BT (2021) Lignin nanoparticles: the next green nanoreinforcer with wide opportunity. Environ Nanotechnol Monit Manag 15(issue):100398
56. Lu Q, Zhu M, Zu Y, Liu W, Yang L, Zhang Y, Zhao X, Zhang X, Zhang X, Li W (2012) Comparative antioxidant activity of nanoscale lignin prepared by a supercritical antisolvent (SAS) process with nonnanoscale lignin. Food Chem 135(issue):63–67

57. Luo T, Wang C, Ji X, Yang G, Chen J, Janaswamy S, Lyu G (2021) Preparation and characterization of size-controlled lignin nanoparticles with deep eutectic solvents by nanoprecipitation. Molecules 26(issue):218
58. Makri SP, Xanthopoulou E, Klonos PA, Grigoropoulos A, Kyritsis A, Tsachouridis K, Anastasiou A, Deligkiozi I, Nikolaidis N, Bikiaris DN (2022) Effect of micro- and nano-lignin on the thermal, mechanical, and antioxidant properties of biobased PLA–lignin composite films. Polymers 14(issue):5274
59. Malcolmson RJ, Embleton JK (1998) Dry powder formulations for pulmonary delivery. Pharm Sci Technol Today 1(issue):394–398
60. Mishra PK, Ekielski A (2019) A simple method to synthesize lignin nanoparticles. Colloids Interfaces
61. Miyashiro D, Hamano R, Umemura K (2020) A review of applications using mixed materials of cellulose, nanocellulose and carbon nanotubes. Nanomaterials 10(issue):186
62. Mojumdar SC, Moresoli C, Simon LC, Legge RL (2011) Edible wheat gluten (WG) protein films. J Therm Anal Calorim 104(issue):929–936
63. Montero B, Rico M, Rodríguez-Llamazares S, Barral L, Bouza R (2017) Effect of nanocellulose as a filler on biodegradable thermoplastic starch films from tuber, cereal and legume. Carbohydr Polym 157(issue):1094–1104
64. Nair SS, Sharma S, Pu Y, Sun Q, Pan S, Zhu JY, Deng Y, Ragauskas AJ (2014) High shear homogenization of lignin to nanolignin and thermal stability of nanolignin-polyvinyl alcohol blends. ChemSusChem 7(issue):3513–3520
65. Pahlevanneshan Z, Deypour M, Kefayat A, Rafienia M, Sajkiewicz P, Esmaeely Neisiany R, Enayati MS (2021) Polyurethane-nanolignin composite foam coated with propolis as a platform for wound dressing: synthesis and characterization. Polymers 13(issue):3191
66. Pang T, Wang G, Sun H, Wang L, Liu Q, Sui W, Parvez AM, Si C (2020) Lignin fractionation for reduced heterogeneity in self-assembly nanosizing: toward targeted preparation of uniform lignin nanoparticles with small size. ACS Sustain Chem Eng 8(issue):9174–9183
67. Parvathy G, Sethulekshmi AS, Jayan JS, Raman A, Saritha A (2021) Lignin based nanocomposites: synthesis and applications. Process Saf Environ Prot 145(issue):395–410
68. Passoni V, Scarica C, Levi M, Turri S, Griffini G (2016) Fractionation of industrial softwood kraft lignin: solvent selection as a tool for tailored material properties. ACS Sustain Chem Eng 4(issue):2232–2242
69. Perera UP, Foo ML, Chew IML (2023) Effect of nanolignin and nanocrystalline cellulose on thermal, mechanical, and water barrier properties of starch composites. In: Foo DCY, Aziz MKTA, Yusup S (eds) Sustainable technologies for the oil palm industry: latest advances and case studies. Springer Nature Singapore, Singapore, pp 117–137
70. Pileni MP (2001) Semiconductor nanocrystals. In Klabunde KJ (ed) Nanoscale materials in chemistry. Wiley, New York, USA, pp 61–84
71. Pylypchuk IV, Lindén PA, Lindström ME, Sevastyanova O (2020) New insight into the surface structure of lignin nanoparticles revealed by $^1$H liquid-state NMR spectroscopy. ACS Sustain Chem Eng 8(issue):13805–13812
72. Qian Y, Deng Y, Qiu X, Li H, Yang D (2014) Formation of uniform colloidal spheres from lignin, a renewable resource recovered from pulping spent liquor. Green Chem 16(issue):2156–2163
73. Rahman OU, Shi S, Ding J, Donglin W, Ahmad S, Yu H (2018) Lignin nanoparticles: synthesis, characterization and corrosion protection performance. New J Chem 42(issue):3415–3425
74. Reverchon E, Adami R (2006) Nanomaterials and supercritical fluids. J Supercrit Fluids 37(issue):1–22
75. Ridho MR, Madyaratri EW, Agustiany EA, Solihat NN, Nawawi DS, Juliana I, Fatriasari W (2022) Isolation and characterization of lignin from black liquor of arecanut leaf sheath (Areca catechu L.) with hydrochloric acid and phosphoric acid. In: Lubis MAR, Hua LS, Antov P, Kristak L (eds) The 2nd international conference of lignocellulose (ICON-LIG) 2022. AIP Publisher, Bogor, Indonesia

76. Sedliaciková M, Stroková Z, Klementová J, Satanová A, Moresová M (2020) Impacts of behavioral aspects on financial decision-making of owners of woodworking and furniture manufacturing and trading enterprises. Acta Fac Xylologiae Zvolen 62(issue):165–176
77. Seidi F, Movahedifar E, Naderi G, Akbari V, Ducos F, Shamsi R, Vahabi H, Saeb MR (2020) Flame retardant polypropylenes: a review. Polymers
78. Shuai L, Saha B (2017) Toward high-yield lignin monomer production. Green Chem 19(issue):3752–3758
79. Sipponen MH, Lange H, Ago M, Crestini C (2018) Understanding lignin aggregation processes. A case study: budesonide entrapment and stimuli controlled release from lignin nanoparticles. ACS Sustain Chem Eng 6(issue):9342–9351
80. Sipponen MH, Smyth M, Leskinen T, Johansson L-S, Österberg M (2017) All-lignin approach to prepare cationic colloidal lignin particles: stabilization of durable Pickering emulsions. Green Chem 19(issue):5831–5840
81. Sohni S, Hashim R, Nidaullah H, Lamaming J, Sulaiman O (2019) Chitosan/nanolignin based composite as a new sorbent for enhanced removal of dye pollution from aqueous solutions. Int J Biol Macromol 132(issue):1304–1317
82. Solihat NN, Sari FP, Falah F, Ismayati M, Lubis MAR, Fatriasari W, Santoso EB, Syafii W (2021) Lignin as an active biomaterial: a review. Jurnal Sylva Lestari 9(issue):1–1
83. Taghiyari HR, Tajvidi M, Taghiyari R, Mantanis GI, Esmailpour A, Hosseinpourpia R (2020) Chapter 19—nanotechnology for wood quality improvement and protection. In: Husen A, Jawaid M (eds) Nanomaterials for agriculture and forestry applications. Elsevier, pp 469–489
84. Tian D, Hu J, Bao J, Chandra RP, Saddler JN, Lu C (2017) Lignin valorization: lignin nanoparticles as high-value bioadditive for multifunctional nanocomposites. Biotechnol Biofuels 10(issue):1–11
85. Tian D, Hu J, Chandra RP, Saddler JN, Lu C (2017) Valorizing recalcitrant cellulolytic enzyme lignin via lignin nanoparticles fabrication in an integrated biorefinery. ACS Sustain Chem Eng 5(issue):2702–2710
86. ur Rahman O, Shi S, Ding J, Wang D, Ahmad S, Yu H (2018) Lignin nanoparticles: synthesis, characterization and corrosion protection performance. New J Chem 42(issue):3415–3425
87. Vahabi H, Brosse N, Latif NHA, Fatriasari W, Solihat NN, Hashim R, Hazwan Hussin M, Laoutid F, Saeb MR (2021) 24—nanolignin in materials science and technology—does flame retardancy matter? In: Kanwar S, Kumar A, Nguyen TA, Sharma S, Slimani Y (eds) Biopolymeric nanomaterials. Elsevier, pp 515–559
88. Villamil Watson DA, Schiraldi DA (2020) Biomolecules as flame retardant additives for polymers: a review. Polymers
89. Wang B, Sun D, Wang H-M, Yuan T-Q, Sun R-C (2019) Green and facile preparation of regular lignin nanoparticles with high yield and their natural broad-spectrum sunscreens. ACS Sustain Chem Eng 7(issue):2658–2666
90. Wang B, Sun D, Wang HM, Yuan TQ, Sun RC (2019) Green and facile preparation of regular lignin nanoparticles with high yield and their natural broad-spectrum sunscreens. ACS Sustain Chem Eng 7(issue):2658–2666
91. Wang M, Leitch M, Xu CC (2009) Synthesis of phenol–formaldehyde resol resins using organosolv pine lignins. Eur Polym J 45(issue):3380–3388
92. World Health Organization, Food, Agriculture Organization of the United United Nations (2020) Safety evaluation of certain food additives: prepared by the eighty-sixth meeting of the joint FAO/WHO expert committee on food additives (JECFA). World Health Organization, Geneva, 572 p
93. Xiong F, Han Y, Wang S, Li G, Qin T, Chen Y, Chu F (2017) Preparation and formation mechanism of size-controlled lignin nanospheres by self-assembly. Ind Crop Prod 100(issue):146–152
94. Yang W, Zhu Y, Wang Q, Ou R, Ma P (2022) Chapter 14—biopolymer/nanolignin films with enhanced toughness. In: Puglia D, Santulli C, Sarasini F (eds) Micro and nanolignin in aqueous dispersions and polymers. Elsevier, pp 433–458

95. Yang W, Qi G, Ding H, Xu P, Dong W, Zhu X, Zheng T, Ma P (2020) Biodegradable poly (lactic acid)-poly (ε-caprolactone)-nanolignin composite films with excellent flexibility and UV barrier performance. Compos Commun 22(issue):100497
96. Yang W, Rallini M, Natali M, Kenny J, Ma P, Dong W, Torre L, Puglia D (2019) Preparation and properties of adhesives based on phenolic resin containing lignin micro and nanoparticles: a comparative study. Mater Des 161(issue):55–63
97. Yang L, Wang X, Cui Y, Tian Y, Chen H, Wang Z (2014) Modification of renewable resources—lignin—by three chemical methods and its applications to polyurethane foams. Polym Adv Technol 25(issue):1089–1098
98. Yang S, Wu J-Q, Zhang Y, Yuan T-Q, Sun R-C (2015) Preparation of lignin-phenol-formaldehyde resin adhesive based on active sites of technical lignin. J Biobased Mater Bioenergy 9(issue):266–272
99. Yang W, Fortunati E, Dominici F, Giovanale G, Mazzaglia A, Balestra GM, Kenny JM, Puglia D (2016) Synergic effect of cellulose and lignin nanostructures in PLA based systems for food antibacterial packaging. Eur Polym J 79(issue):1–12
100. Yang W, Kenny JM, Puglia D (2015) Structure and properties of biodegradable wheat gluten bionanocomposites containing lignin nanoparticles. Ind Crop Prod 74(issue):348–356
101. Yang W, Owczarek JS, Fortunati E, Kozanecki M, Mazzaglia A, Balestra GM, Kenny JM, Torre L, Puglia D (2016) Antioxidant and antibacterial lignin nanoparticles in polyvinyl alcohol/chitosan films for active packaging. Ind Crop Prod 94(issue):800–811
102. Yearla SR, Padmasree K (2016) Exploitation of subabul stem lignin as a matrix in controlled release agrochemical nanoformulations: a case study with herbicide diuron. Environ Sci Pollut Res 23(issue):18085–18098
103. Younesi-Kordkheili H, Pizzi A (2022) Improving the properties of phenol-lignin-glyoxal as a wood adhesive by lignin nanoparticles. Eur J Wood Wood Prod
104. Zhang C-W, Nair SS, Chen H, Yan N, Farnood R, Li F-Y (2020) Thermally stable, enhanced water barrier, high strength starch biocomposite reinforced with lignin containing cellulose nanofibrils. Carbohydr Polym 230(issue):115626
105. Zhang Q, Li H, Guo Z, Xu F (2020) High purity and low molecular weight lignin nanoparticles extracted from acid-assisted MIBK pretreatment. Polymers 12(issue):378
106. Zhang Z, Terrasson V, Guénin E (2021) Lignin nanoparticles and their nanocomposites. Nanomaterials
107. Zhao X, Zhang Z, Pang J, Su L (2022) Study on the preparation of epoxy resin materials from nanolignin polyols. Ind Crop Prod 185(issue):115158

# Sustainable and Eco-Friendly Starch-Based Adhesives for High-Performance Wood-Based Composites

**Muhammad Adly Rahandi Lubis, Apri Heri Iswanto, and Muhammad Iqbal Maulana**

**Abstract** Starch-based adhesives have emerged as a promising sustainable alternative for wood-based composites due to the necessity to substitute oil-derived adhesives and lessen environmental impact. The renewable, biodegradable, and abundant nature of starch provides a cost-effective and eco-friendly raw material for adhesive production. Recent progressions have concentrated on enhancing the water resistance and bonding strength of starch-based adhesives through various modifications and treatments. Nevertheless, challenges such as inadequate water resistance and weak adhesion in wet conditions are encountered. To tackle these issues, researchers have investigated chemical modifications like esterification, which notably enhance the water resistance of starch-based adhesives, rendering them suitable for applications in paper, plywood, and wood composites. The strategic alteration of starch through gelatinization, oxidation, and the integration of compatibilizers has additionally boosted the mechanical properties and processability of starch-based adhesives, making them comparable to synthetic materials. Additionally, the advancement of bio-based adhesives from avocado seed starch reinforced with clay particles has exhibited promising outcomes, with optimized shear strength and favorable interaction among the adhesive components. The increasing interest in bio-based wood adhesives is also mirrored in the industrial and research emphasis on developing non-toxic, low-emission wood composites with enhanced properties, capitalizing on the renewability and strong adhesion properties of starch. The transition towards starch-based adhesives signifies a sustainable, cost-effective, and eco-friendly approach to manufacturing high-performance wood-based composites.

**Keywords** Adhesives · Starch · Compatibilizer · Esterification · Gelatinization · Oxidation

M. A. R. Lubis (✉) · M. I. Maulana
Research Center for Biomass and Bioproducts, National Research and Innovation Agency, Bogor, Indonesia
e-mail: muha142@brin.go.id

A. H. Iswanto
Department of Forest Product, Faculty of Forestry, Universitas Sumatera Utara, Medan, Indonesia

 
P. Antov et al. (eds.), *Nanomaterials Additives in Bioadhesives for Wood Composites*, Smart Nanomaterials Technology, https://doi.org/10.1007/978-981-96-3400-2_5

# 1 Introduction

The increasing environmental concerns and the push towards sustainable development have driven significant interest in the development of eco-friendly adhesives for wood-based composites. Traditional wood composites often rely on synthetic, formaldehyde-based adhesives, which, despite their effectiveness, pose serious health and environmental risks due to the emission of hazardous volatile organic compounds (VOCs), including carcinogenic formaldehyde [4, 7, 17]. In response to these challenges, researchers have been exploring alternative adhesive solutions that are both effective and environmentally benign. Among these alternatives, starch-based adhesives have emerged as a promising candidate due to their biodegradability, renewability, and low-cost [16, 20].

Starch is the mixture of two distinct polysaccharide fractions of amylose and amylopectin which both are composed of glucose with different sizes and shapes. The ratio of these two components varies based on the botanical origin of the starch. Amylose has a relatively long linear α-glucan structure composed of approximately 99% $(1 \rightarrow 4)$-α-linkages and slight $(1 \rightarrow 6)$-α-linkages [11, 14, 15]. The molecular weight of amylose is around 105–106 Da with the degree of polymerization (DP) around 500–5,000. Amylose structure has the branch number 9–20 per molecule and each chain is built of 200–700 glucose units. However, amylopectin is a highly branched polysaccharide that mainly contains 95% $(1 \rightarrow 4)$-α-linkages and 5–6% $(1 \rightarrow 6)$-α-linkages. The molecular weight of amylopectin varies in the range of 107–109 Da with a DP of around 9600–15,900 [28, 35]. The molecular chain of amylopectin is made up of 18–25 units, which is shorter than the amylose chain.

Starch, a natural polymer, can be modified and combined with other materials to enhance its adhesive properties. The morphology of starch is the mixture of loose amorphous regions in highly crystalline regions due to the hydrogen bonds between the starch molecules [24, 30, 31, 38, 41]. The crystalline regions inhibit the penetration of water and chemical components into the starch structure resulting in a higher gelation temperature and a lower reactivity of the starch. Therefore, it has been recommended to perform some modifications on the crystalline region of starch or to reduce the size of crystalline segments. Various modification methods have been proposed to decrease the starch crystallinity, including chemical treatments (oxidation, esterification, etherification, and cationization), physical treatments (mechanical activation, microwave, ultrasonic degradation, heat-moisture treatment), and enzymatic treatment. For instance, starch-stabilized polyvinyl acetate (PVAc) emulsions have shown improved water resistance and bonding strength, making them suitable for wood applications [38]. Additionally, the incorporation of other natural materials, such as dextrin and Arabic gum, has been found to significantly enhance the adhesive characteristics, including pull-off adhesion and lap shear strength, while also improving thermal and electrical insulation properties. Furthermore, innovative approaches such as the synthesis of waterborne lignin-based epoxy resin emulsions (WLEPs) and their curing with polyamide have demonstrated the potential to create formaldehyde-free wood adhesives with high performance, including excellent storage stability and enhanced thermal and mechanical properties [10, 25]. Moreover, the use of oxidized starch in the preparation of biomass-derived aldehyde solutions for wood adhesives has shown promising results in terms of bonding performance and water resistance, offering a viable alternative to conventional urea–formaldehyde resins [22].

This review aims to provide a comprehensive overview of the advancements in starch-based adhesives for wood-based composites, examining the various formulations, modifications, and applications that contribute to their effectiveness and sustainability. By highlighting the latest research and technological developments, this article seeks to underscore the potential of starch-based adhesives in promoting eco-friendly practices within the wood industry, aligning with the broader goals of reducing environmental impact and enhancing material sustainability.

## 2 History of Starch

Starch is a natural polymer and is widely used in several applications including food, paper-making, additives, and adhesives [12, 29]. Some developments are covered in the unrecorded past. Strips of Egyptian papyrus, cemented together with a starchy adhesive and dated to the pre-dynastic period of 3500–4000 B.C., give evidence of the early use of starch [33, 39]. However, a description of starch and its application is not found until much more recent times. The historian and philosopher, Caius Plinius Secundus described documents of 130 B.C. made by sizing papyrus with modified wheat starch to produce a smooth surface. The adhesive was made from fine ground wheat flour, boiled with a dilute solution of vinegar. The paste was spread over strips of papyrus; the strips were beaten with a mallet and further strips lapped over the edges to give a broader sheet. Chinese paper documents from about 312 A.D. are reported to contain starch size. Later, Chinese documents were first coated with a high-fluidity starch to provide resistance to ink penetration and then covered with powdered starch to provide weight and thickness. Starches from rice, wheat, and barley were commonly used at that time.

Starch was introduced into Northern Europe to stiffen linen, possibly early in the fourteenth century. Colored and uncolored starches were used as cosmetics. Uncolored starch was used principally as a hair powder. Blue starch was employed by the Puritans until its use was banned by Queen Elizabeth in 1596. Yellow starch was quite fashionable until a notorious woman prisoner wearing a bright yellow-starched ruffle was publicly executed. Red starch cosmetics remained in fashion for many years. As starch became a more important industrial commodity, work was done on its modification. This included the great discovery of Kirchoff in 1811 that sugar could be produced from potato starch by acid-catalyzed hydrolysis. Then, there was the accidental discovery of the torrefaction method for producing dextrins, now termed British gums. In 1821, a fire occurred in a Dublin textile mill that utilized starch as a size. After the blaze was extinguished, one of the workmen noticed that some of the starch had been turned brown by the heat and dissolved easily in water to produce a thick, adhesive paste. The roasting of new starch was repeated and the product was shown to have the useful properties previously observed. In this way, heat dextrinization became known and subsequently elaborated into wide usage [27, 33, 39].

The first American wheat starch plant was started by Gilbert at Utica, New York in 1807, but was converted to one producing com starch in 1849. The change from wheat to com starch began with Thomas Kingsford's development in 1842 of a manufacturing process in which crude com starch was purified by alkaline treatment [27]. Potato starch manufacture began in 1820 in Hillsborough County, New Hampshire. Usage of potato starch grew rapidly until in 1895, sixty-four factories were in operation, of which forty-four were in Maine. They produced 24 million pounds of starch per year during approximately 3 months of operation. Most of the starch was sold to the textile industry. Rice starch manufacture was begun in 1815 using caustic treatment of rice. However, production did not increase significantly and the little rice starch later used was mainly imported. By 1890 the number of American starch plants had increased to eighty, producing 240 million pounds of starch per year. The United Starch Company and the National Starch Manufacturing Company joined forces to form the National Starch Company of New Jersey in 1902. The Corn Products Company, representing 80% of the corn starch industry with a daily grind of 1800 tons, was formed by a union of the National Starch Company of New Jersey, the Glucose Sugar Refining Company, the Illinois Sugar Refining Company, and the Charles Pope Glucose Company [1, 2, 8].

Most of the starch used in the U.S. (95%) and worldwide (77%) is derived from corn (Kirby 1992). Corn starch usage in the U.S. in 1995 was approximately 3.6 × 109 kg with a growth rate of about 2% a year. This represents about 3% of the annual production of corn in the USA. The price of corn starch has averaged $0.2–0.3/kg over the past 10 years. Current prices, however, are higher due to poor weather in 1995 and high current export demand. Potato starch, most of which is imported from Europe, costs approximately $ 0.65/kg in the U.S. A variety of starches are available that have been modified by chemical, physical, or genetic means to better satisfy the requirements of different applications. These are priced in the range of about $0.50–2.20/kg [33].

## 3 Source of Starch

Starch is a polysaccharide produced by most plants as a means of storing energy. It is stored intracellularly in the form of spherical granules 2–100 μm in diameter. Most commercially available starches are isolated from grains such as corn and wheat and tubers such as potato and tapioca. These cereals and tubers contain large quantities of starch, usually 60–90% on a dry weight basis [33, 39]. The ready availability of corn at relatively low and steady prices, its storability from season to season, and its high starch content led naturally to the development of commercial processes for the recovery of corn starch. From the early nineteenth century, when Yankee inventors first discovered that corn starch was fairly easily recovered by grinding the soaked grain, the process has gradually evolved into today's highly sophisticated automated process, which produces a multitude of useful food and industrial products. Mechanical innovations developed by trial and error during full-scale operations stimulated much of the early growth of the corn starch industry [33]. Today, process and product improvements more commonly follow research and engineering studies and thorough pilot plant evaluations.

Cassava (*Manihot utilissima* Pohl) is typically grown in tropical climates while corn, wheat, and potatoes are grown in colder climates [33]. Manufacturing plants for cassava starch are located close to root growing areas to minimize root transport costs and, more importantly, to enable the processing of tubers in the shortest time. The roots are delivered to the factory and stored in wooden or concrete bunkers. Bunker filling and emptying must be closely supervised to ensure that the roots that are harvested first are consumed first. The roots are usually delivered to a washing station by a belt conveyor. After washing, the outer skin is removed. The inner part of the peel, or cortex, is not removed because it has starch recoverable in modern processes. The washer is usually a U-shaped trough with paddles that carry the roots to the peeler. The peeler can be an integral part of the washer or a separate unit. The roots are peeled by the abrasion of one against another and against the walls and paddles of the washing and peeling device. To recover the starch, all cell walls must be ruptured. This has been accomplished, at times, by mild fermentation of the roots, then grinding them to a pulp with starch recovered by screening and washing, or by centrifugation. The fermentation process does not give good starch yields, and the quality of the starch is generally inferior [1, 2].

Rice (*Oryza sativa* L.) is the staple food of East, Southeast, and South Asia, where 90% of the world's rice crop is produced and consumed [8]. Commercial preparation of starch from rice is limited owing to the high cost of brewer's rice relative to other cereals and tubers. In the European Economic Community, only 7000 tons of rice starches are produced annually in factories in Belgium, Germany, The Netherlands, and Italy. Factories in Egypt and Syria are also producing rice starch, but it has not been produced in the United States since 1943 [27]. Sodium hydroxide is used for purifying milled rice starch because at least 80% of the protein of milled rice is an alkali-soluble protein called glutelin. Commercial preparation of rice starch has been described by Hogan, and there has been little change in the process since the 1960s. The process consists of steeping broken rice in 0.3–0.5% sodium hydroxide solution, wet milling the steeped grain, removal of cell walls, extraction of the protein with sodium hydroxide solution, washing, and drying. Trash and dirt are removed by preliminary cleaning. Steeping softens the grain and aids in extracting the protein. Steeping temperature varies from room temperature to 50 °C, and the steeping period is usually 24 h. Wet grinding is preferred to dry grinding, as starch damage is more extensive in dry milling, resulting in greater starch losses through dissolution in alkali.

Wheat starch and wheat gluten are important and valuable co-products of the wet-processing of wheat flour [1, 2, 8]. These two products are entering a new development phase, mainly due to the preponderance of value-added products that are currently being commercialized because of their sustainability. In the wheat starch market, the products available for food and industrial uses are large-granule starch, low-moisture starch, resistant starch, and cook-up and pre-gelatinized versions of unmodified starch. There are six general techniques of wheat starch and gluten separation: dough; batter; aqueous dispersion; chemical dispersion; wet-milling of kernels; and non-aqueous separation [27]. The last three techniques are not used because of poor product quality, high operating cost, effluent problems, and poor efficiency.

The total production of potato starch is small compared to the corn starch produced in the world. Potato starch production was only 2.5 million tons per year whereas corn starch reached 45.8 million tons/year in 2005 [1, 2]. For effective production of potato starch, potatoes should contain as much starch as possible. Therefore, in modern potato starch factories in Europe, only special species of industrial potatoes are used. These potatoes are not very tasty due to the high amount of starch and, therefore, are not consumed as food. The potatoes are harvested and processed in Europe between August and April; this period is referred to as the starch campaign. Culled food potatoes are not normally used because of their low dry matter content and relatively small starch granules, which are more difficult to process. When the price of food potatoes is low due to overproduction, food potatoes are sometimes processed between the starch campaigns in Europe. Reclaimed potato starch, which is starch recovered from the process waters from other potato processing industries (French fry, chip/crisp, and potato puree manufacturers) is processed on a small scale in Europe and North America.

## 4 Modification of Starch

The increasing environmental concerns and the depletion of petroleum resources have driven the adhesive industry to seek sustainable and eco-friendly alternatives to traditional petroleum-based adhesives. Starch-based adhesives have emerged as a promising solution due to their renewability, biodegradability, and relatively low cost. However, the inherent hydrophilicity of starch poses challenges in terms of water resistance and bonding strength, necessitating modifications to enhance its performance. Recent advancements have focused on various modification techniques, including chemical cross-linking, enzymatic treatments, and the incorporation of other bio-based materials to improve the adhesive properties of starch. For instance, the development of a double cross-linked structure by reacting dialdehyde starch (DAS) with hyperbranched polyamide (HBP) and glucose has significantly enhanced water resistance and thermal stability, achieving wet strengths of up to 1.26 MPa [40]. Similarly, wheat gluten modified with transglutaminase (TGase) and epichlorohydrin (ECH) has shown improved dry and wet shear strengths, meeting the Chinese National Standard for plywood adhesives [34]. The exploration of protein-rich elements, such as agricultural and animal byproduct proteins, as sustainable alternatives to formaldehyde-based adhesives further underscores the potential of bio-based adhesives in the wood industry. Additionally, starch/$Ca^{2+}$ gels have demonstrated excellent flame retardancy and stable adhesive properties, particularly in high-amylose corn starch variants, making them suitable for paper and wood products [21].

The shift towards bio-adhesives is also driven by the need to comply with stringent environmental standards, with wood and wood composites being promising substrates due to their high strength, stiffness, and formability. The use of natural binders such as polylactic acid (PLA), polycaprolactone (PCL), and thermoplastic starch (TPS) in high-density fiberboards (HDF) has shown that increasing binder content can significantly improve mechanical properties, although challenges remain in terms of water absorption [13]. Overall, the development of starch-based adhesives represents a significant step towards sustainable and eco-friendly wood-based composites, offering a viable alternative to traditional synthetic adhesives while addressing environmental and health concerns.

## *4.1 Chemical Modifications*

Owing to low solid content, the very slow setting time, and the frequently observed "retrogradation", i.e. tendency to a rise of viscosity, such starch adhesives are not used in modern manufacturing processes. By modifying native starch of different origins, a broad range of adhesive raw materials of starch origin can be prepared. The surface energy of starch is such that it is usually impossible to achieve bonding to plastic materials. Since starch adhesives are always provided as an aqueous, colloidal solution to achieve good wetting, sufficient setting time is only guaranteed if at least one of the substrates can absorb water (e.g. wood or paper). Starch can be modified as shown in Fig. 1. Esterification is a typical chemical modification of starch by converting hydroxyl groups into esters to improve the hydrophobicity of starch. The water absorptivity and water solubility of the esterified starch depend on the chain length of the esterification agent and the degree of esterification (DS). An esterified corn starch was obtained by reacting with maleic anhydride and then crosslinking with a poly-isocyanate pre-polymer [3]. Esterification, involving the introduction of ester groups into the starch molecule, significantly improves water resistance and bonding strength.

Various methods, including transesterification, alkylation, acetylation, and succinylation, have been employed to produce esterified starches suitable for adhesive applications [38]. For instance, $K_3PO_4$-catalyzed transesterification of cassava starch with methyl laurate has been shown to enhance the hydrophobicity of starch laurates, resulting in superior shear strength for bonding paper substrates compared to commercial adhesives [8]. Additionally, the incorporation of polyphenols, such as tannin, into starch-based formulations can further enhance water resistance, as demonstrated by starch-tannin adhesives achieving satisfactory performance in underwater dipping conditions [32, 36].

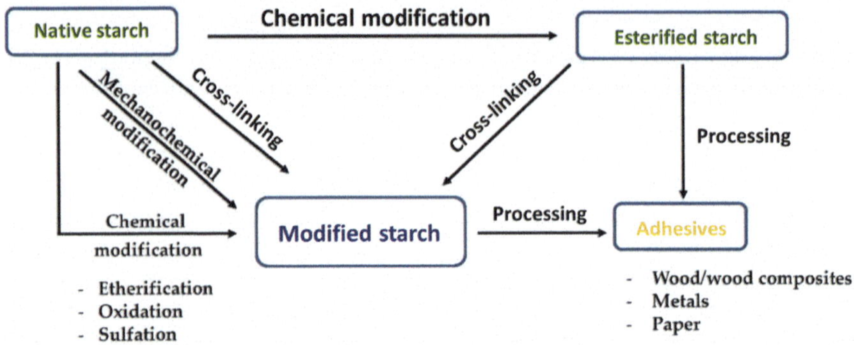


**Fig. 1** Illustration of chemical modifications of starch-based adhesives [37]. Open Access CC BY-NC-ND 4.0

Grafting copolymerization of the synthetic polymers onto the starch backbone enhanced the bonding properties of the starch. A study tried to modify a starch-based adhesive by the addition of blocked isocyanate and an auxiliary agent. The dry and wet bonding strength reached the peaks when the mixing ratio of starch and blocked isocyanate was 100/20 and 100/25, respectively. Furthermore, bonding the isocyanate to the starch-based adhesive could reduce the viscosity of the adhesive. Introducing additive bentonite could thicken the adhesive and improve its water resistance [9].

Etherification, a chemical modification process, involves introducing ether groups into the starch molecule, which enhances its adhesive properties, particularly in terms of water resistance and bonding strength (Fig. 2). Etherification process is less common compared to other modifications like esterification and oxidation but is gaining attention for its unique benefits [6, 19]. Etherified starches are produced through reactions such as alkylation, which can significantly improve the adhesive's performance by altering the starch's molecular structure to enhance its interaction with other materials [19]. For instance, the use of etherified starch in adhesives has been shown to improve the distribution of the adhesive at the bonding interface, which is crucial for achieving strong and durable bonds in wood products [3, 23]. Additionally, etherified starch adhesives can be combined with other modifications, such as crosslinking with polyisocyanates, to further enhance their thermal stability and bonding performance [9]. The development of etherified starch-based adhesives also aligns with the growing consumer demand for non-toxic, biodegradable, and renewable materials in industrial applications. Research has demonstrated that these adhesives can achieve high adhesion strength and good film-forming performance, making them suitable for various wood products, including plywood, fiberboard, and particleboard [18]. Moreover, the use of etherified starches in adhesives can be optimized by adjusting the amylose/amylopectin ratio, as studies have shown that amylopectin-rich starches tend to perform better in terms of viscosity and bonding strength under different conditions.

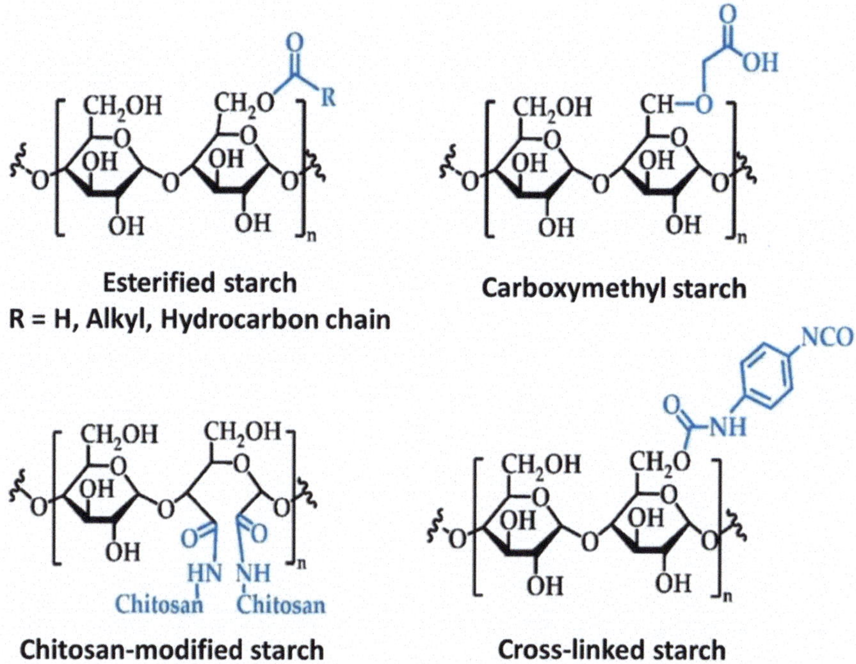


**Fig. 2** Example of modified starch-based adhesives [37]. Open Access CC BY-NC-ND 4.0

Regarding the chemical modification of starch, an oxidation process forms a more reactive wood adhesive (Fig. 3). Starch is oxidized to obtain low-viscosity, high-solids dispersions, and resistance to viscosity increases or gelling in aqueous dispersion. The oxidation causes de-polymerization, which results in lower viscosity dispersion and introduces carbonyl and carboxyl groups, which minimize retrogradation of amylose, thus giving viscosity stability [19]. A study prepared a cornstarch adhesive through an oxidation-gelatinization process and then characterized its rheological properties [16]. The apparent viscosity was measured as a function of several parameters such as temperature, starch-to-water ratio, and shear rate. It was found that the apparent viscosity reached a maximum value at 10 °C and then reduced with increasing the temperature; and, the viscosity decreased slightly on increasing the shear rate from 6 to 60 RPM. The starch adhesive behaved as a pseudo-plastic fluid. In general, oxidized starches are sensitive to heat, tending to yellow or brown when exposed to high temperatures. This yellowing tendency during drying has been related to the aldehyde content. With increasing aldehyde content, the oxidized starch becomes increasingly yellow on storage. Yellowing of oxidized starch dispersed in water by cooking or by alkali is also related to the aldehyde content. An innovative starch-based adhesive was formulated by the addition of a silane coupling agent ($CH_2 = CH–Si(OC_2H_5)_3$), an olefin monomer (butyl acetate and vinyl acetate, as a co-monomer), and hydrogen peroxide (as an oxidant) [40]. They reported that the

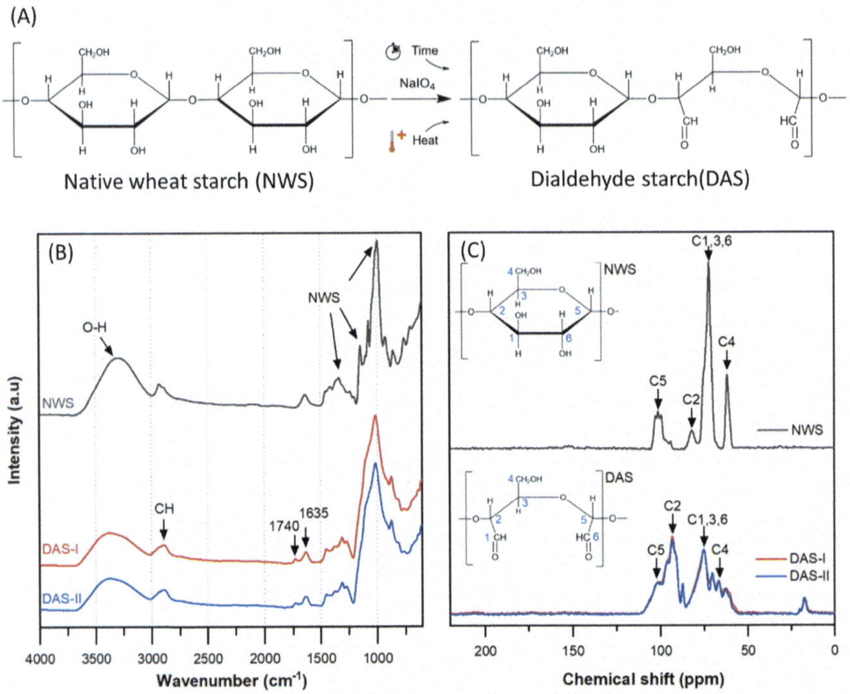


**Fig. 3** The oxidation reaction of starch with NaIO4 (**a**), FTIR spectra (**b**), and 13C-NMR spectra of dialdehyde starch [26]. Open Access CC BY-NC-ND 4.0

hydroxyl groups of starch were converted into carboxyl and aldehyde groups in the presence of the oxidant agent. The graft copolymerization improved the thermal stability, water resistance, and bonding strength of the prepared adhesive. The optimization process suggested that the addition of 9 wt % of coupling agent and 3 wt % of $NaIO_2$ produce a modified starch-based adhesive with dry and wet bonding strengths at approximately 7.88 and 4.09 MPa, respectively.

## *4.2 Physical Modifications*

Treatment with ultrasounds (sound waves at or above a frequency range of 15–20 kHz) is one of the physical methods of modification of starch leading to its depolymerization [5]. This process in contrast to chemical methods is considered to be environmentally friendly as it reduces the usage of chemicals, waste production, and energy consumption e.g. ultrasonication shortened the time of octenyl-succinylation of carboxymethylated (OC-CMS) potato starch from even 24 h conventionally spent to a few minutes without the use of p-toluenesulphonic acid. The impact of ultrasonic radiation on granular starch or starch gels has been studied for years [37].

According to many reports, ultrasounds cause physical degradation of granules with visible fissures and pores on the surface (Fig. 4) but no alterations in granules' shape and size have been observed [5]. In addition, prolonged exposure of polysaccharide solutions to high-energy ultrasound causes an apparent drop in viscosity due to macromolecular chain degradation.

Ultrasonic treatment resulted in changes in granular structure, swelling power and solubility, gelatinization transition temperature, syneresis, and pasting properties [5]. A study showed that ultrasonic treatment degraded preferentially the amorphous regions and more easily attacked linear amylose than highly branched amylopectin molecules. The extent of the changes induced by ultrasonic treatment depended not only on the crystal structure of starch but also on its amylose content. It is also shown that physical treatments allowed the OS-CMS derivatives can be prepared using the assistance of microwave or ultrasound irradiation, which permits to shorten the esterification time from 24 h of the previously reported conventional heating to several minutes.

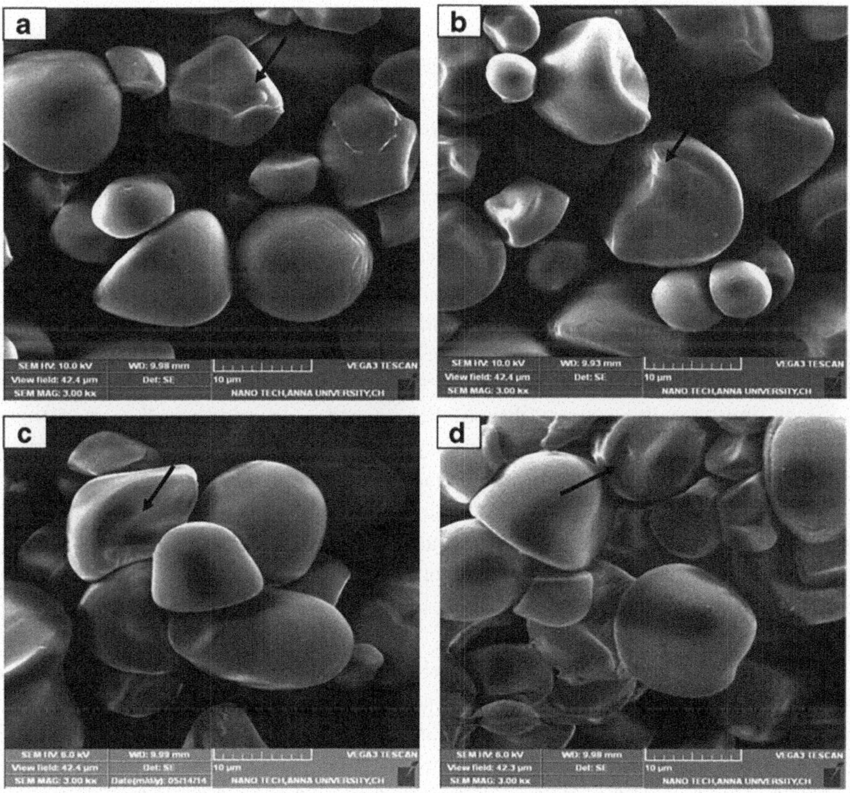


**Fig. 4** SEM micrographs of native and modified starch granules [5] Open Access CC BY-NC-ND 4.0

Starch paste clarity is very important for many food applications, especially for such products as jellies or fruit fillings. It depends on the associative bonds between the starch molecules in the granule. If opacity is an unacceptable feature of the product, usually waxy starches or modified starches are used to avoid it. As is seen in Fig. 4, pastes from native starches show rather low clarity, except potato starch. Sonication in ethanol caused a small decrease in paste clarity for rice and wheat starches and did not change it in the case of corn and potato starches. Modification in water influenced significantly the clarity of potato starch paste causing its increase but for the other starches, it did not cause a noticeable effect. A study reported that the clarity of sonicated starch pastes increased as the effect of the disruption of swollen granules during the process [36]. However, there is a lack of data concerning this determination for starches sonicated in the form of granules.

### 4.3 *Enzymatic Treatments*

Modification of starch using the enzyme Porcine pancreatic (alpha-amylase) was studied by published work [39]. When starch granules in water are exposed to heat, the inter- and intra-molecular hydrogen bonds between starch chains are disrupted, allowing the granules to swell and then disintegrate. The availability of starch chains to the digestive enzymes is thus increased as gelatinization progresses. The partially gelatinized starch samples, however, showed differences in enzymatic action mainly in the initial stage of digestion, but little at the maximum level of digestion. Both pre-gelatinized and native starch samples showed similar values of maximum hydrolysis which was lower (~by 5%) than that of gelatinized. This implies that there were enzyme-resistant starch residues in amounts that were similar among the native and partially gelatinized samples.

The residual crystallites may also affect the enzymatic digestibility by imposing physical limits on the accessibility of the enzymes. In addition, the crystalline morphology caused differences in the enzymatic hydrolysis pattern. The effects of the residual crystallites, however, may change the enzymatic hydrolysis (digestion) pattern, mainly in the intermediate and late stages of digestion. As indicated by the hydrolysis patterns, the formation of the amorphous matrix, which in these experiments was composed exclusively of amylopectin, played a major role in the differing digestion behaviors among the partially gelatinized and retrograded waxy rice starch samples [39].

The degree of gelatinization is positively correlated with in vivo plasma glucose and insulin responses [39]. The starches in some processed food products are not fully gelatinized, so that the level of native or partially gelatinized starch in the foods is of particular importance from the view of nutrition. The consumption of native or minimally processed cereal products as replacements for staple meals has been increasing in Korea and some Asian countries, because of their health-promoting effects. Incomplete gelatinization effectively retards glucose release from starch during the enzymatic digestion. The digestion rate was reduced as the melting enthalpy increased.

The amounts of slowly digestible starch and resistant starch correlated positively with the relative melting enthalpy of the partially gelatinized or retrograded starch samples.

## 5 Conclusion

High-performance starch-based adhesives are emerging as viable alternatives to traditional petrochemical adhesives, driven by their renewability and environmental benefits. Recent studies highlight various modifications to enhance their properties. For instance, the optimization of starch-tannin formulations has demonstrated significant improvements in water resistance and adhesion strength, achieving D2 classification under EN 204 standards. Additionally, the development of a zero formaldehyde starch-based adhesive has effectively addressed issues of formaldehyde emissions while enhancing water resistance and adhesion through innovative modifications with konjac glucomannan and milk protein. Furthermore, the use of oxidative cross-linking via Fenton's reagent has resulted in adhesives with superior hydrophobicity and mechanical strength, showcasing a 94% increase in wet shear strength. Overall, these advancements indicate that starch-based adhesives can meet high-performance requirements while promoting sustainability in adhesive technology. However, ongoing research is essential to further optimize these formulations for broader industrial applications.

## References

1. Abbott AP, Ballantyne AD, Conde JP et al (2012) Salt modified starch: sustainable, recyclable plastics. Green Chem 14:1302. https://doi.org/10.1039/c2gc16568f
2. Abbott AP, Palazuela Conde J, Davis SJ, Wise WR (2012) Starch as a replacement for urea-formaldehyde in medium density fibreboard. Green Chem 14:3067. https://doi.org/10.1039/c2gc36194a
3. Ačkar D, Babić J, Jozinović A et al (2015) Starch modification by organic acids and their derivatives: A review. Molecules 20:19554–19570. https://doi.org/10.3390/molecules201019554
4. Ayrilmis N, Lee Y-K, Kwon JH et al (2016) Formaldehyde emission and VOCs from LVLs produced with three grades of urea-formaldehyde resin modified with nanocellulose. Build Environ 97:82–87. https://doi.org/10.1016/j.buildenv.2015.12.009
5. Babu AS, Parimalavalli R, Jagannadham K, Rao JS (2015) Chemical and structural properties of sweet potato starch treated with organic and inorganic acid. J Food Sci Technol 52:5745–5753. https://doi.org/10.1007/s13197-014-1650-x
6. Bayazeed A, Farag S, Shaarawy S, Hebeish A (1998) Chemical modification of starch via etherification with methyl methacrylate. Starch/Staerke 50:89–93. https://doi.org/10.1002/(SICI)1521-379X(199803)50:2/3%3c89::AID-STAR89%3e3.0.CO;2-B
7. Böhm M, Salem MZM, Srba J (2012) Formaldehyde emission monitoring from a variety of solid wood, plywood, blockboard and flooring products manufactured for building and furnishing materials. J Hazard Mater 221–222:68–79. https://doi.org/10.1016/j.jhazmat.2012.04.013

8. Buléon A, Colonna P, Planchot V, Ball S (1998) Starch granules: Structure and biosynthesis. Int J Biol Macromol 23:85–112. https://doi.org/10.1016/S0141-8130(98)00040-3
9. Chen X, Sun C, Wang Q et al (2022) Preparation of glycidyl methacrylate grafted starch adhesive to apply in high-performance and environment-friendly plywood. Int J Biol Macromol 194:954–961. https://doi.org/10.1016/j.ijbiomac.2021.11.152
10. Duval A, Lawoko M (2014) A review on lignin-based polymeric, micro- and nano-structured materials. React Funct Polym 85:78–96. https://doi.org/10.1016/j.reactfunctpolym.2014.09.017
11. Ferdosian F, Pan Z, Gao G, Zhao B (2017) Bio-based adhesives and evaluation for wood composites application. Polymers (Basel) 9:70. https://doi.org/10.3390/polym9020070
12. Goel C, Semwal AD, Khan A et al (2020) Physical modification of starch: changes in glycemic index, starch fractions, physicochemical and functional properties of heat-moisture treated buckwheat starch. J Food Sci Technol 57:2941–2948. https://doi.org/10.1007/s13197-020-04326-4
13. Gumowska A, Kowaluk G (2023) Physical and mechanical properties of high-density fiberboard bonded with bio-based adhesives. Forests 14. https://doi.org/10.3390/f14010084
14. Islam MN, Rahman F, Das AK, Hiziroglu S (2021) An overview of different types and potential of bio-based adhesives used for wood products. Int J Adhes Adhes 112:102992. https://doi.org/10.1016/j.ijadhadh.2021.102992
15. Islam MN, Rahman F, Das AK, Hiziroglu S (2022) An overview of different types and potential of bio-based adhesives used for wood products. Int J Adhes Adhes 112:102992. https://doi.org/10.1016/j.ijadhadh.2021.102992
16. Iswanto AH, Lubis MAR (2022) Technological properties of formaldehyde free adhesives based on oxidized starch mixed with different crosslinkers for plywood. Proc Int Conf. https://doi.org/10.1007/978-981-19-0308-3_4
17. Kristak L, Antov P, Bekhta P et al (2023) Recent progress in ultra-low formaldehyde emitting adhesive systems and formaldehyde scavengers in wood-based panels: A review. Wood Mater Sci Eng 18:763–782. https://doi.org/10.1080/17480272.2022.2056080
18. Lamaming J, Heng NB, Owodunni AA et al (2020) Characterization of rubberwood particleboard made using carboxymethyl starch mixed with polyvinyl alcohol as adhesive. Compos Part B Eng 183:107731. https://doi.org/10.1016/j.compositesb.2019.107731
19. Li W, Wu L, Xu Z, Liu Z (2020) Adhesion-to-fibers and film properties of etherified–oxidized cassava starch/polyvinyl alcohol blends. Iran Polym J 29:331–339. https://doi.org/10.1007/s13726-020-00798-5
20. Li Z, Wang J, Cheng L et al (2014) Improving the performance of starch-based wood adhesive by using sodium dodecyl sulfate. Carbohydr Polym 99:579–583. https://doi.org/10.1016/j.carbpol.2013.08.062
21. Liu P, Ling J, Mao T et al (2023) Adhesive and flame-retardant properties of starch/Ca2+ gels with different amylose contents. Molecules 28. https://doi.org/10.3390/molecules28114543
22. Lubis MAR, Park B-D (2018) Modification of urea-formaldehyde resin adhesives with oxidized starch using blocked pMDI for plywood. J Adhes Sci Technol 32:2667–2681. https://doi.org/10.1080/01694243.2018.1511075
23. Maulana MI, Lubis MAR, Febrianto F et al (2022) Environmentally friendly starch-based adhesives for bonding high-performance wood composites: A review. Forests 13:1614. https://doi.org/10.3390/f13101614
24. Meimoun J, Wiatz V, Saint-Loup R et al (2017) Modification of starch by graft copolymerization. Starch—Stärke 1600351:1600351. https://doi.org/10.1002/star.201600351
25. Nasiri A, Wearing J, Dubé MA (2020) The use of lignin in emulsion-based pressure-sensitive adhesives. Int J Adhes Adhes 100. https://doi.org/10.1016/j.ijadhadh.2020.102598
26. Neitzel N, Hosseinpourpia R, Adamopoulos S (2023) A dialdehyde starch-based adhesive for medium-density fiberboards. BioResources 18:2155–2171
27. Radley JA (1976) Industrial uses of starch and its derivatives, 1st edn. Applied Science Publishers Ltd., London

28. Rindlav-Westling Å, Gatenholm P (2003) Surface composition and morphology of starch, amylose, and amylopectin films. Biomacromol 4:166–172. https://doi.org/10.1021/bm0256810
29. Santoso D, Wardana AA, Surono IS (2019) Resistant starch content, pasting properties, and morphology of taro (Colocasia esculenta L. Schott) flour modified by heat-moisture treatment. J Phys Conf Ser 1363. https://doi.org/10.1088/1742-6596/1363/1/012007
30. Sethy PK, Prusty K, Mohapatra P, Swain SK (2019) Nanoclay decorated polyacrylic acid/ starch hybrid nanocomposite thin films as packaging materials. Polym Compos 40:229–239. https://doi.org/10.1002/pc.24636
31. Sheikhi A, van de Ven TGM (2018) Colloidal starch and cellulose nanocrystals unite to improve the mechanical properties of paper: From enhanced coatings to reinforced nanocomposites. ACS Appl Nano Mater 1:1841–1852. https://doi.org/10.1021/acsanm.8b00251
32. Shirmohammadli Y, Efhamisisi D, Pizzi A (2018) Tannins as a sustainable raw material for green chemistry: A review. Ind Crops Prod 126:316–332. https://doi.org/10.1016/j.indcrop.2018.10.034
33. Shogren RL (1998) Starch: Properties and materials applications. Biopolym Renew Resour 30–46. https://doi.org/10.1007/978-3-662-03680-8_2
34. Solt P, Konnerth J, Gindl-Altmutter W et al (2019) Technological performance of formaldehyde-free adhesive alternatives for particleboard industry. Int J Adhes Adhes 94:99–131. https://doi.org/10.1016/j.ijadhadh.2019.04.007
35. Vanier NL, El Halal SLM, Dias ARG, da Rosa ZE (2017) Molecular structure, functionality and applications of oxidized starches: A review. Food Chem 221:1546–1559. https://doi.org/10.1016/j.foodchem.2016.10.138
36. Wang S, Li C, Copeland L et al (2015) Starch retrogradation: A comprehensive review. Compr Rev Food Sci Food Saf 14:568–585. https://doi.org/10.1111/1541-4337.12143
37. Watcharakitti J (2022) Modified starch-based adhesives: A review. Polymers (Basel) 14
38. Watcharakitti J, Win EE, Nimnuan J, Smith SM (2022) Modified starch-based adhesives: A review. Polymers (Basel) 14:2023. https://doi.org/10.3390/polym14102023
39. Whistler RL, Bemiller JN, Paschall E (1984) Starch: Chemistry and technology. Academic press, London
40. Zhang S, Liu F, Peng H et al (2015) Preparation of novel c-6 position carboxyl corn starch by a green method and its application in flame retardance of epoxy resin. Ind Eng Chem Res 54:11944–11952. https://doi.org/10.1021/acs.iecr.5b03266
41. Zhou J, Tong J, Su X, Ren L (2016) Hydrophobic starch nanocrystals preparations through crosslinking modification using citric acid. Int J Biol Macromol 91:186–1193. https://doi.org/10.1016/j.ijbiomac.2016.06.082

# Multifunctional Nanotechnology Application for Wood Properties Enhancement: Adhesive and Coating

**Ria Aniza, Anelie Petrissans, and Mathieu Petrissans**

**Abstract** In the rapidly evolving world of materials science, the integration of nanotechnology into various applications is revolutionizing industries across the board. One such groundbreaking development is the utilization of nanotechnology in enhancing the characteristics of wood adhesive and wood coating by synthesizing the functional material in nano-scale units. This innovative approach combines the natural and sustainable properties of wood with the developed performance and durability provided by nanoparticles. Their unique characteristics and properties hold great potential in improving the strength, flexibility, and resistance of wood materials. Nanoparticles typically are classified into two variations according to the materials including organic and inorganic nanoparticles. These nanoparticles can be manipulated at the molecular level to enhance specific traits, such as fire resistance, antimicrobial, UV protector, water repellency, durability, and dimensional stability. This makes nanoparticle-enhanced technology ideal for indoor and outdoor applications on wood. Incorporating these nanoparticles into wood can result in highly functional and sustainable materials that surpass traditional wood products in terms of performance and longevity.

**Keywords** Nanotechnology · Wood adhesive · Wood coating · Organic-inorganic nanomaterials

R. Aniza (✉) · A. Petrissans · M. Petrissans
Université de Lorraine, INRAE, LERMAB, Epinal F88000, France
e-mail: ria.aniza@univ-lorraine.fr

R. Aniza
International Doctoral Degree Program in Energy Engineering, National Cheng Kung University, Tainan 701, Taiwan

Department of Environmental Engineering, National Cheng Kung University, Tainan 701, Taiwan

Department of Environmental Engineering, Institut Teknologi Sepuluh Nopember, East Java Surabaya 60111, Indonesia

 
P. Antov et al. (eds.), *Nanomaterials Additives in Bioadhesives for Wood Composites*, Smart Nanomaterials Technology, https://doi.org/10.1007/978-981-96-3400-2_6

## 1 Introduction

The world of construction and manufacturing is constantly evolving, driven by the need for sustainable and environmentally friendly solutions. One such innovation that is making waves in this industry is nanotechnology on adhesive wood composite and coating-impregnated material. By incorporating cutting-edge nanotechnology with renewable materials, these composites and coating offer a breakthrough solution for enhancing the strength, durability, and versatility of wood-based products [7] as well as offer advanced benefits such as protecting the wood surface from UV light, fireproof, bacteria, and fungus attacks [10]. Enter the multifunctional nanoparticle employment adhesive and coating for the wood, thus, this groundbreaking technology promises to reform the way some researchers approach wood composite production and coating technology, offering a more efficient and eco-friendly alternative [18].

Currently, nanotechnology-based methods for producing novel wood derivatives have also substantially taken over the wood and paper industries. The manufacture of outstanding wood-based products has been produced practicable by cellulosic nanofibers, nanosheets, nanogels, nanomembranes, nanocoatings, and nanopolymeric compounds of pectin and lignin [15]. However, nanoparticle technology is being applied in wood industries for some in adhesive production and widely in coating production. Rajendran et al. [23] studied the life cycle and techno-economic analysis of cellulose nanocrystals (CNCs) production from wood pulp using $H_2S$ treatment, the study indicated that the CNC manufactured using the acid recovery technique was more cost-effective than the one generated without it, with higher project net present values. However, the greater capital requirements led to a longer period of repayment and a worse return on the capital spent, for CNC manufactured with and without acid recovery, the anticipated minimum selling prices were $4.69/kg and $4.89/kg, respectively. Furthermore, Khanjanzadeh et al. [14] investigated the application of cellulose nanocrystals (CNCs) to improve the performance of urea–formaldehyde (UF) adhesive used in wood-based composite for medium-density fireboard (MDF) production. According to the findings, UF adhesives' solid content, density, viscosity, and free formaldehyde reduced while the gel time increased when improved CNC was added to the system. The formaldehyde emission values of the panels ranged from 11 to 17% lower than those of the panels manufactured from plain UF. The study also noted that modified CNC could improve the physical and mechanical properties of MDF.

In the matter of wood coating, nanotechnology leads the advancement of the wood preservation technique. Most of the purpose of the nanomaterial added in the wood coating is to protect the wood from environmental factors such as UV radiation, microbial attacks, and antifire. Zikeli et al. [36] studied the behavior of the essential oil of thyme containing lignin nanoparticles (EO-LNPs) and polylactic glycolic acid nanoparticles (EO-PLGA NPs) on wood spray coating to substitute the traditional fossil fuel-based wood coating and improve the longevity of wood. The study indicated that despite having bigger average particle diameters and a more uneven particle size distribution, lignin nanoparticles were more successful than

PLGA nanoparticles at capturing and holding onto essential oils. Moreover, Lazari and Elmi [16] investigated the hybrid coating of superhydrophobic chitosan/silica nanocomposite for wood in a seawater environment. For about 6 months of immersion in the seawater environment, the study indicated that the wood with the coating material of superhydrophobic chitosan/silica hybrid nanocomposite exhibited about 50% less degradation than the control (uncoated wood). These results show that nanotechnology may improve the properties of the wood material regardless of the challenges of environmental types [10, 21], such as common ambient air or seawater exposure.

According to the primary holistic literature review, the study of nanotechnology is scattered and hard to implement due to various parameters, therefore, this study aims to summarize the recent application of nanotechnology on wood. The discussion is classified into two majors: nanotechnology in wood adhesive and wood coating. This study has some benefits in providing comprehensive information on nanotechnology in wood for the researcher and sharing the same interest with the industrial practitioner in improving the quality of wood products.

## 2 Bibliometric Analysis and Fundamentals of Nanotechnology

There has been a significant increase in interest in the use of wood as a manufacturing material for the future. In terms of nanotechnology in wood, according to the platform of Web of Science (WoS), using the main keywords of "Nanotechnology" and "Wood", there are about 1,103 types of scientific studies. The tree map chart (Fig. 1) exhibits the five major categories of publications focus on the nanotechnology application in wood, including (1) Material science and multidisciplinary (410 publications); (2) Nanoscience nanotechnology (329 publications); (3) Chemistry multidisciplinary (216 publications); (4) Physics applied (213 publications); and (5) Materials science paper wood (101 publications).

In wood production, nanotechnology is gradually increased to be applied to improve wood's properties. According to the historical record in WoS, nanotechnology was introduced to wood applications in 1996. The study focusing on nanotechnology for wood has been gradually increasing for the past 10 years (2014–2024). The WoS record (Fig. 2) data revealed that the highest number of publications appeared in 2020 (99 publications) and significantly decreased in 2021 (61 publications). It might be due to the tremendous movement restrictions in the world as consequence of the pandemic COVID-19. However, the phase of the new normal leads to the study of nanotechnology for wood progressively increasing by the year (2022, 84 publications).

Nanotechnology is the branch of science and engineering that focuses on the design, analysis, and application of the material in a "nanoscale" size. Nano-sized materials typically range from 1 to 100 nm (nm) [21]. The main motive behind the

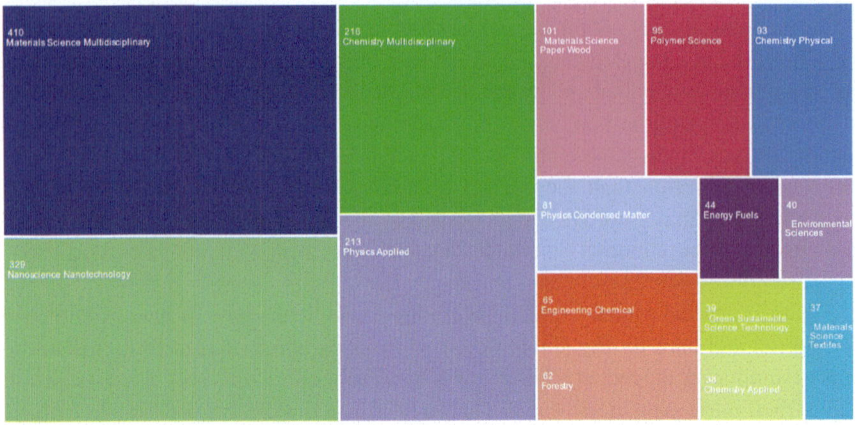


**Fig. 1** Treemap chart of nanotechnology in wood according to WoS

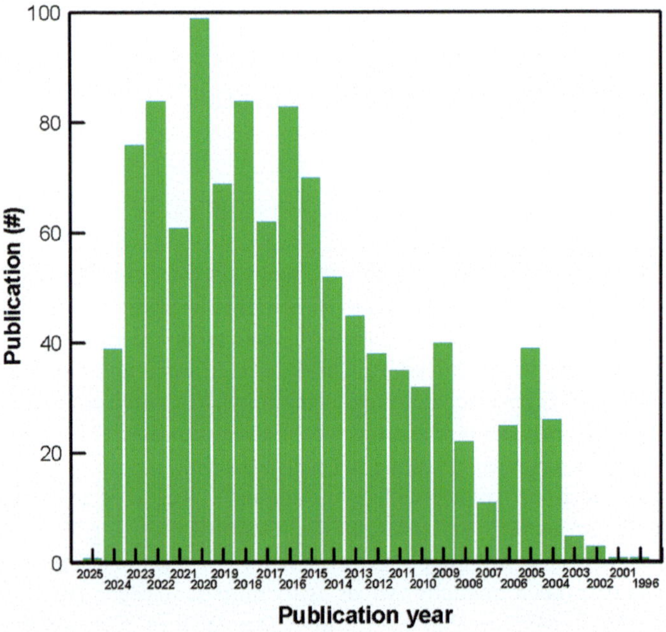


**Fig. 2** Scientific publication of nanotechnology for wood

application of nanotechnology referred to as wood science and engineering is the special ability of nano-based materials to effectively and profoundly penetrate the wood material, it may change the surface chemistry of the substrate in the process, and improve the characteristics of the wood as a result. Two major pathways are currently

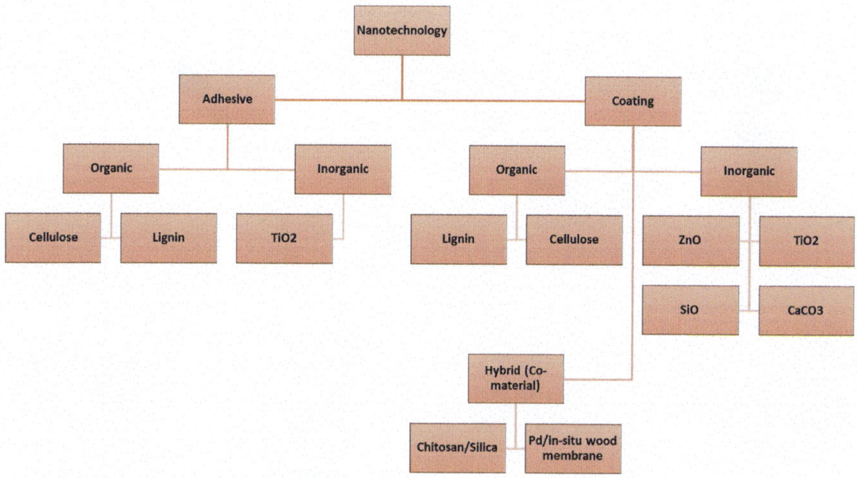


**Fig. 3** Nanoparticle organic and inorganic material in wood products: adhesive and coating

employed to apply nanotechnology on wood materials, including (1) adhesive and (2) coating.

The nanomaterials for applying nanotechnology in wood may be classified into several categories (Fig. 3). According to the performance of the nanomaterials, wood adhesives may commonly be used in two categories, including organic (cellulose- and lignin-based substrate) and inorganic (metal such as $TiO_2$) nanomaterials. Meanwhile, wood coating may applied in three categories, including organic (cellulose and lignin), inorganic (ZnO, SiO, $TiO_2$, and $CaCO_3$), and hybrid (Chitosan/Silica and Pd/in-situ wood membrane). In terms of the in-situ nanotechnology application of wood material, the nanoparticle mat is derived from the wood by polymerization process (Fig. 4) [19]. The modification processes involve a structure degradation scale from the macroscale to the nanoscale. This process may be used to obtain cellulose- or lignin-based nanoparticles.

## 3 Nanotechnology for Wood Adhesive

### 3.1 Overview

Nano bio-adhesive refers to the integration of nanoparticle materials into a bio-based adhesive such as cellulose- and lignin-based, creating a high-performance bonding agent. Fundamentally, the nano-sized materials are dispersed within the wood adhesive matrix, reinforcing its structure at a molecular level [9]. This results in improved mechanical properties, such as increased strength, stiffness, and toughness [20]. The bio-based nature of the adhesive component is what sets nano bio-adhesive

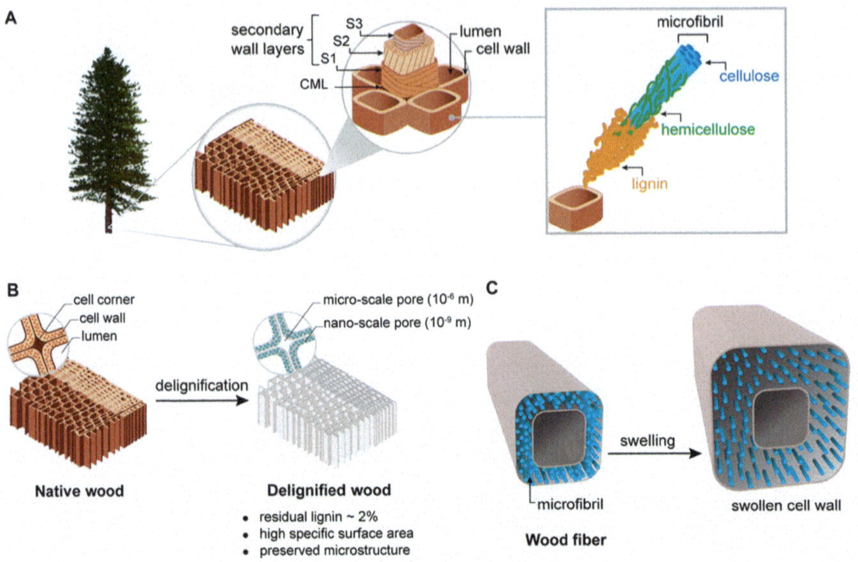


**Fig. 4** A representation of the primary components of the cell wall in softwood, which serves as a base for nanostructural alteration [19]

apart from traditional adhesives. Instead of relying on synthetic chemicals derived from fossil fuels, nano bio-adhesives are derived from renewable resources such as plant-based polymers or proteins [15]. This makes them more environmentally friendly and sustainable compared to their counterparts.

Zhu et al. [35] introduced a novel approach for producing eco-friendly lignin-based wood adhesives using aminated lignin-Cu nanoparticles. The study revealed that without utilizing conventional solvents such as aldehydes, phenols, or organic solvents, the bioadhesive properties of water resistance and mildew resistance were significantly improved. Additionally, the bonding strength and debonding work increased by 175% and 533%, respectively. The graphic diagram of the lignin nanoparticle bioadhesive is illustrated in Fig. 5. Additionally, another organic–inorganic hybrid material for adhesive wood may derive from the soybean waste and the $Al(OH)_3$ nanoparticle (Fig. 6) to establish a dense, noncovalent bonding network based on coordination and hydrogen bonding, improving the soybean waste adhesive's ability to attach to surfaces and withstand water [3]. The durability of the organic–inorganic hybrid adhesive of soybean waste to Al nanoparticle increased to 11.2 $MJ/m^3$ (a 133.3% improvement).

Among the natural plant-based polymers (starch, plant proteins, tannins, chitosan, vegetable oils, cellulose, and lignin) on the planet for bio-based adhesives for wood composite, however, most of the studies prefer to utilize cellulose- or lignin-based materials for employing nanotechnology in wood. Among cellulose and lignin, cellulose has been recognized for having the best performance in terms of biocompatibility, decomposition, and safety [24]. However, more development is still required

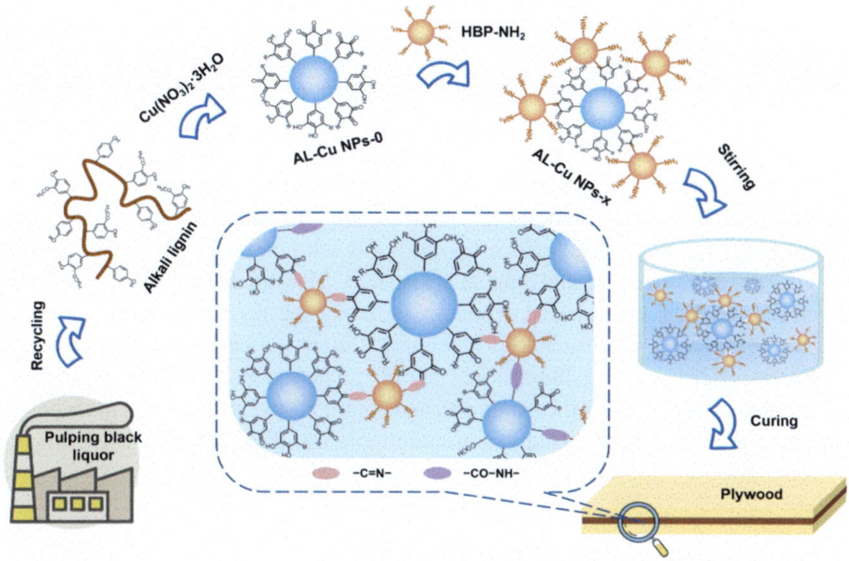


Fig. 5 Alkali lignin and the production of eco-friendly biomass adhesives [35]

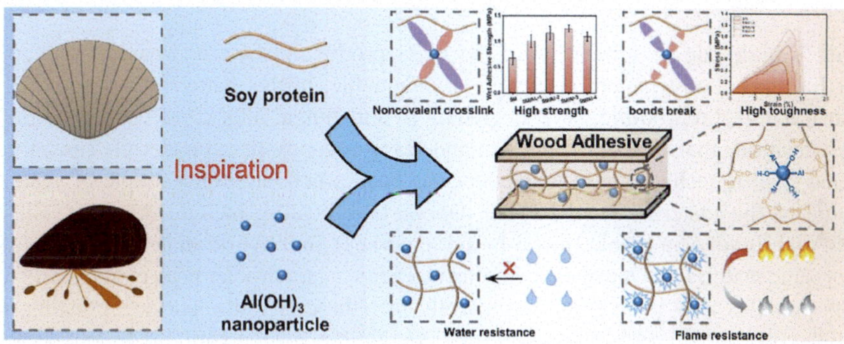


Fig. 6 Organic–inorganic hybrid adhesive for wood [3]

before they can be utilized in a variety of applications, such as hydrogels, adhesives, and composite membranes, owing to the high inter- or intramolecular interactions and intrinsic hydrophilic bonds in the cellulose chain structure [8, 17]. As a result, employing cellulose as an adhesive and attaining acceptable bonding qualities continue to be quite difficult.

### 3.2 Characterization

The integration of nano bio-adhesives into wood composites brings forth a multitude of advantages. Firstly, these composites exhibit exceptional resistance to deformation, moisture, and fire [26]. The reinforced wood matrix prevents warping, cracking, and swelling, making it an ideal material for high-stress applications such as structural components in construction [32]. Additionally, the introduction of nanoparticles enhances the mechanical properties of the wood composite, including tensile strength, flexural strength, and impact resistance. This allows for the creation of lighter and stronger wood-based products, reducing the need for excess materials and contributing to overall material efficiency. Likewise, nano bio-adhesive wood composites offer improved dimensional stability. The reduced moisture absorption and desorption rates prevent the expansion and contraction of the material in response to changes in humidity, resulting in a more reliable and durable product.

## 4 Nanotechnology for Wood Coating

### 4.1 Overview

With a high rate of urbanization, the mass production of wood-based products for furniture, construction, or energy is undeniable. In the matter of furniture and construction, fireproofing, slow degradation, anti-fungal attack, and mites are the specifications that are commonly demanded for a woody-based material. Thus, the wood coating technology in woodwork has been a headline for a couple of years. Some of the materials used for the nanoparticle coating include organic (cellulose and lignin), inorganic (metal-based materials), and hybrid materials (organic/inorganic materials). One of the common inorganic nanoparticles used is silica. Silica nanoparticles may penetrate intermittently into the cell walls' surface, creating a rough covering [2]. In addition to a small weight percentage gain, the hydrophobic silica nanoparticles permanently increase the hydrophobicity of wood, improving the dimensional stability [16]. Additionally, the treatment dramatically reduced the water intake and equilibrium moisture content [22]. This mechanism, therefore, improves the dimensional stability of wood surfaces, which display noticeably larger water contact angles (see Fig. 7).

Another study by Huang et al. [10] inspired by snail shells to produce the fire retardant for wood using nano-coating tannic acid (TA)-hexagonal boron nitride nanosheets (TA-BNNS) and polyvinyl alcohol (PVA). Then such a nanocoating was applied for the surface treatment of rigid polyurethane foam (RPUF). The study showed that TA/PVA@BNNS coated RPUF could reduce the peak heat release rate (PHRR) and peak smoke production rate (PSPR) by 52.4 and 53.4%, respectively. Moreover, Akhtari and Ganjipour [1] studied and compared the effects of nanosilver, nanocopper, and nanozincoxide on Paulownia (*Paulownia fortunei*) wood's ability to

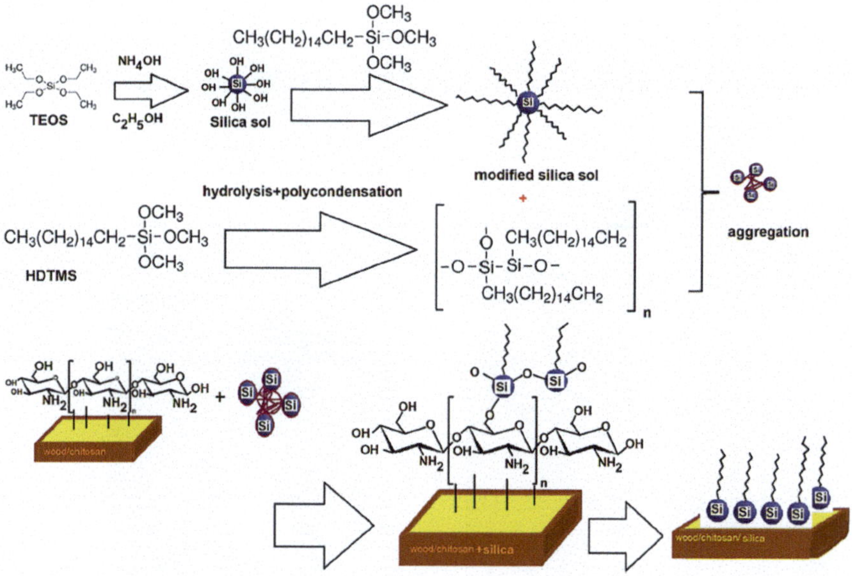


**Fig. 7** Hydrophobic mechanism of Si-nanoparticle on wood surface [16]

withstand the fungal disease that causes white rot (*Coriolus versicolor*). The results proven that nanosilver, nanocopper, and nanozinc oxide significantly improved the decay endurance of Paulownia against *C. versicolor* by decreasing the weight loss of wood from 28 to 2%. Likewise, nanoparticles can also provide additional functionalities to the wood composite. For example, the incorporation of antimicrobial nanoparticles can help prevent the growth of fungi and bacteria, making the composite suitable for use in humid environments [4]. Similarly, the addition of flame-retardant nanoparticles can enhance the fire resistance of the wood composite, making it safer for certain applications. In the wastewater treatment plant, nanotechnology has been used to treat methylene blue (MB, $C_{16}H_{18}N_3ClS$) with a high removal efficiency >99.8% (Fig. 8) [6].

Additionally, carbon nanotubes (CNTs) are also commonly used in biocomposite wood applications. CNTs are known for their exceptional mechanical properties, such as high tensile strength and stiffness [13]. By incorporating CNTs into biocomposite wood, the material's overall strength and durability can be significantly improved, allowing for a broader range of applications, and exceptional mechanical properties, such as high tensile strength and stiffness. In this regard, the nanotechnology added to the coating materials consider full of benefit to be applied in the wood industry.

In the hybrid coating material, such as cellulose wood/chitosan, there is a lot of spectrum overlap in the absorption bands of the two materials observed during the FTIR test (Fig. 9). It may be due to chitosan and cellulose being both polysaccharides. When silica is applied to wood or chitosan, the distinctive Si–O absorption bands

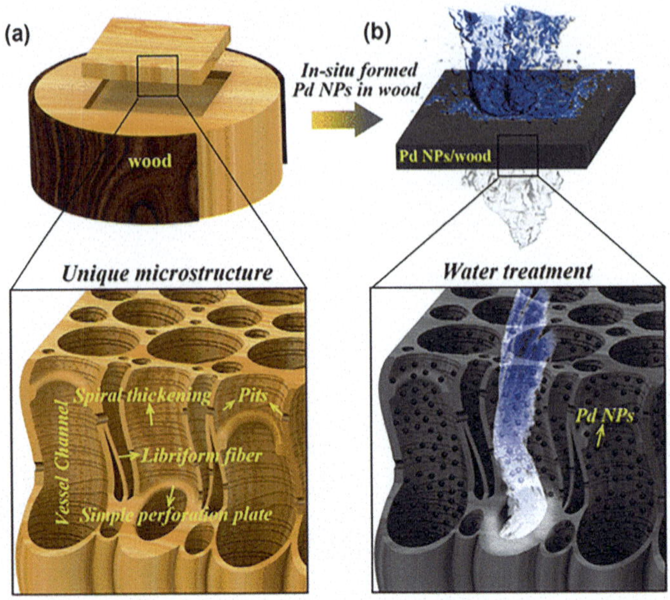


**Fig. 8** **a** Schematic of a 3D wood membrane decorated with Pd NPs for water treatment and **b** In situ formed Pd NPs within the wood where lignin acts as the reducing agent [6]

coincide with the C–O stretching vibrations in cellulose and chitosan in the 1000–1130 $cm^{-1}$ range. Peaks 2852 and 2922 $cm^{-1}$ are associated with the stretching vibrations of the alkyl groups in the hydrophobic chain of hexadecyl trimethoxysilane (HDTMS) on the wood surface, specifically $CH_2$ and $CH_3$ [16].

## 4.2 Characterization

To better understand the characteristics, distribution, and efficacy of nanoparticles in strengthening wood coatings, a variety of approaches and methodologies are used in the characterization of nanoparticles in wood coatings. To characterize the nanoparticle in wood coating, several factors are considered such as particle size and morphology, chemical composition and structure, dispersion and distribution in the coating, mechanical-thermal-optical properties, barrier properties, and durability or weathering. In terms of applications, there are four major classifications of nanotechnology that can be applied in wood coating including (1) UV radiation protection; (2) Antimicrobial; (3) Enhanced mechanical strength; and (4) Improved aesthetic. Zinc oxide (ZnO) and titanium dioxide ($TiO_2$) nanoparticles are efficient UV blockers [11, 29]. Added to wood coatings, they prevent UV light from damaging the coating and the wood fibers by absorbing and scattering it [25]. Some of the studies, incorporated ZnO- or $TiO_2$-based nanoparticle wood coating not only for UV radiation protection

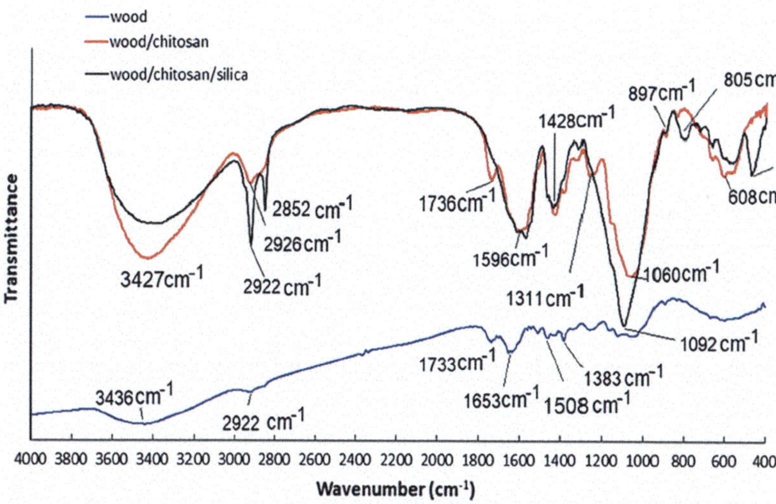


**Fig. 9** FTIR test for wood surface before and after Si-nanoparticle dopping [16]

but also for moisture protection [11], antiflaming [34], and antifungus attack [29]. Additionally, the fabrication of nanoparticles on the surface of the wood may improve the mechanical strength of the wood [28]. Some studies revealed that choosing cellulose nanocrystals [12], lignin/titanium oxide [27], or polydimethylsiloxane-Si nanoparticles [5] for the nanofiller is considered an excellent process in wood coatings.

## 5 Comparing Traditional to Advanced Wood Adhesive and Coating: Challenge Identification

When comparing nanotechnology invented to wood adhesive and coating with traditional ways, several fundamental differences emerge. Traditional wood composites, such as plywood or particleboard, typically rely on synthetic adhesives such as urea–formaldehyde or phenol–formaldehyde resins [33]. These adhesives often contain harmful chemicals and emit (volatile organic compounds) VOCs, posing health risks to both manufacturers and end-users as well as hazard to the environment. In contrast, nano bio-adhesive wood composites utilize bio-based adhesives derived from renewable resources, eliminating the need for toxic chemicals and reducing the emission of harmful substances [30]. This makes them a more sustainable and environmentally friendly option. Additionally, the incorporation of nanoparticles into the wood matrix improves the mechanical properties of the composite, resulting in a stronger and more durable material compared to traditional wood composites. Nano bio-adhesive wood composites also exhibit enhanced resistance to deformation, moisture, and fire, making them a more reliable choice for various applications. While nano bio-adhesive wood composites offer numerous advantages, there are also challenges and limitations to consider. One of the main challenges is the cost of production. The synthesis of nano bio-adhesives and the incorporation of nanoparticles into the wood matrix can be expensive, making these composites more costly compared to traditional wood materials [31].

Compared to traditional wood coating, nanoparticle-enhanced wood coatings improve the properties of the wood, as a consequence, the life span of the wood may be longer than the wood with the traditional treatment [10, 15]. Traditional wood coating typically relies on organic absorbers that may be considered can degrade over time, leading to a decrease in the effectiveness and lifespan of the wood, however, wood coatings are scientifically improved by exploring formulation cooperation with nanotechnology.

The challenge that currently appears for the nanotechnology application in wood adhesive and coating is the scalability of the manufacturing process. Currently, the production of nanoparticle adhesive and coating are limited to small-scale operations, making it difficult to meet the demands of large-scale construction projects or mass production in other industries. Further research and development are needed to optimize the manufacturing process and improve its efficiency. Furthermore, the long-term durability and performance of nanoparticle adhesive and coating under various environmental conditions need to be thoroughly evaluated. Factors such as UV exposure, temperature fluctuations, and moisture content can impact the stability and integrity of the composite over time. Continued research and testing are necessary to ensure the reliability and longevity of these composites.

## 6 Conclusions

By offering creative solutions that improve the material's qualities and performance, nanotechnology's application in wood products has completely changed the industry's trajectory. Wood adhesives are now more robust, long-lasting, and immune to external conditions including heat, moisture, and biological deterioration, thanks to the use of nanoparticles. This improvement broadens the use of wood products under a variety of difficult circumstances while simultaneously enhancing their durability and structural integrity. A more sustainable approach to the production of wood products is also encouraged by the development of eco-friendly nanoadhesives, which help allay health and environmental worries related to conventional adhesives. Additionally, nanotechnology has significantly improved the preservation and upkeep of wood coatings' surfaces. While preserving the wood's original beauty, nanocoatings offer remarkable toughness, scratch resistance, and UV protection. By adding nanoparticles, surfaces that are antimicrobial, antifungal, and self-cleaning can be created, which lessens the need for routine maintenance and increases the lifespan of wood goods. These developments add to the general sustainability of wood-based materials while also enhancing the useful qualities of wood. The wood industry can drive innovation and increase the use of wood in a variety of sectors by utilizing nanotechnology to create more robust, adaptable, and environmentally friendly products.

## References


1. Akhtari M, Ganjipour M (2013) Effect of nano-silver and nano-copper and nano-zinc oxide on Paulownia wood exposed to white-rot fungus. Paper presented at the 44th annual meeting of the international research group on wood protection, Çesme, Turkiye
2. Bak M, Takács D, Rákosa R, Németh ZI, Németh R (2023) One-step process for the fabrication of hydrophobic and dimensional stable wood using functionalized silica nanoparticles. Forests 14(3). https://doi.org/10.3390/f14030651
3. Bian Y, Bai M, Cao J, Li J (2023) A strong soybean meal adhesive enhanced by aluminum hydroxide nanoparticles via a low-cost and simple organic-inorganic hybrid strategy. Int J Adhes Adhes 125:103442. https://doi.org/10.1016/j.ijadhadh.2023.103442
4. Cai X, Riedl B, Wan H, Zhang S (2010) Montmorillonite nanoparticle distribution and morphology in melamine-urea-formaldehyde resin-impregnated wood nanocomposites. Wood Fiber Sci 285–291
5. Chang H, Tu K, Wang X, Liu J (2015) Fabrication of mechanically durable superhydrophobic wood surfaces using polydimethylsiloxane and silica nanoparticles. RSC Adv 5(39):30647–30653. https://doi.org/10.1039/C5RA03070F
6. Chen F, Gong AS, Zhu M, Chen G, Lacey SD, Jiang F, Li Y, Wang Y, Dai J, Yao Y, Song J, Liu B, Fu K, Das S, Hu L (2017) Mesoporous, three-dimensional wood membrane decorated with nanoparticles for highly efficient water treatment. ACS Nano 11(4):4275–4282. https://doi.org/10.1021/acsnano.7b01350
7. Dunky M (2003) Adhesives in the wood industry. In: Handbook of adhesive technology. ABD, New York, pp 872–941
8. Gadhave RV, SK V, Mahanwar PA, Gadekar PT (2021) Effect of addition of boric acid on thermo mechanical properties of microcrystalline cellulose/polyvinyl alcohol blend and applicability as wood adhesive. J Adhes Sci Technol 35(10):1072–1086
9. Guild FJ, Tsang WL, Taylor AC (2022) Silica nano-particle filled polymers: debonding and microstructure. Compos Sci Technol 218:109202. https://doi.org/10.1016/j.compscitech.2021.109202
10. Huang J, Yang L, Wang W, Yu B, Cao C, Bi Z, Wang J, Shen J, Song L (2024) Biomimetic, highly adhesive, supramolecular nano-coatings for fire-safe polymeric foams. Prog Org Coat 186:107995. https://doi.org/10.1016/j.porgcoat.2023.107995
11. Jnido G, Ohms G, Viöl W (2020) One-step deposition of polyester/$TiO_2$ coatings by atmospheric pressure plasma jet on wood surfaces for UV and moisture protection. Coatings 10(2). https://doi.org/10.3390/coatings10020184
12. Kaboorani A, Auclair N, Riedl B, Landry V (2017) Mechanical properties of UV-cured cellulose nanocrystal (CNC) nanocomposite coating for wood furniture. Prog Org Coat 104:91–96. https://doi.org/10.1016/j.porgcoat.2016.11.031
13. Kaymakci A, Ayrilmis N, Gulec T, Tufan M (2016) Preparation and characterization of high-performance wood polymer nanocomposites using multi-walled carbon nanotubes. J Compos Mater 51(9):1187–1195. https://doi.org/10.1177/0021998316674265
14. Khanjanzadeh H, Behrooz R, Bahramifar N, Pinkl S, Gindl-Altmutter W (2019) Application of surface chemical functionalized cellulose nanocrystals to improve the performance of UF adhesives used in wood based composites—MDF type. Carbohyd Polym 206:11–20. https://doi.org/10.1016/j.carbpol.2018.10.115

15. Kumar A, Nguyen TA, Sharma S, Bhat R (2022) Chapter 1—nanotechnology in paper and wood engineering: an introduction. In: Bhat R, Kumar A, Nguyen TA, Sharma S (eds) Nanotechnology in paper and wood engineering. Elsevier, pp 3–13. https://doi.org/10.1016/B978-0-323-85835-9.00015-5
16. Lazari M, Elmi F (2024) Structural study of coated wood with superhydrophobic chitosan/silica hybrid nanocomposite in seawater. Prog Org Coat 186:108076. https://doi.org/10.1016/j.porgcoat.2023.108076
17. Mariana M, Alfatah T, HPS AK, Yahya EB, Olaiya N, Nuryawan A, Mistar E, Abdullah C, Abdulmadjid S, Ismail H (2021) A current advancement on the role of lignin as sustainable reinforcement material in biopolymeric blends. J Mater Res Technol 15:2287–2316
18. Monroy Y, Rivero S, García MA (2019) Sustainable panels design based on modified cassava starch bioadhesives and wood processing byproducts. Ind Crops Prod 137:171–179. https://doi.org/10.1016/j.indcrop.2019.04.062
19. Montanari C, Olsén P, Berglund LA (2021) Sustainable wood nanotechnologies for wood composites processed by in-situ polymerization. Front Chem 9
20. Ni K, Du G, Liu C, Wu Y, Yang H, Yin C, Su H, Ran X, Gao W, Yang L (2023) Cross-linked entanglement of aldehyde and amine-functionalized nanocellulose reinforced with biomineralization to produce an all-bio-based adhesive. Chem Eng J 465:142888. https://doi.org/10.1016/j.cej.2023.142888
21. Papadopoulos AN, Taghiyari HR (2019) Innovative wood surface treatments based on nanotechnology. Coatings 9(12):866
22. Qu L, Rahimi S, Qian J, He L, He Z, Yi S (2021) Preparation and characterization of hydrophobic coatings on wood surfaces by a sol-gel method and post-aging heat treatment. Polym Degrad Stab 183:109429. https://doi.org/10.1016/j.polymdegradstab.2020.109429
23. Rajendran N, Runge T, Bergman RD, Nepal P, Houtman C (2023) Techno-economic analysis and life cycle assessment of cellulose nanocrystals production from wood pulp. Biores Technol 377:128955. https://doi.org/10.1016/j.biortech.2023.128955
24. Rinaudo M (2008) Main properties and current applications of some polysaccharides as biomaterials. Polym Int 57(3):397–430
25. Salla J, Pandey KK, Srinivas K (2012) Improvement of UV resistance of wood surfaces by using ZnO nanoparticles. Polym Degrad Stab 97(4):592–596. https://doi.org/10.1016/j.polymdegradstab.2012.01.013
26. Siahkamari M, Emmanuel S, Hodge DB, Nejad M (2022) Lignin-glyoxal: a fully biobased formaldehyde-free wood adhesive for interior engineered wood products. ACS Sustain Chem Eng 10(11):3430–3441
27. Song X, Guo W, Zhu Z, Han G, Cheng W (2024) Preparation of uniform lignin/titanium dioxide nanoparticles by confined assembly: a multifunctional nanofiller for a waterborne polyurethane wood coating. Int J Biol Macromol 258:128827. https://doi.org/10.1016/j.ijbiomac.2023.128827
28. Sow C, Riedl B, Blanchet P (2011) UV-waterborne polyurethane-acrylate nanocomposite coatings containing alumina and silica nanoparticles for wood: mechanical, optical, and thermal properties assessment. J Coat Technol Res 8(2):211–221. https://doi.org/10.1007/s11998-010-9298-6
29. Wang Y, Wu X, Wang Y, Tian Y, Mu H, Li J (2021) Hydrophobic and UV-resistant properties of environmentally friendly nano-ZnO-coated wood 75(2):138–147. https://doi.org/10.1515/hf-2019-0312
30. Xiong J, Chen F, Sun L, Yu X, Zhao J, Hu Y, Wang Y (2019) Characterization of VOC emissions from composite wood furniture: Parameter determination and simplified model. Build Environ 161:106237. https://doi.org/10.1016/j.buildenv.2019.106237
31. Yue H, Xu C, Yao J, He M, Yin G, Cui Y, Yang C, Guo J (2022) Characterization and properties of plywood bioadhesive derived from cottonseed protein and sawdust cellulose. Cellulose 29(10):5869–5881. https://doi.org/10.1007/s10570-022-04611-9
32. Zhang B, Pizzi A, Petrissans M, Petrissans A, Baptiste C (2023) Self-cross-linked tannin-aminated tannin surface coatings for particleboard. J Renew Mater 11(12):4097–4121

33. Zhang B, Pizzi AP, Pétrissans M, Petrissans A, Colin B (2023) A melamine-dialdehyde starch wood particleboard surface finish without formaldehyde. J Renew Mater 11. https://doi.org/10.32604/jrm.2023.028888
34. Zhou L, Fu Y (2020) Flame-retardant wood composites based on immobilizing with chitosan/sodium phytate/nano-$TiO_2$-ZnO coatings via layer-by-layer self-assembly. Coatings 10(3). https://doi.org/10.3390/coatings10030296
35. Zhu Y, Bian R, Yu Y, Li J, Li C, Lyu Y, Li X, Luo J, Li J (2024) Aminated alkali lignin nanoparticles enabled formaldehyde-free biomass wood adhesive with high strength, toughness, and mildew resistance. Chem Eng J 494:152914. https://doi.org/10.1016/j.cej.2024.152914
36. Zikeli F, Jusic J, Palocci C, Mugnozza GS, Romagnoli M (2024) Spray coating of wood with nanoparticles from lignin and polylactic glycolic acid loaded with thyme essential oils. Polymers 16(7):947

# Potential of Nanomaterials as Formaldehyde Scavengers in Wood-Based Composites

**Petar Antov, Jakub Kawalerczyk, Seng Hua Lee, Muhammad Adly Rahandi Lubis, and Viktor Savov**

**Abstract** Conventional wood-based composites are commonly produced using synthetic, formaldehyde-based resins, derived from petroleum-based constituents, like urea, phenol, and melamine. However, these thermoset adhesives have certain downsides, including the emission of harmful volatile organic compounds, notably formaldehyde, from the finished wood products. The advancements of nanotechnology have created many possibilities for the adhesive sector to innovate and produce a new wave of novel, high-performance adhesives for bonding wood-based composites. Using high aspect ratio nanomaterials as reinforcing agents has shown significant promise in improving the properties of wood-based composites. Therefore, the aim of this chapter is to discuss the recent advances in the use of inorganic and organic nanomaterials as formaldehyde scavengers used in the production of environmentally-friendly, high-performance wood-based composites with enhanced characteristics for a wide range of value-added applications. The challenges and future perspectives of using nanomaterials as formaldehyde scavengers in wood-based composites have also been discussed.

**Keywords** Wood adhesives · Wood-based composites · Free formaldehyde emission · Formaldehyde scavengers · Nanoparticles · Nanocellulose · Nanolignin · Nanoclay

---

P. Antov (✉) · V. Savov
Department of Mechanical Technology of Wood, Faculty of Forest Industry, University of Forestry, Sofia, Bulgaria
e-mail: p.antov@ltu.bg

J. Kawalerczyk
Department of Mechanical Wood Technology, Faculty of Forestry and Wood Technology, Poznań University of Life Sciences, Poznan, Poland

S. H. Lee
Department of Wood Industry, Faculty of Applied Sciences, Universiti Teknologi MARA, Cawangan Pahang Kampus Jengka, Bandar Tun Razak, Malaysia

M. A. R. Lubis
Research Center for Biomaterials, Indonesian Institute of Sciences, Cibinong, West Java, Indonesia

 
P. Antov et al. (eds.), *Nanomaterials Additives in Bioadhesives for Wood Composites*, Smart Nanomaterials Technology, https://doi.org/10.1007/978-981-96-3400-2_7

## 1 Introduction

In recent years, the wood-based panel industry is undergoing a crucial transformation towards a greener and digital bioeconomy. This shift is vital for the industry's sustainable evolution, aiming to reduce carbon emissions and adopt a circular economy model. By transforming linear industrial value chains and processes into circular ones, the industry seeks minimize pollution and waste generation, fostering sustainable growth and job creation. Embracing this green transition offers a chance to address climate change, lessen the reliance on fossil-based materials, enhance resource efficiency, and bolster security within the industry [1]. A key aspect of the circular economy is to use natural wood resources more efficiently, especially as wood demand is projected to triple from 2010 to 2050 [51]. The recent legislative requirements related to waste management and the increased environmental consciousness are the main driving forces for shifting the current wood-based industry practices to a low-carbon, circular bioeconomy [14, 73, 129]. Designing and creating sustainable, eco-friendly materials is now crucial for producing high-value products with a reduced environmental impact across diverse industries [16, 101, 159, 230]. The wood-based panel industry has recently undergone substantial transformations with new materials and technologies, becoming one of the most rapidly expanding manufacturing sectors globally [120, 153, 199, 246]. With rising demand for wood and its derivatives, global wood-based composite production has been growing steadily and is projected to increase from 403.87 million $m^3$ in 2024 to 474.36 million $m^3$ by 2029, with an expected compound annual growth rate (CAGR) of over 3% (Wood-based panel market 2024).

Adopting circular economy practices in the wood-based composites industry, connected with the cascading use of wood raw materials, and the increased global demand for sustainable, eco-friendly wood-based products, have created certain challenges to both industry and academia, related to recycling and reusing wood composites [39, 79, 92, 137, 138, 195, 232], development of innovative wood-based composites [9, 12, 33, 66, 155, 185, 208], optimal utilization of the available lignocellulosic feedstocks [7, 95, 102, 116, 166, 167], and search for alternative raw materials for their manufacturing [6, 28, 78, 119, 145, 170, 181, 206, 207, 211, 243].

Adhesives are crucial for the performance of wood-based composites, as the bond quality and material properties largely depend on the adhesive's type and characteristics [126, 160, 176, 239]. Selecting the appropriate adhesive for bonding wood-based composites depends on several factors, such as the desired properties of the panels, production conditions, the acceptable production costs, environmental aspects, and country-specific regulations [216]. Adhesives used in the wood industry can be categorized by their chemical composition, physical characteristics, or specific applications such as in construction, furniture making, or wood-composite production. Based on their chemical composition, formaldehyde-based adhesives, e.g., phenol–formaldehyde (PF), urea–formaldehyde (UF), and melamine-fortified urea–formaldehyde (MUF), resorcinol–formaldehyde resin (RF), phenol-resorcinol–formaldehyde resin (PRF), synthesized from petrochemical components, such as urea, phenol, melamine, etc. [119, 142, 176, 196, 201, 228]. These conventional thermosetting adhesives constitute about 95% of all adhesives utilized in the production of wood-based composites, including particleboards, medium-density fiberboards (MDF), and plywood [126, 142]. Urea–formaldehyde resins, produced through the polymeric condensation of formaldehyde and urea, are the primary type of thermosetting adhesives used in the manufacture of wood-based composites, with an estimated global annual consumption of approximately 11 million tons [126]. The synthesis of UF resins consists of two steps: an alkaline methylolation step followed by an acidic condensation reaction. UF adhesives constitute about 80% of the total global production volume of amino resins, followed by melamine–formaldehyde (MF) resins, along with smaller amounts of resins derived from other amino compounds or aldehydes [126, 165]. UF resins, composed of a mixture of linear or branched oligomers and polymers, invariably contain some unreacted monomers. Unreacted urea is purposefully included to attain specific effects, such as improved storage stability. Once UF resins have hardened, they form insoluble, three-dimensional networks, making the cured resins incapable of being melted or thermoformed subsequently. However, during the application stage, UF resins are either water-soluble, dispersed in water, or exist as spray-dried powders. Typically, these powders are dissolved and dispersed again in water before being used in various applications [126].

The extensive use of formaldehyde-based adhesives in wood industry can be attributed to their versatile characteristics, e.g., high chemical reactivity, solubility and dispersibility, compatible and adaptable to a variety of curing conditions, microorganism and abrasion resistance, and relatively low cost [59, 68, 142]. Despite their numerous beneficial characteristics, UF adhesives have certain significant disadvantages including the release of formaldehyde from the finished wood-based panels and their lack of durability in moist conditions. This is particularly true when heat is present, due to the aminonethylene link reversibility, accounting for the low resistance of UF resins to water and moisture, particularly at elevated temperatures, as well as their hydrolysis during composite manufacturing and usage [26, 50, 119]. These shortcomings have restricted the use of UF adhesive-bonded panels in indoor applications, as the formaldehyde released from these panels is a primary contributor to the deterioration of indoor air quality. It is also a key factor in causing the so-called "sick building syndrome" within indoor environments [118, 163]. Formaldehyde is one of the most common and significant indoor pollutants [32, 130, 135]. Prolonged exposure to formaldehyde may result in numerous health hazards, such as damage to the central nervous system, immune system issues, asthma, respiratory discomfort, and even cancer. Due to these health risks associated with formaldehyde exposure, the International Agency for Research on Cancer (IARC), adhering to a recommendation by the World Health Organization (WHO), has categorized formaldehyde as a human carcinogen (Group 1) [90]. This led to the implementation of more stringent formaldehyde emission regulations in different countries worldwide, e.g., in Europe, the USA, and Japan, which have raised the industrial focus on creating new wood adhesives that emit extremely low levels of formaldehyde and possess improved properties [119, 202]. The European (EN 120) standard regulation allows for formaldehyde levels in wood-based panels to be up to 8 mg/100 g for E1 class, and above 8 mg/100 g but not exceeding 30 mg/100 g for E2 class. The Japanese standard (JIS A 1460) imposes the strictest limits, permitting formaldehyde levels ranging from less than 0.3 mg $L^{-1}$ to a maximum of 5.0 mg $L^{-1}$, depending on the panel class [119]. These regulations have driven the wood-based panel industry to strive for developing products with close-to-zero formaldehyde emissions [9, 12, 88, 89]. Markedly, decreasing the formaldehyde content in adhesives like UF resin can negatively impact the physical and mechanical characteristics of wood-based products. This presents an ongoing challenge for manufacturers to produce panels that meet formaldehyde regulatory standards while maintaining high-performance characteristics.

Over the past decades, significant advancements have been made in reducing formaldehyde emissions from wood-based panels bonded with formaldehyde-based synthetic adhesives. These include reformulating the adhesive or partially replacing formaldehyde with other aldehydes, substituting formaldehyde-emitting adhesives with different adhesive types, and creating adhesives from renewable resources. The adverse effects of formaldehyde emissions from wood composites can be mitigated by employing isocyanate adhesives, specifically polymeric 4,4'-diphenyl methane diisocyanate (pMDI), which do not contain formaldehyde [67, 174, 176]. However, their broader industrial use as wood adhesives is hindered by the higher costs of pMDI compared to traditional formaldehyde-based adhesives. Additionally, pMDI's lower wetting angle relative to water-based condensation resins may lead to insufficient glue coverage, necessitating adjustments in application methods [85]. Other viable strategies to reduce the free formaldehyde emissions from wood-based panels involve incorporating a range of organic, inorganic, and mineral substances into the adhesive systems to capture formaldehyde, such as lignin, tannins, rice husk, hemp, wheat flour, bark, modified pine needles, pulp and paper sludge, urea, phosphates, nanoparticles, charcoal, pozzolan, wollastonite, and zeolites [8, 28, 36, 40, 41, 48–50, 52, 53, 57, 63, 107–109, 117, 144, 182, 208–210], adjusting the hot pressing parameters, such as temperature and duration [30, 105, 177, 220], treating or post-treating the surface of finished wood-based panels [31, 81, 184], and employing eco-friendly, biobased wood adhesives that are free from formaldehyde, are additional methods to reduce the harmful formaldehyde emissions [8–10, 14, 17, 82, 86, 157, 193, 194, 215, 224, 240]. A multitude of research has been dedicated to creating innovative, eco-friendly biobased adhesives for wood-based panel production, utilizing renewable resources and effectively repurposing waste and by-products. This includes technical lignins, e.g., lignosulfonates, hydrolysis, kraft and organosolv lignins [11, 12, 20, 37, 47, 62, 74, 88, 94, 173, 192, 197], modified condensed and hydrolyzed tannins [45, 60, 87, 151, 161, 172, 175, 217, 226], soy derivatives [15, 16, 44, 69, 70, 250], starch [15, 75, 87, 141, 143, 204], and a combination of various natural feedstocks, such as corn flour with lignin [242], mimosa tannin and corn starch [161], shrimp protein with corn starch and mimosa tannin [38], soybean modified with mimosa tannin [64], etc. Producing wood-based panels without additional binders is another feasible alternative, considering that wood naturally contains biopolymeric components such as cellulose, lignin, and hemicelluloses [65, 96].

**Fig. 1** Natural feedstocks for the synthesis of biobased wood adhesives [15, 16] Copyright from Elsevier with License number 5800801433854

Figure 1 visually depicts the primary categories of natural raw materials employed in formulating biobased adhesives for manufacturing wood-based panels.

The suitability of other biosourced materials in creating biobased wood adhesives has also been assessed, with studies showing encouraging results for the use of carbohydrates such as polysaccharides, oligomers, gums, and simple sugars. Notably, chitosans, which are abundant biodegradable polysaccharides sourced from crustacean shells, are considered a promising and sustainable option due to their reactive amine and hydroxyl groups [2, 91, 112, 219]. Additionally, furanic-aldehyde resin made from furfuryl alcohol and the non-toxic aldehyde glyoxal has exhibited strong water resistance and adhesive properties in plywood production [234]. Another promising technique for developing wood adhesive involves liquefying wood or similar lignocellulosic materials using substances like ethylene glycol, sulfuric acid, or phenol [128, 132]. Numerous studies have also highlighted the effectiveness of citric acid as a crosslinking agent or hardener to improve adhesive strength, as well as its use as a primary eco-friendly bonding agent [131, 190, 191, 200, 205]. Numerous studies have demonstrated that renewable resources, e.g., lignin, tannin, and soy protein isolate, can be utilized to produce environmentally-friendly, non-isocyanate polyurethanes (NIIPUs for wood adhesive applications [19, 20, 46, 192]. Additionally, the feasibility of using unsaturated vegetable oils, such as seed oil derivatives, has been explored for adhesive applications [61, 213]. Successful efforts have also been made to use cashew nut shell liquid, a significant agricultural byproduct of cashew nut production, mostly composed of anacardic acid, cardanol, and cardol, in the development of wood adhesives [97, 174, 194]. To note, the majority of research on developing fully biobased, zero-formaldehyde wood adhesive systems has been conducted at the laboratory level, targeted on the chemical characterization and modification of biosourced materials, as well as evaluating the exploitation properties and performance of the resulting wood-bases composites. The primary drawbacks for developing 100% biobased wood adhesives include enhancing the chemical reactivity of natural feedstocks through additional modifications, improving the mechanical characteristics and dimensional stability of the resulting wood composites, and adjusting the technological processes used, Currently, tannin-based bioadhesives are the only ones with significant industrial adoption [169, 253].

As ecological concerns rise with the widespread use of petroleum-based materials, it becomes increasingly vital to create innovative and sustainable wood adhesives offering improved performance, affordability, and a smaller environmental footprint [218].

Nanomaterials are widely employed across various sectors to improve the mechanical, chemical, and thermal properties of materials [100, 222]. In the field of composites, incorporating nanomaterials has been shown to boost mechanical strength, elasticity, and longevity [121, 146]. Biobased nanomaterials, derived from abundant and renewable lignocellulosic feedstocks, provide a sustainable and eco-friendly alternative, thanks to their natural properties like low density, biodegradability, robust mechanical and thermal characteristics, dimensional stability, and superior reinforcement capabilities [209].

Integrating nanomaterials into wood adhesive formulations can greatly diminish reliance on chemicals derived from fossil fuels. These nanomaterials present a sustainable, economical substitute for traditional synthetic adhesives based on formaldehyde, serving dual roles as fillers and reinforcements of adhesives used in the production of wood composites [223]. Notably, adding nanomaterials has been shown to reduce the release of harmful free formaldehyde from the finished wood-based panels and improve their mechanical characteristics [98, 103, 134, 212, 214]. Hence, the purpose of this chapter is to explore and assess the potential uses of nanomaterials, specifically nanocellulose, nanolignin, and nanoclay, as formaldehyde scavengers and reinforcing agents in various wood adhesives, and evaluate their effects on the overall adhesive performance and characteristics of wood-based composites [27, 162, 168].

## 2 Application of Nanomaterials as Formaldehyde Scavengers in Wood-Based Composites

The advent of nanotechnology has provided numerous possibilities for the adhesive sector to create novel types of high-performance adhesives [99]. Therefore, turning to nanotechnology and employing nanoparticles presents an alternative approach to overcoming the limitations of commercial wood adhesives. Previous studies have demonstrated that the addition of nanofillers can significantly enhance the properties of wood adhesives. Indeed, incorporating fillers/extenders can mitigate the constraints of polymer chemistry, as the adhesive's performance can be bolstered by integrating various nanomaterials [133, 221]. Nanomaterial fillers can be divided into two main categories, i.e., inorganic and organic. Inorganic nanoparticles, including nano-aluminum oxide, $SiO_2$, $TiO_2$, carbon nanotubes, and nanoclay, have been used in the composition of wood adhesives [42, 43, 99, 124]. To note, the application of these inorganic nanomaterials in wood adhesive formulations is connected with certain human health hazards and their renewability [100].

Alumina nanoparticles have demonstrated the capacity to absorb free formaldehyde in UF resin [53, 56]. Adding these nanoparticles reduces formaldehyde emissions in wood-based panels, although it may affect key features such as the peak curing temperature [122]. Furthermore, the addition of $Al_2O_3$ nanoparticles into UF resin can enhance heat distribution during the hot pressing of medium-density finerboards (MDF) and improve key mechanical characteristics of the composites, such as internal bond strength and modulus of elasticity [123]. Another study explored the effects of adding small amounts of $Al_2O_3$ nanoparticles to UF-bonded MDF panels on its rheology, morphology, thermodynamic mechanics and thermal kinetics [53]. The incorporation of $Al_2O_3$, particularly in larger quantities, altered the viscosity of UF resin, resulting in shear thinning behaviour (Fig. 2). Additionally, the thermodynamic analysis indicated a positive effect of $Al_2O_3$ addition on UF characteristics, including reduced formaldehyde emission during UF resin curing reactions at high temperatures, and a reduction of up to 14% in MDF panels at room temperature.

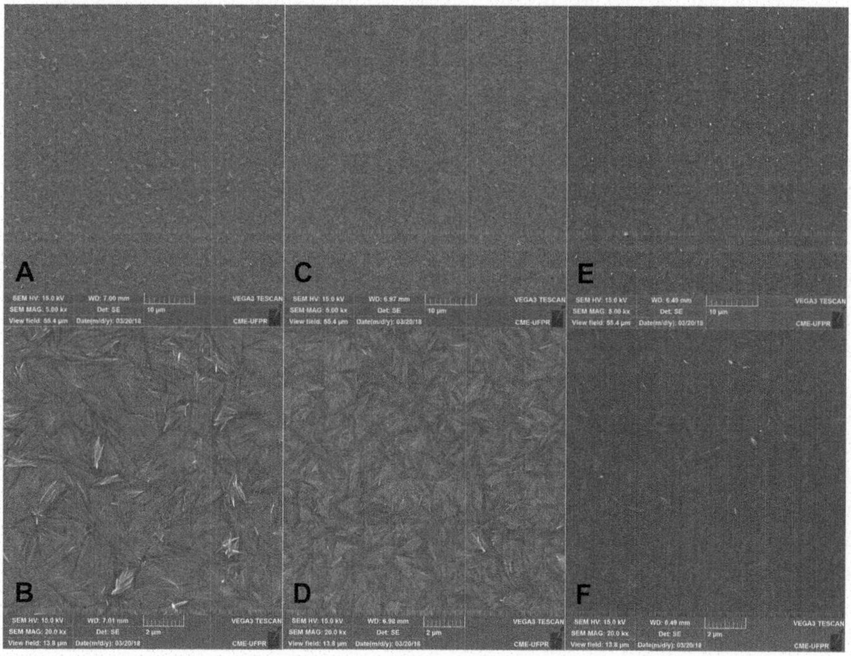


**Fig. 2** Scanning electron microscopy (SEM) images of unmodified and modified UF resins: **a** and **b** Unmodified UF at 5kx and 20kx; **c** and **d** UF + 0.5% Al2O3 at 5kx and 20kx; **e** and **f** UF + 2% $Al_2O_3$ at 5kx and 20kx [53]. Copyright from Elsevier with License number 5801770209418

The effects of incorporating nano-$SiO_2$ on the properties of UF resin used for bonding wood-based panels (plywood, particleboard and MDF) was also studied by Lin et al. [136]. The authors reported that the addition of nano-$SiO_2$ content up to 1.5% resulted in increased viscosity and bonding strength without extending the curing time, and decreased free formaldehyde emission levels, surpassing the requirements of the respective national standard and achieving E1 emission category.

Roumeli et al. [186] investigated the impact of adding different amounts of nano-$SiO_2$ on the properties of UF resin and the main physical, mechanical, and thermal characteristics of particleboards [186]. The panels exhibited slightly improved mechanical characteristics, such as internal bond strength and modulus of rupture, and enhanced thickness swelling values. However, only a small reduction of free formaldehyde content was reported, attributed to the relatively small (up to 3.5 wt%) nanoparticle content and their uneven distribution (Fig. 3).

The positive effect of incorporating nano-$SiO_2$ into UF resin on the reduction of free formaldehyde emission from wood-based panels was also confirmed for oriented strand board (OSB) manufactured from underutilized paulownia species (*Paulownia fortunie*) [189], and for plywood [235].

The application of graphene oxide (GO) as an effective formaldehyde scavenger in plywood panels containing phenol–formaldehyde (PF) resins has also been explored [34]. Adding just 1% GO significantly improved mechanical properties, including a 45% increase in shear strength, 35% in modulus of elasticity, and 25% in modulus of rupture compared to reference panels. Additionally, the moisture resistance of the panels improved, with a 25% increase in shear strength after 24 h in cold water and a 37% increase after 12 h of immersion in boiling water. Markedy, GO also demonstrated exceptional formaldehyde capture efficiency, reducing formaldehyde levels by over 60% compared to the control panels. The positive effect of incorporating graphene nanoplatelets (GNPs) in PF resin used for bonding particleboard panels on their formaldehyde emission was confirmed in the study by Łukawski et al. [140]. The authors reported an optimal weight percentage of GNPs of 18.5%, which resulted in decreased formaldehyde emission and satisfactory mechanical properties of the panels, allowing their use in furniture and interior design applications.

Another study also confirmed the potential of GO, synthesized using the improved Hummers' method, to effectively scavenge formaldehyde in the UF resin [187]. The addition of small amounts of GO did not have a significant effect on formaldehyde emission due to the limited GO adsorption sites (Fig. 4a). By adding 0.20 wt% GO to the UF resin, the UF/GO composite resin achieved a significant reduction in formaldehyde emissions, reaching $0.22 \pm 0.03$ mg $L^{-1}$—a remarkable 81.5% decrease compared to the control UF resin (Fig. 4b). However, higher GO concentrations (above 0.20 wt%) led to increased formaldehyde emissions, approaching levels similar to pure UF resin ($1.19 \pm 0.36$ mg $L^{-1}$). These results were attributed to the increased acidity of the composite which accelerated the curing reaction of UF resin, impacting emission levels (Fig. 4c).

The application of multiwalled carbon nanotubes (MWCNT) as nanofillers in UF resins for bonding fiberboard panels was investigated by Kumar et al. [124, 125]. A significant improvement of the mechanical properties, i.e., modulus of

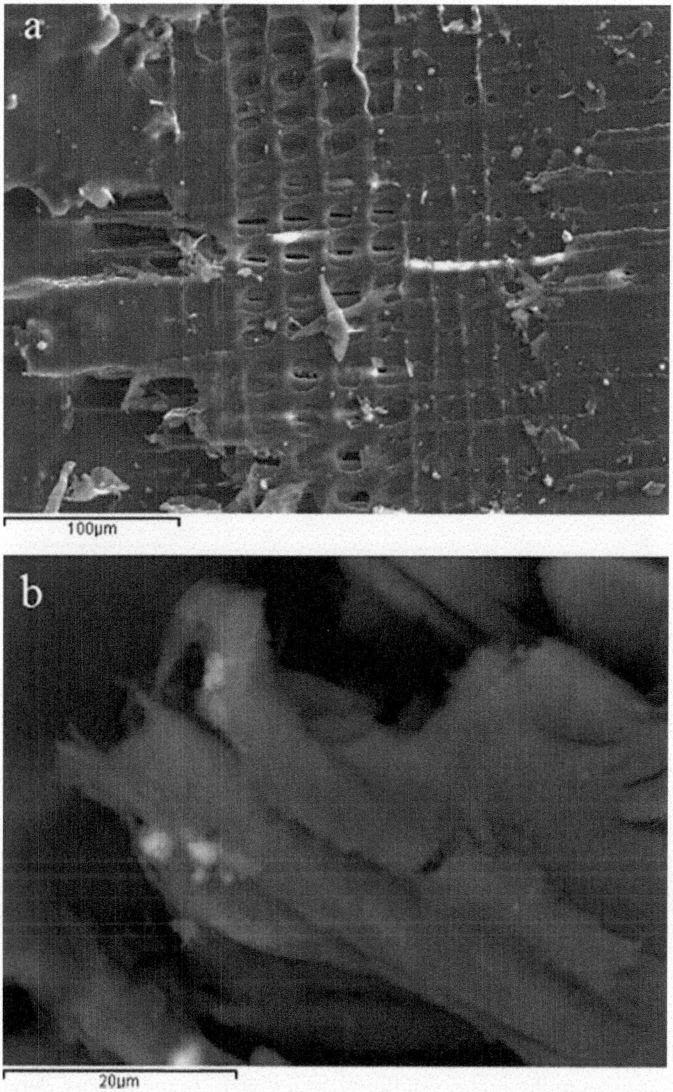


**Fig. 3** SEM of wood particles mixed with UF resin wood particles mixed with UF resin (**a**) and aggregations of nanoparticles (**b**) [186]. Copyright from Elsevier with License number 5801340607888

rupture and modulus of elasticity, of the panels bonded with UF resin, modified with MWCNT. Field emission scanning electron microscopy (FESEM) images of MWCNT, unmodified UF resin, of UF composite with the addition of MWCNT, and of a UF-MWCNT-bonded MDF panel are shown on Fig. 5.

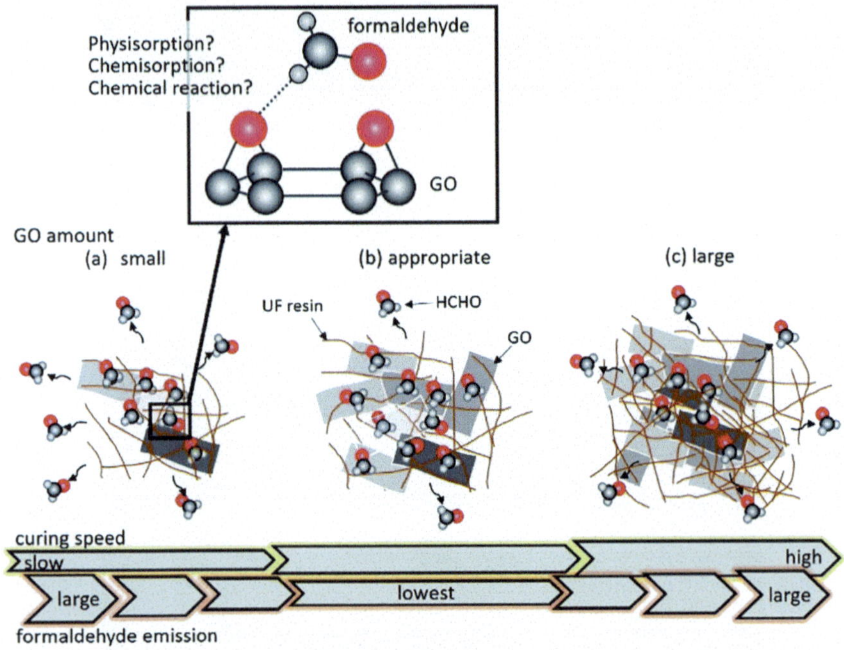


**Fig. 4** Formaldehyde emission from UF resin containing different wt% of GO as formaldehyde scavenger: **a** 0.10, 0.15 wt%, **b** 0.20 wt%, and **c** 0.40, 1.0, 1.9 wt% GO [187] (Open access)

A schematic representation of the UF-MWCNT nanocomposite formulation is given in Fig. 6. The authors achieved the presented morphology by applying ultrasonication. Under these conditions, the carboxylic groups (C = O) of the MWCNTs likely interact covalently with hydroxyl groups (OH) present in the UF adhesive during subsequent cross-linking and polymerization processes. The result is the formation of UF-nano MWCNT composites with enhanced properties [125].

A very interesting example of formaldehyde scavengers are the mineral nanoparticles defined as the layered mineral silicates also called nanoclays, such as e.g. montmorillonite, wollastonite and sepiolite. Due to the near-surface atomic structure, surface topography and crystallinity which were found to vary depending on the particles' size, they exhibit completely different properties than their bulk counterparts [84]. Besides the low toxicity, they are characterized by the large specific surface area and a great adsorption capacity may be very desirable for use as a formaldehyde scavenger [3, 83, 233].

Yadav et al. [238] performed interesting research on the use of pristine nanoclay and nanoclay modified with the transition metal ion as a modifier for UF resin in plywood production. Interestingly, nanoparticles were not added in the most common way as a filler, but were introduced during the synthesis of the resin. Authors found that the addition of nanoclay, regardless of the applied modification and the moment when the nanoparticles were added, contributed to the significant reduction

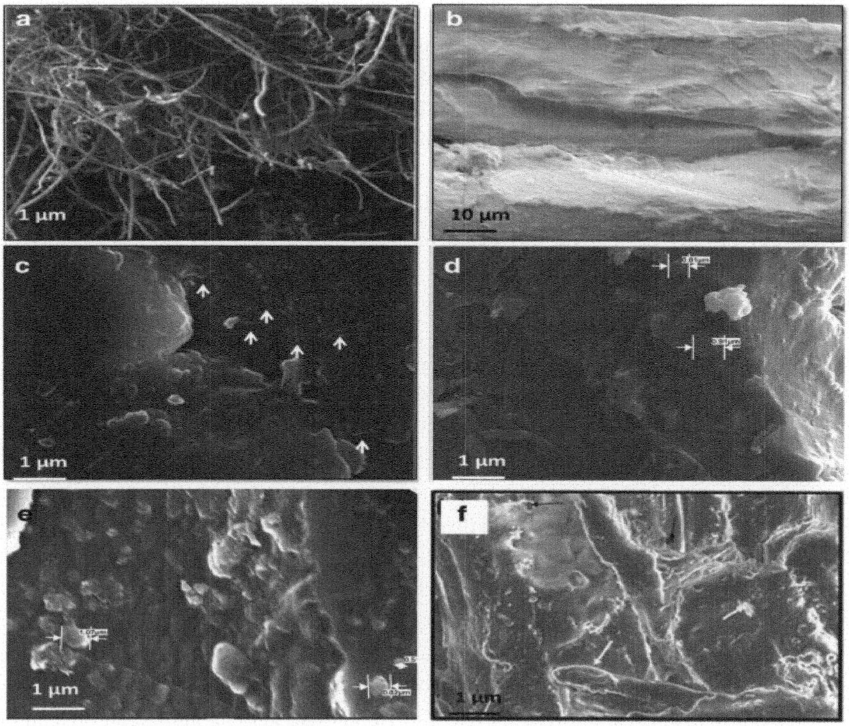


**Fig. 5** FESEM pictures of **a** MWCNT, **b** unmodified UF, **c** UF-MWCNT 1%, **d** UF-MWCNT 3.5%, **e** UF-MWCNT 7%, and **f** MDF panel bonded with UF-MWCNT [54]. Copyright from Elsevier with License number 5802540563610

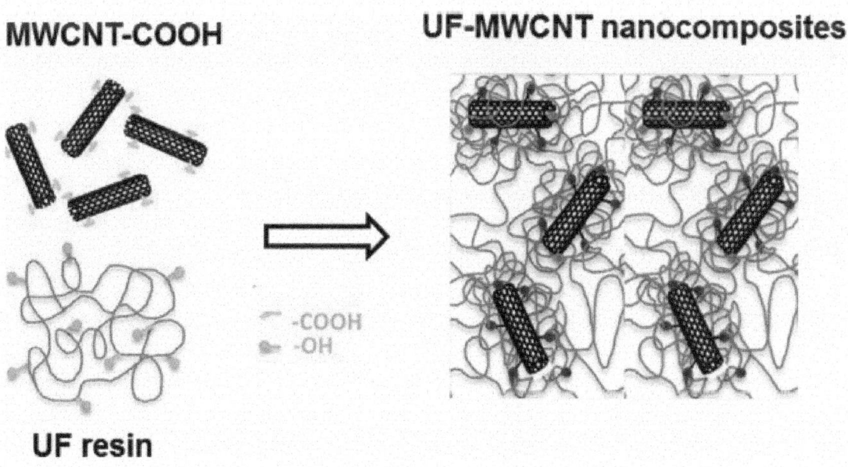


**Fig. 6** Formulation of UF- MWCNT nanocomposite [125]. Copyright from Elsevier with License number 5802540563610

in formaldehyde emission. What is also important, the modification also resulted in the improvement in bonding quality. It increases the chance of potential implementation, because modifiers that significantly reduce the strength of bond lines usually also demonstrate a limited applicability. The authors explained the observed phenomenon with the gas-phase absorption of formaldehyde which nanoclays are capable of when a good level of dispersion is achieved [189, 248]. Continuation of the research investigating the effect of the nanoclay modified with copper (II) chlorine addition during the alkaline stage of UF resin synthesis showed a positive effect on the formaldehyde emission from plywood as well [237]. Recorded emission values dropped from 1.351 even to 0.98 mg/L. In addition to gas-phase adsorption, authors also explained the outcomes as the consequence of the shielding effect, intramolecular interactions between nanoclay and UF resin and progressing reactions with the functional groups of UF resin leading to a decrease in free formaldehyde content [13, 229, 248]. Investigations conducted by Lei et al. [133] confirm this formaldehyde-scavenging effect even at low quantities of a modifier. Even the addition of 1% of nanoparticles reduced the formaldehyde emission from plywood, measured with the desiccator method, from 0.905 to 0.827 mg/L. This observation may be important considering the tendency of nanoparticles to agglomerate and dispersion-related issues occurring at higher loadings [127, 247]. In addition to the most common interior-grade plywood bonded with UF resin, the performed research also showed a positive impact of adhesive reinforcement with nanoclay in the case of water-resistant plywood [58, 225], plywood made of wood grown on tropical plantations [150] and plywood glued with MUF resin used in the conditions of increased humidity [55, 254].

Nanoclay is an interesting modifier of formaldehyde-based resins not only in the production of plywood, but also other wood-based materials such as particleboards, MDF boards or OSB boards. It was found that it contributes to the improvement in mechanical properties (modulus of rupture, bending strength and internal bond) and reduction in thickness swelling and water absorption of manufactured particleboards [93, 149]. Moreover, research performed by [236] has shown that the formaldehyde content was significantly reduced due to the addition of different nanoclays. The best results were obtained in the case of the 2% addition of wollastonite (Fig. 7b) and 1.5% montmorillonite (Fig. 7a), and the HCHO content was reduced from 6.85 to 3.20 mg/100 g and from 5.90 to 2.48 mg/100 g, respectively.

Interestingly, in the case of montmorillonite, a further increase in the content of nanoparticles resulted in an increase in the formaldehyde content, which, according to the authors, resulted from the progressive agglomeration of nanoparticles [53, 203]. Furthermore, the use of UF resin reinforced with various amounts of nanoclay (1, 3 and 5%) in the production of OSB boards from underutilized low-quality paulownia wood was also investigated [188, 189]. In the case of the most favorable UF adhesive formulation containing 5% nanoclay, the recorded formaldehyde emission decreased from 1.130 to 0.645 mg/L. According to authors, this phenomenon can be attributed to a tremendous barrier properties of nanoclay introduced into a polymer biostructures. Moreover, the reinforcing effect of nanoclay was also noted in the case of medium density fiberboard (MDF board). The outcomes of conducted analysis have shown that such modification of UF adhesive with the well-dispersed, exfoliated nanoclay

**Fig. 7** SEM images of: **a** montmorillonite provided by [252]; **b** wollastonite provided by [198] (Open access)

can considerably influence the thermal-oxidative stability, thermal conductivity, air permeability, modulus of rupture, modulus of elasticity, internal bond and thickness swelling of resultant panels in a positive way [23, 244, 245]. The addition of 5% nanoparticles also resulted in a significant decrease in formaldehyde emissions from 0.60 to 0.38 mg/L, which was associated with large specific surface area, high aspect ratio and the availability of hydroxyl groups to react with free formaldehyde [227]. An interesting concept that has not been thoroughly explored before is the use of functionalized nanoclay. This is the so-called two-stage modification of the resin, where in the first stage the nanoparticle is modified, and in the second stage the functionalized nanoparticle is added to the resin in order to improve its properties. For example, the functionalization of the nanoclay by a silane grafting is widely investigated in the production of bio-composites [4, 22, 72]. Silanization of nanoclays is usually performed to increase the interfacial interactions and compatibility between the hydrophobic polymer matrix and hydrophilic nanoparticles. Moreover, it usually also leads to improved dispersion of nanoparticles within the matrix and, what is important from the perspective of formaldehyde-scavenging abilities of nanoclay, the attachment of certain specific functional groups may also add some functions to such nano-modifier [76, 156, 179]. The results of the recent studies carried out by Khorramabadi et al. [115] and Khorramabadi et al. [114], showed that the surface functionalization of the nanoclay with 3-aminopropyltriethoxysilane (APTES) and L-lysine increased its potential in terms of reducing the formaldehyde emission. The use of unmodified nanoclay reduced formaldehyde emissions from MDF boards by up to 25%, however, the reductions recorded for APTES- and lysine-functionalized nanoparticles were reaching up to 61 and 50%, respectively. This phenomenon was most likely caused by the attachment of amino group, which has the ability to chemically bond both free and hydrolyzed formaldehyde [71, 171]. Although no other functionalization methods have been studied so far for nanoclay used as a UF resin modifier, these type of solutions certainly have great potential.

The example of nanomaterial widely distributed through the nature is nanocellulose (NC). It is the most abundant biopolymer occurring in the biosphere. Nano-sized cellulose is characterized by renewability, biocompatibility, great availability and biodegradability [104]. According to Gupta and Shukla [77], the possible advanced applications for nanocellulose include, for example, biomedical applications, energy conversion and harvesting devices, nanocomposite materials, sorbent matrix for liquid materials and food industry. To obtain it, chemical, biological and mechanical methods are usually used and some of them are presented in Fig. 8.

Research shows that the modification of formaldehyde-based resins with nanocellulose leads to a significant improvement in the properties of wood-based materials. Nano-sized cellulosic particles were previously found to exhibit reinforcing effect on many properties of veneer-based layered composites such as plywood and laminated veneer lumber (LVL). It contributed to the improvement in bonding quality [24], flexural strength [111], reduction of pressing parameters [110], reduction of adhesive consumption [106] etc. Furthermore, the introduction of NC to UF resin may improve the performance of particleboard, OSB board and MDF board by increasing their strength and reducing the water uptake [21, 148, 221]. Overall, the enhancement in these properties is attributed to the increased ductility, fracture energy, fracture toughness of the bond line and facilitated transfer of stresses which could be possible due to the interfacial interactions with the matrix, the large specific surface area and the high aspect ratio of NC [147, 154, 222]. However, the effect of unmodified

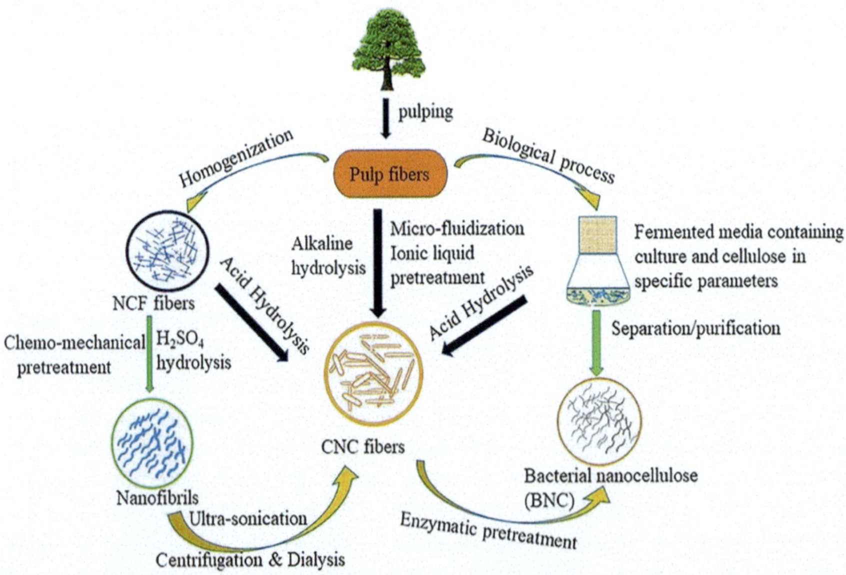


**Fig. 8** The nanocellulose production process [77, 180] (Open access)

nanocellulose on formaldehyde emission from wood-based materials still remains inconclusive.

Artner et al. [21], investigated the effect of nanocellulose addition on the properties of MDF boards. The applied modification using 1% NC caused a reduction in formaldehyde emission, determined with desiccator method, from 5.9 mg/L to 4.1 mg/L. Moreover, Moslemi et al. [148], used cellulosic nanoparticles extracted from rice straw as a modifier of UF resin in MDF board production. Authors found that in the case of 2% NC addition, the recorded formaldehyde emission decreased from 7.85 mg/L to 5.95 mg/L. In both studies, authors indicate that the reasons for the observed formaldehyde-scavenging effect may be the shielding effect, great specific surface area and availability of hydroxyl groups for the reaction with the active groups of UF resin. However, the investigations of formaldehyde emission from plywood and LVL manufactured with the use of NC-reinforced UF adhesive did not confirm this effect and showed no statistically significant changes [25, 108]. Therefore, more research is needed to evaluate the influence of numerous factors, related to e.g. the nanocellulose itself and the wood-based materials production technology, on the ability of NC to reduce the formaldehyde emission.

The results also indicate that, similarly as in the case of the nanoclay, the effect on the reduction of formaldehyde from wood-based materials could be more pronounced in the case of the surface-functionalized nanocellulose addition. The examples of methods which can be used for that purpose are summarized in Fig. 9.

In the case of studies on the modification of UF resin, only functionalization with APTES has been investigated so far. Overall, this is the most common method

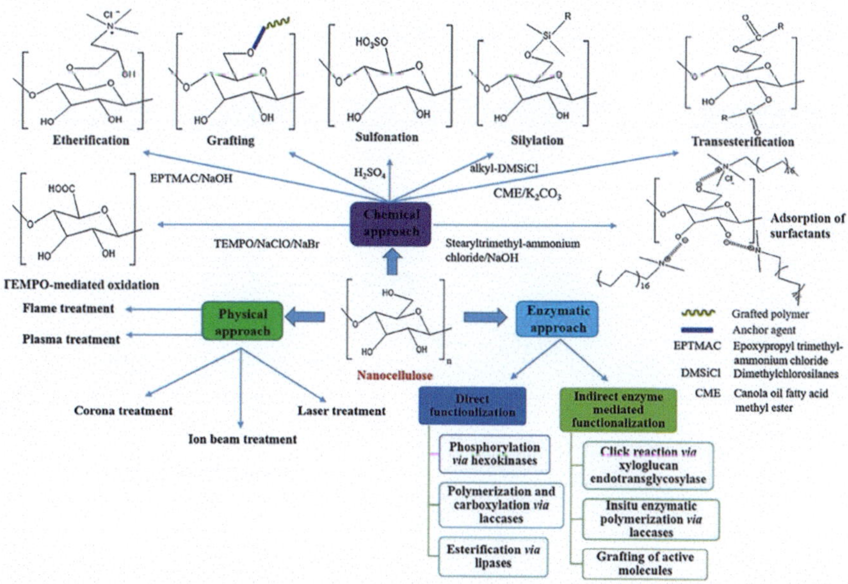


**Fig. 9** Selected methods used for nanocellulose functionalization [158] (Open access)

used to modify the cellulose nanoparticles intended for the use is the production of bio-composites [152]. Zhang et al. [251], used APTES-functionalized nanocellulose to enhance the adhesive behavior and improve the properties of plywood. Authors found that the addition of 1.5% NC resulted in the most significant reduction in formaldehyde emission from 0.47 to 0.22 mg/L. Similar results were obtained in the case of MDF boards, where formaldehyde emission decreased from 2.3 to 2.0 mg/100 g with the addition of 1.5% APTES-NC [249]. The positive effect of NC modified with APTES on the formaldehyde emission from MDF boards was confirmed by Khanjanzadeh et al. [113], who determined that the applied nano-modification decreased the HCHO emission from 7.03 to 5.83 mg/L. The functionalized cellulosic nanoparticles were also effective in the case of particleboard and strawboard made from rape straw particles. The emission recorded using a gas chamber analysis was reduced from 4.9 to 3.1 mg/m$^2$h and from 4.8 to 3.1 mg/m$^2$h in the case of particleboard and strawboard, respectively [107]. The reason for such a noticeable effect is probably the amino group occurring at the top of the APTES molecule (Fig. 10). Amino groups can easily react with both hydrolyzed and free formaldehyde and, as a result, it leads to the formation of methylol groups which can form methylene bridges due to further reactions [35, 80, 164]. Such promising results obtained for functionalized nanocellulose indicate a great need to conduct further research in this field.

Another example of a nanoparticle that can be obtained from lignocellulosic biomass is nanolignin. It is estimated that approximately 50–70 million tons of lignin are produced annually in the world. Despite its interesting chemical structure, the vast majority of lignin is intended for combustion, while at least some of it could be used for the production of nanomaterials [88, 89]. The advantages of nanolignin include, for example, biodegradability, no toxicity, great compatibility, antibacterial features

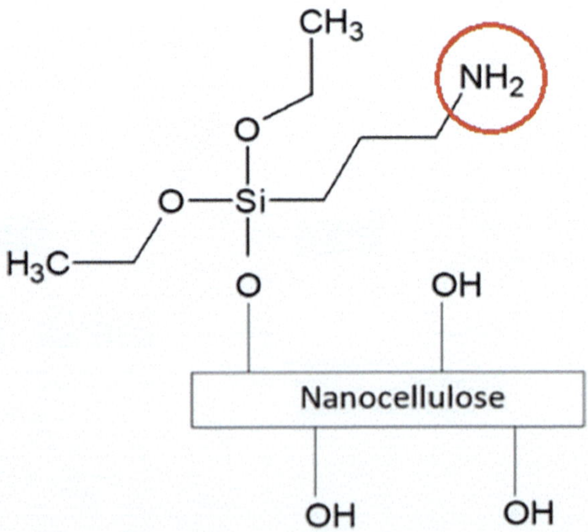


**Fig. 10** Schematic presentation of APTES-modified nanocellulose [107] (Open access)

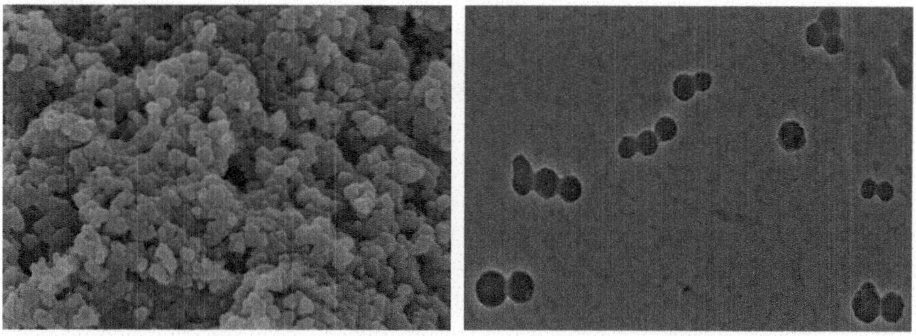


**Fig. 11** The appearance of lignin nanoparticles: **a** SEM images, **b** TEM images [139] (Open access)

etc. [5]. For this reason, the addition of nanolignin has been investigated in applications such as food additives, biocomposites, food packing, pharmaceutical and cosmetic products, and many others [178]. The appearance of lignin nanoparticles is shown in images taken using SEM and TEM presented in Fig. 11. The number of works on nanolignin used as a modifier of formaldehyde-based resins is very scarce, much lower than, for example, in the case of nanocellulose or metal nano-oxides.

Ridho et al. [183], examined the suitability of nanolignin obtained from the areca leaf sheath for the modification of the ultra-low molar ratio UF resin. The results have shown the acceleration in adhesive curing, reduced crystallinity of resin in cured state and improvement in tensile strength of plywood. Moreover, the formaldehyde emission was reduced from 0.078 to 0.065 mg/L due to the introduction of 1% nanolignin. Furthermore, Younesi-Kordkheili [241] conducted investigations aimed at determining the effect of nanolignin obtained by acidic processing of bulk lignin introduced to UF resin on the properties of plywood. Based on the outcomes, it was found that the mechanical properties were improved and the formaldehyde emission was reduced. The obtained emission values decreased from 4.3 to 2.4 mg/100 g due to the addition of maleated nanolignin. According to authors, the reason for lower formaldehyde emission was more crosslinked structure and availability of active sites of nanolignin for the reaction with free formaldehyde. Taking into account the promising results presented in recently published works, it can be concluded that research in this direction demonstrates a great potential.

## 3 Summary and Future Challenges

Nanotechnology as a science is developing at a very fast pace by implementing the achievements from many different fields. Very high progress is noticeable in the production of various types of composites modified with nanomaterials, which are used in many industrial branches. Materials engineering focuses on both the most efficient and cost-effective production of nanoparticles, their modification and finding their potential applications. One of the possibilities is for sure the addition of nanomaterials as formaldehyde scavengers to formaldehyde-based resins commonly used in wood-based materials manufacturing. This is caused by reduction in the level of allowable emissions and growing awareness of potential health threats caused by the exposure to HCHO. It has been shown that nanoparticles such as, for example, nano-$Al_2O_3$, nano-$SiO_2$, nano-graphene oxide, multiwalled carbon nanotubes, nanoclay, nanocellulose and nanolignin introduced to the UF resin contribute to a decrease in formaldehyde emission. In most cases, the formaldehyde-scavenging effect is attributed to the large specific surface area, high aspect ratio, high reactivity, shielding effect, significant adsorption capacity and the availability of hydroxyl groups to react with free formaldehyde. However, further studies are needed to fully understand this complex issue. To achieve this, three directions of further research can be listed: (i) the effect of nanoparticles' characteristics on the formaldehyde-scavenging ability, (ii) the effect of nanoparticle functionalization on the formaldehyde-scavenging ability and (iii) health risks associated with the use of nanomaterials. As shown in this work, the effect of even the same type of nanoparticle on formaldehyde emission from wood-based materials can vary in a significant way. It is known that the properties of nanoparticles may depend on many factors, such as their crystallinity, size, agglomeration, chemical composition, state, shape etc., which may also affect their formaldehyde-scavenging ability. However, no comprehensive analyzes in this area have been carried out yet. It has also been shown that the functionalization of nanoparticles may significantly increase their potential in reducing formaldehyde emissions. So far, only functionalization using APTES has been studied, but many methods exhibiting even greater potential can be used for this purpose. Furthermore, the toxicity of nanoparticles varies greatly depending on their features. The studies conducted in vitro and in vivo, in some cases present different outcomes in terms of potential threats, which indicates the need for further research. Thus, there is also an urgent need to develop precautions to protect the users of nanomaterials and protocols for conducting research on the nano-sized particles.

**Acknowledgements** This research was supported by the project "Development, Exploitation Properties and Application of Eco-Friendly Wood-Based Composites from Alternative Lignocellulosic Raw Materials", project no. НИС-Б-1290/19.10.2023, carried out at the University of Forestry, Sofia, Bulgaria.

## References

1. Antov P, Lee SH, Lubis MAR, Kristak L, Réh R (2023) Advanced eco-friendly wood-based composites II. Forests 14:826
2. Abdelmoula M, Ben Hlima H, Michalet F, Bourduche G, Chavant JY, Gravier A, Delattre C, Grédiac M, Mathias JD, Abdelkafi S, Michaud P, de Baynast H (2021) Chitosan-based adhesive: optimization of tensile shear strength in dry and wet conditions. Polysaccharides 2(1):110–120

3. Abdelnour SA, Alagawany M, Hashem NM, Farag MR, Alghamdi ES, Hassan FU, Bilal RM, Elnesr SS, Dawood MA, Nagadi SA (2021) Nanominerals: fabrication methods, benefits and hazards, and their applications in ruminants with special reference to selenium and zinc nanoparticles. Animals 11(7):1916
4. Abeywardena SBY, Perera S, Nalin De Silva KM, Tissera NP (2017) A facile method to modify bentonite nanoclay with silane. Int Nano Lett 7(3):237–241. https://doi.org/10.1007/s40089-017-0214-2
5. Abraham B, Syamnath VL, Arun KB, Zahra PF, Anjusha P, Kothakotta A, Chen Y-H, Ponnusamy VK, Nisha P (2023) Lignin-based nanomaterials for food and pharmaceutical applications: Recent trends and future outlook. Sci Total Environ 881:163316
6. Akinyemi BA, Kolajo TE, Adedolu O (2022) Blended formaldehyde adhesive bonded particleboards made from groundnut shell and rice husk wastes. Clean Technol Environ:1–10
7. Anggini AW, Lubis MAR, Sari RK, Papadopoulos AN, Antov P, Iswanto AH, Lee SH, Mardawati E, Kristak L, Juliana I (2023) Cohesion and adhesion performance of tannin-glyoxal adhesives at different formulations and hardener types for bonding particleboard made of Areca (*Areca catechu*) leaf sheath. Polymers 15:3425
8. Antov P, Jivkov V, Savov V, Simeonova R, Yavorov N (2020a) Structural application of eco-friendly composites from recycled wood fibres bonded with magnesium lignosulfonate. Appl Sci 10:7526
9. Antov P, Savov V, Krišt'ák L, Réh R, Mantanis GI (2021a) Eco-friendly, high-density fiberboards bonded with urea-formaldehyde and ammonium lignosulfonate. Polymers 13:220
10. Antov P, Savov V, Neykov N (2020b) Reduction of formaldehyde emission from engineered wood panels by formaldehyde scavengers—A review. In: Proceedings of the 13th international scientific conference wood ema 2020 and 31st international scientific conference ICWST 2020 sustainability of forest-based industries in the global economy, Vinkovci, Croatia, pp 7–11
11. Antov P, Savov V, Neykov N (2020c) Sustainable bio-based adhesives for eco-friendly wood composites—A review. Wood Res 65(1):51–62
12. Antov P, Savov V, Trichkov N, Krišťák Ľ, Réh R, Papadopoulos AN, Taghiyari HR, Pizzi A, Kunecová D, Pachikova M (2021b) Properties of high-density fiberboard bonded with urea-formaldehyde resin and ammonium lignosulfonate as a bio-based additive. Polymers 13(16):2775
13. Arafa IM, Fares MM, Barham AS (2004) Sol–gel preparation and properties of interpenetrating, encapsulating and blend silica based urea formaldehyde hybrid composite materials. Eur Polymer J 40(7):1477–1487
14. Arias A, Entrena-Barbero E, Feijoo G, Maria Teresa Moreira MT (2022) Sustainable non-isocyanate polyurethanes bio-adhesives for engineered wood panels are revealed as promising candidates to move from formaldehyde-based alternatives J Environ Chem Eng 10:107053
15. Arias A, Feijoo G, Moreira MT (2021a) Evaluation of starch as an environmental-friendly bioresource for the development of wood bioadhesives. Molecules 26:4526
16. Arias A, Feijoo G, Moreira MT (2023) Biorefineries as a driver for sustainability: Key aspects, actual development and future prospects. J Clean Prod 418(3):137925
17. Arias A, González-García S, Feijoo G, Moreira MT (2021b) Environmental benefits of soy-based bio-adhesives as an alternative to formaldehyde-based options. Environ Sci Pollut Res 28:29781–29794
18. Arias A, González-Rodríguez S, Vetroni Barros M, Salvador R, de Francisco AC, Piekarski CM, Moreira MT (2021) Recent developments in bio-based adhesives from renewable natural resources. J Clean Prod 314:127892
19. Aristri MA, Lubis MAR, Iswanto AH, Fatriasari W, Sari RK, Antov P, Gajtanska M, Papadopoulos AN, Pizzi A (2021a) Bio-based polyurethane resins derived from tannin: Source, synthesis, characterisation, and application. Forests 12(11):1516
20. Aristri MA, Lubis MAR, Yadav SM et al (2021b) Recent developments in lignin- and tannin-based non-isocyanate polyurethane resins for wood adhesives—A review. Appl Sci 11:4242
21. Artner MA, de Cademartori PHG, Avelino F, Lomonaco D, Magalhães WLE (2021) A novel design for nanocellulose reinforced urea–formaldehyde resin: a breakthrough in amino resin synthesis and biocomposite manufacturing. Cellulose 28(6):3435–3450

22. Asgari M, Abouelmagd A, Sundararaj U (2017) Silane functionalization of sodium montmorillonite nanoclay and its effect on rheological and mechanical properties of HDPE/clay nanocomposites. Appl Clay Sci 146:439–448
23. Ashori A, Nourbakhsh A (2009) Effects of nanoclay as a reinforcement filler on the physical and mechanical properties of wood-based composite. J Compos Mater 43(18):1869–1875
24. Ayrilmis N, Kwon J-H, Lee S-H, Han T-H, Park C-W (2016a) Microfibrillated-cellulose-modified urea-formaldehyde adhesives with different F/U molar ratios for wood-based composites. J Adhes Sci Technol 30(18):2032–2043
25. Ayrilmis N, Lee Y-K, Kwon JH, Han T-H, Kim H-J (2016) Formaldehyde emission and VOCs from LVLs produced with three grades of urea-formaldehyde resin modified with nanocellulose. Build Environ 97:82–87
26. Baharoglu M, Nemli G, Sarı B, Bardak S, Ayrılmıs N (2012) The influence of moisture content of raw material on the physical and mechanical properties, surface roughness, wettability, and formaldehyde emission of particleboard composite. Compos B Eng 43:2448–2451
27. Bandara N, Esparza Y, Wu J (2017) Exfoliating nanomaterials in canola protein derived adhesive improves strength and water resistance. RSC Adv 7:6743–6752
28. Barbu MC, Lohninger Y, Hofmann S, Kain G, Petutschnigg A, Tudor EM (2020) Larch bark as a formaldehyde scavenger in thermal insulation panels. Polymers 12(11):2632
29. Barbu MC, Sepperer T, Tudor EM, Petutschnigg A (2020) Walnut and hazelnut shells: Untapped industrial resources and their suitability in lignocellulosic composites. Appl Sci 10:6340
30. Bekhta P, Sedliačik J, Bekhta N (2020) Effect of veneer-drying temperature on selected properties and formaldehyde emission of birch plywood. Polymers 12(3):593
31. Bekhta P, Sedliačik J, Jones D (2018) Effect of short-term thermomechanical densification of wood veneers on the properties of birch plywood. Eur J Wood Wood Prod 76(2):549–562
32. Bekhta P, Noshchenko G, Réh R, Kristak L, Sedliačik J, Antov P, Mirski R, Savov V (2021a) Properties of eco-friendly particleboards bonded with lignosulfonate-urea-formaldehyde adhesives and pMDI as a crosslinker. 14(17): 4875
33. Bekhta P, Sedliačik J, Noshchenko G, Kačík F, Bekhta N (2021b) Characteristics of beech bark and its effect on properties of UF adhesive and on bonding strength and formaldehyde emission of plywood panels. Eur J Wood Prod 79:423–433
34. Benhamou AA, Salim MH, Boussetta A, Mennani M, Essifi K, Kassab Z, El Achaby M, Moubarik A (2023) Graphene oxide a promising formaldehyde scavenger of phenolic resin-bonded plywood: From elaboration to evaluation. J Appl Polym Sci. https://doi.org/10.1002/app.55093
35. Boran S, Usta M, Gümüşkaya E (2011) Decreasing formaldehyde emission from medium density fiberboard panels produced by adding different amine compounds to urea formaldehyde resin. Int J Adhes Adhes 31(7):674–678
36. Boran S, Usta M, Ondaral S, Gümüşkaya E (2012) The efficiency of tannin as a formaldehyde scavenger chemical in medium density fiberboard. Compos Part B Eng 43(5):487–2491
37. Borrero-López AM, Valencia C, Domínguez G, Eugenio ME, Franco JM (2021) Rheology and adhesion performance of adhesives formulated with lignins from agricultural waste straws subjected to solid-state fermentation. Ind Crop Prod 171:113876
38. Boussetta A, Benhamou AA, Ihammi A, Ablouh EH, Barba FJ, Boussetta N, Grimi N, Moubarik A (2022) Shrimp waste protein for bio-composite manufacturing: Formulation of protein-cornstarch-mimosa-tannin wood adhesives. Ind Crop Prod 187:115323
39. Bütün Buschalsky FY, Mai C (2021) Repeated thermo-hydrolytic disintegration of medium density fibreboards (MDF) for the production of new MDF. Eur J Wood Prod 79:1451–1459
40. Buyuksari U, Ayrilmis N, Avci E, Koc E (2010) Evaluation of the physical, mechanical properties and formaldehyde emission of particleboard manufactured from waste stone pine (*Pinus pinea* L.) cones. Bioresour Technol 101:255–259
41. Camlibel O (2020) Mechanical and formaldehyde-related properties of medium density fiberboard with zeolite additive. BioResources 15(4):7918–7932

42. Candan Z, Akbulut T (2013) Developing environmentally friendly wood composite panels by nanotechnology. BioResources 8(3):3590–3598
43. Candan Z, Akbulut T (2015) Physical and mechanical properties of nanoreinforced particleboard composites. Maderas Ciencia Y Tecnologia 17(2):319–334
44. Chen N, Zheng P, Zeng Q, Lin Q, Rao J (2017) Characterization and performance of soy-based adhesives cured with Epoxy resin. Polymers 9(10):514
45. Chen X, Pizzi A, Fredon E, Gerardin C, Zhou X, Zhang B, Du G (2022) Low curing temperature tannin-based non-isocyanate polyurethane (NIPU) wood adhesives: Preparation and properties evaluation. Int J Adhes Adhes 112:103001
46. Chen X, Pizzi A, Xi X, Zhou X, Fredon E, Gerardin C (2021) Soy protein isolate Non-Isocyanates Polyurethanes (NIPU) wood adhesives. J Renew Mater 9(6):1045–1057
47. Chen X, Xi X, Pizzi A, Fredon E, Du G, Gerardin C, Amirou S (2020) Oxidized demethylated lignin as a bio-based adhesive for wood bonding. J Adhes 97:873–890
48. Costa NA, Ohlmeyer M, Ferra J, Magalhães F, Mendes A, Carvalho L (2014) The influence of scavengers on VOC emissions in particleboards made from pine and poplar. Eur J Wood Prod 72:117–121
49. Costa NA, Pereira J, Ferra J, Cruz P, Martins J, Magalhães FD, Mendes A, Carvalho LH (2013b) Sodium metabisulphite as a scavenger of air pollutants for wood-based building materials. Int Wood Prod J 4(4):242–247
50. Costa NA, Pereira J, Ferra J, Cruz P, Martins J, Magalhães FD, Mendes A, Carvalho LH (2013a) Scavengers for achieving zero formaldehyde emission of wood-based panels. Wood Sci Technol 47:1261–1272
51. Dammer L, Bowyer C, Breitmayer E, Eder A, Nanni S, Allen B, Carus M, Essel R (2016) Mapping study on cascading use of wood products. WWF Technical Report. https://wwfeu.awsassets.panda.org/downloads/wwf_mondi_cascading_use_of_wood_final_web.pdf (accessed on 6 Mar 2024)
52. Darmawan S, Sofyan K, Pari G, Sugiyanto K (2010) Effect of activated charcoal addition on formaldehyde emission of medium density fibreboard. J For Res 7(2):100–111
53. de Cademartori PHG, Artner MA, de Freitas RA, Magalhães WLE (2019) Alumina nanoparticles as formaldehyde scavenger for urea-formaldehyde resin: Rheological and in-situ cure performance. Compos Part B: Eng 176:107281
54. Dorieh A, Selakjani PP, Shahavi MH, Pizzi A, Movahed SG, Pour MF, Aghaei R (2022) Recent developments in the performance of micro/nanoparticle-modified urea-formaldehyde resins used as wood-based composite binders: A review. Int J Adhes Adhes 114:103106
55. Doosthoseini K, Zarea-Hosseinabadi H (2010) Using Na+ MMT nanoclay as a secondary filler in plywood manufacturing. J Indian Acad Wood Sci 7:58–64
56. Dudkin BN, Krivoshapkin VP, Krivoshapkina EF (2006) Effect of aluminum oxide nanoparticles on the properties of urea-formaldehyde resin. Russ J Appl Chem 79(9):1522–1525
57. Dukarska D, Kawalerczyk J, Kmieciak J (2024) Modified pine needles as a formaldehyde scavenger for urea-formaldehyde resin in plywood production. Eur J Wood Prod 82:147–158
58. Dukarska D, Kawalerczyk J, Włodarczyk J (2022) Properties of the water-resistant plywood glued with pf resin with the addition of tannin filler and nanoclay. Wood Research 67:241–253
59. Dunky M (2003) Adhesives in the wood industry. Chapter 47. In: Pizzi A, Mittal KL (eds) Handbook of adhesive technology. Marcel Dekker, New York, pp 872–941
60. Dunky M (2020) Wood adhesives based on natural resources: A critical review Part III. Tannin- and Lignin-based adhesives. Rev Adhes Adhes 8:379–525
61. Dunky M (2021) Wood adhesives based on natural resources: A critical review: Part IV. Special Topics. In: Mittal KL (ed) Progress in adhesion and adhesives, vol 6. https://doi.org/10.1002/9781119846703.ch18
62. El Mansouri NE, Pizzi A, Salvadó J (2006) Lignin-based wood panel adhesives without formaldehyde. Holz Roh Werkst 65:65
63. Eom YG, Kim JS, Kim S, Kim JA, Kim HJ (2006) Reduction of formaldehyde emission from particleboards by bio-scavengers. Mokchae Konghak 34(5):29–41

64. Esposito L, Ciannamea EM, Solaberrieta I, Piter JC, Ruseckaite RA, Stefani PM (2022) Tannin-modified soybean protein concentrate for wood adhesive. J Appl Res Technol Eng 3(1):1–7
65. Ferrandez-Garcia A, Ferrandez-Garcia MT, Garcia-Ortuño T, Ferrandez-Villena M (2022) Influence of the density in binderless particleboards made from sorghum. Agronomy 12:1387
66. Foti D, Voulgaridou EE, Karastergiou S, Taghiyari HR, Papadopoulos AN (2022) Physical and mechanical properties of eco-friendly composites made from wood dust and recycled polystyrene. J Renew Mater 10(1):75
67. Frazier CE (2003) Isocyanate wood binders. In: Pizzi A, Mittal KL (eds) Handbook of adhesive technology, 2nd ed, Chapter 33. Marcel Dekker, New York, NY, USA, pp 681–694
68. Frihart CR (2015) Wood adhesives: Past, present, and future. For Prod J 65:4–8
69. Frihart CR, Satori H (2013) Soy flour dispersibility and performance as wood adhesive. J Adhes Sci Technol 27:2043–2052
70. Ghahri S, Pizzi A (2018) Improving soy-based adhesives for wood particleboard by tannins addition. Wood Sci Technol 52:261–279
71. Ghani A, Ashaari Z, Bawon P, Lee SH (2018) Reducing formaldehyde emission of urea formaldehyde-bonded particleboard by addition of amines as formaldehyde scavenger. Build Environ 142:188–194
72. Ghosh SK, Das TK, Ganguly S, Nath K, Paul S, Ganguly D, Das NC (2023) Silane functionalization of sodium montmorillonite and halloysite (HNT) nanoclays by 'grafting to' method to improve physico-mechanical and barrier properties of LLDPE/clay nanocomposites. Polym Bull 80(4):4307–4335
73. Gillela S, Yadav SM, Sihag K, Lubis MAR, Wibowo ES, Negi A, Iswanto AH, Antov P, Kristak L (2022). A review on Lantana camara lignocellulose fiber-reinforced polymer composites Biomass Conv Bioref. https://doi.org/10.1007/s13399-022-02402-7
74. Gonçalves S, Ferra J, Paiva N, Martins J, Carvalho LH, Magalhães FD (2021) Lignosulphonates as an alternative to non-renewable binders in wood-based materials. Polymers 13:4196
75. Gu Y, Cheng L, Gu Z, Hong Y, Li Z, Li C (2019) Preparation, characterization and properties of starch-based adhesive for wood-based panels. Int J Biol Macromol 134:247–254
76. Guo F, Aryana S, Han Y, Jiao Y (2018) A review of the synthesis and applications of polymer–nanoclay composites. Appl Sci 8(9):1696
77. Gupta GK, Shukla P (2020) Lignocellulosic biomass for the synthesis of nanocellulose and its eco-friendly advanced applications. Front Chem 8
78. Güler C, Sarikaya A, Sertkaya AA, Canli E (2024) Almond shell particle containing particleboard mechanical and physical properties. Constr Build Mater 431:136565
79. Hagel S, Joy J, Cicala G, Saake B (2021) Recycling of waste MDF by steam refining: evaluation of fiber and paper strength properties. Waste Biomass Valor 12:5701–5713
80. Hassannejad H, Shalbafan A, Rahmaninia M (2018) Reduction of formaldehyde emission from medium density fiberboard by chitosan as scavenger. J Adhes
81. Hematabadi H, Behrooz R, Shakibi A, Arabi M (2012) The reduction of indoor air formaldehyde from wood-based composites using urea treatment for building materials. Constr Build Mater 28:743–746
82. Hemmilä V, Adamopoulos S, Karlsson O, Kumar A (2017) Development of sustainable bioadhesives for engineered wood panels—a review. RSC Adv 7:8604–38630
83. Ho TA, Greathouse JA, Lee AS, Criscenti LJ (2018) Enhanced Ion adsorption on mineral nanoparticles. Langmuir 34(20):5926–5934
84. Hochella MF, Lower SK, Maurice PA, Penn RL, Sahai N, Sparks DL, Twining BS (2008) Nanominerals, mineral nanoparticles, and earth systems. Science 319(5870):1631–1635
85. Hornus M, Via KB, Gallagher T, Peresin MS (2020) Partial substitution of pMDI with lignin containing cellulose nanofibrils: Low density oriented strand board. Wood Mater Sci Eng 16(6):391–396
86. Hosseinpourpia R, Adamopoulos S, Mai C, Taghiyari HR (2019) Properties of medium-density fiberboards bonded with dextrin-based wood adhesives. Wood Res 64:185–194

87. Hou M, Lei H, Zhou X, Du G, Pizzi A, Essawy H, Zhang Q, Wu D, Yan R, Xi X Preparation and characterization on the eco-friendly corn starch based adhesive of with salient water resistance, mildew resistance. Int J Biol Macromol 269:132043
88. Hussin MH, Appaturi JN, Poh NE, Abd Latif NH, Brosse N, Ziegler-Devin I, Vahabi H, Syamani FA, Fatriasari W, Solihat NN (2022) A recent advancement on preparation, characterization and application of nanolignin. Int J Biol Macromol
89. Hussin MH, Latif NHA, Hamidon TS, Idris NN, Hashim R, Appaturi JN, Brosse N, Ziegler-Devin I, Chrusiel L, Fatriasari W, Syamani FA, Iswanto AH, Lee SH, Osman Al Edrus SSA, Lum WC, Antov P, Savov V, Lubis MAR, Kristak L, Reh R, Sedliačik J (2022) Latest advancements in high-performance bio-based wood adhesives: A critical review. J Mater Res Technol 21:3909–3946
90. IARC (2006) Monographs on the evaluation of carcinogenic risks to humans: Formaldehyde, 2-butoxyethanol and 1-tert-butoxypropan-2-ol, vol 8, p 280. https://monographs.iarc.fr/ENG/Monographs/vol88/mono88.pdf (accessed on 15 May 2024)
91. Ilyas RA, Aisyah HA, Nordin AH, Ngadi N, Zuhri MYM, Asyraf MRM, Sapuan SM, Zainudin ES, Sharma S, Abral H, Asrofi M, Syafri E, Sari NH, Rafidah M, Zakaria SZS, Razman MR, Majid NA, Ramli Z, Azmi A, Bangar SP, Ibrahim R (2022) Natural-fiber-reinforced chitosan, chitosan blends and their nanocomposites for various advanced applications. Polymers 14(5):874
92. Irle M, Privat F, Couret L, Belloncle C, Déroubaix G, Bonnin E, Cathala B (2018) Advanced recycling of post-consumer solid wood and MDF. Wood Mat Sci Eng 11:1–5
93. Ismita N, Lokesh C (2017) Effects of different nanoclay loadings on the physical and mechanical properties of Melia composita particle board. Bois & For Trop 334:7–12
94. Iswanto AH, Lubis MAR, Sutiawan J, Al-Edrus SSO, Lee SH, Antov P, Kristak L, Reh R, Mardawati E, Santoso A, Kusumah SS (2023) Latest advancements in the development of high-performance lignin- and tannin-based non-isocyanate polyurethane adhesive for wood composites. Polymers 15:3864
95. Iždinský J, Vidholdová Z, Reinprecht L (2020) Particleboards from recycled wood. Forests 11:1166
96. Jerman M, Böhm M, Dušek J, Černý R (2022) Effect of steaming temperature on microstructure and mechanical, hygric, and thermal properties of binderless rape straw fiberboards. Build Environ 223:109474
97. Jial P, Song F, Li Q, Xia H, Shu X, Zhou Y (2019) Recent development of cardanol based polymer materials: A review. J Renew Mater 7:601–619
98. Jiang W, Haapala A, Tomppo L, Pakarinen T, Sirviö JA, Liimatainen H (2018) Effect of cellulose nanofibrils on the bond strength of polyvinyl acetate and starch adhesives for wood. BioResources 13:2283–2292
99. Kaboorani A, Riedl B (2011) Effects of adding nano-clay on performance of polyvinyl acetate (PVA) as awood adhesive. Comp Part A, Appl Sci Man 42:1031–1039
100. Kaboorani A, Riedl B, Blanchet P, Fellin M, Hosseinaei O, Wang S (2012) Nanocrystalline Cellulose (NCC): A renewable nano-material for Polyvinyl Acetate (PVA) Adhesive. Eur Polym J 48:1829–1837
101. Kargarzadeh H, Huang J, Lin N, Ahmad I, Mariano M, Dufresne A, Thomas S, Galeski A (2018) Recent developments in nanocellulose-based biodegradable polymers, thermoplastic polymers, and porous nanocomposites. Prog Polym Sci 87:197–227
102. Karliati T, Lubis MAR, Dungani R, Maulani RR, Hadiyane A, Rumidatul A, Antov P, Savov V, Lee SH (2024) Performance of particleboard made of agroforestry residues bonded with thermosetting adhesive derived from waste styrofoam. Polymers 16:543
103. Karthäuser J, Biziks V, Mai C, Militz H (2021) Lignin and lignin-derived compounds for wood applications—a review. Molecules 26(9):2533
104. Kaur P, Sharma N, Munagala M, Rajkhowa R, Aallardyce B, Shastri Y, Agrawal R (2021) Nanocellulose: Resources, physio-chemical properties, current uses and future applications. Front Nanotechnol 3:747329

105. Kawalerczyk J, Antov P, Dziurka D, Mirski R, Lee SH (2023) The effect of pressing parameters and hardener content on the properties of plywood bonded with propylamine-UF adhesive. Wood Mater Sci Eng 19(3):710–717
106. Kawalerczyk J, Dziurka D, Mirski R, Siuda J (2021) The reduction of adhesive application in plywood manufacturing by using nanocellulose-reinforced urea-formaldehyde resin. J Appl Polym Sci 138(7):49834
107. Kawalerczyk J, Siuda J, Mirski R, Dziurka D (2020) Hemp flour as a formaldehyde scavenger for melamine-urea-formaldehyde adhesive in plywood production. BioResources 15:4052–4064
108. Kawalerczyk J, Dziurka D, Mirski R, Szentner K (2020b) Properties of plywood produced with urea-formaldehyde adhesive modified with nanocellulose and microcellulose. Drv Ind 71:61–67
109. Kawalerczyk J, Walkiewicz J, Sedliačik J, Dukarska D, Woźniak M, Mirski R (2024) Modified buckwheat husk as a filler for urea-formaldehyde resin in plywood production. Polymers 16:1350
110. Kawalerczyk J, Dziurka D, Mirski R (2022a) The effect of a phenol-formaldehyde adhesive reinforcement with nanocellulose on the pressing parameters of plywood. Wood Research 67(5):796–808
111. Kawalerczyk J, Walkiewicz J, Dziurka D, Mirski R, Brózdowski J (2022b) APTES-modified nanocellulose as the formaldehyde scavenger for UF adhesive-bonded particleboard and strawboard. Polymers 14(22):5037
112. Khalaf Y, El Hage P, Mihajlova J, Bergeret A, Lacroix P, El Hage R (2021) Influence of agricultural fibers size on mechanical and insulating properties of innovative chitosan-based insulators. Constr Build Mater 287:123071
113. Khanjanzadeh H, Behrooz R, Bahramifar N, Pinkl S, Gindl-Altmutter W (2019) Application of surface chemical functionalized cellulose nanocrystals to improve the performance of UF adhesives used in wood based composites-MDF type. Carbohyd Polym 206:11–20
114. Khorramabadi LA, Behrooz R, Kazemi S (2023) Effects of nanoclay modification with amino-propyltriethoxysilane (APTES) on the performance of urea–formaldehyde resin adhesives. BioResources 18(3):5417
115. Khorramabadi LA, Behrooz R, Najafi SK (2023) Reduction of formaldehyde emission from medium density fiberboard using nanoclay modified with 3-aminopropyltriethoxysilane and l-Lysine as additives to urea-formaldehyde adhesive. Int J Adhes Adhes 125:103426
116. Koynov D, Antov P, Valyova M, Savov V, Dochev I, Lee SH (2024) Properties of hybrid plywood produced by utilisation of peeler cores. Forests 15:582
117. Kim S (2009) Environment-friendly adhesives for surface bonding of wood-based flooring using natural tannin to reduce formaldehyde and TVOC emission. Bioresour Technol 100:744–748
118. Kim WJ, Terada N, Nomura T, Takahashi R, Lee SD, Park JH, Konno A (2002) Effect of formaldehyde on the expression of adhesion molecules in nasal microvascular endothelial cells: the role of formaldehyde in the pathogenesis of sick building syndrome. Clin Exp Allergy 32(2):287–295
119. Kristak L, Antov P, Bekhta P, Lubis MAR, Iswanto AH, Reh R, Sedliacik J, Savov V, Taghiyari HR, Papadopoulos AN, Pizzi A, Hejna A (2023) Recent progress in ultra-low formaldehyde emitting adhesive systems and formaldehyde scavengers in wood-based panels: a review. Wood Mater Sci Eng 18(2):763–782
120. Krišťáková S, Neykov N, Antov P, Sedliačiková M, Reh R, Halalisan AF, Hajdúchová I (2021) Efficiency of wood-processing enterprises—evaluation based on DEA and MPI: A comparison between Slovakia and Bulgaria for the period 2014–2018. Forests 12:1026
121. Köse K, Mavlan M, Youngblood JP (2020) Applications and impact of nanocellulose based adsorbents. Cellulose 27:2967–2990
122. Kumar A, Gupta A, Sharma KV, Gazali SB (2013) Influence of aluminum oxide nanoparticles on the physical and mechanical properties of wood composites. BioResources 8(4):6231–6241

123. Kumar A, Gupta A, Sharma KV, Nasir M (2013) Use of aluminum oxide nanoparticles in wood composites to enhance the heat transfer during hot-pressing. Eur J Wood Wood Prod 71(2):193–198
124. Kumar A, Gupta A, Sharma KV (2015) Thermal and mechanical properties of urea-formaldehyde (UF) resin combined with multiwalled carbon nanotubes (MWCNT) as nanofiller and fiberboards prepared by UF-MWCNT. Holzforschung 69(2):199–205
125. Kumar A, Sharma KV, Gupta A, Tywoniak J, Hajek P (2017) Optimization of processing parameters of medium density fiberboard using response surface methodology for multiwalled carbon nanotubes as a nanofiller. Eur J Wood Wood Prod 75:203–213
126. Kumar RN, Pizzi A (2019) Urea-formaldehyde resins. Adhesives for wood and lignocellulosic materials. Wiley-Scrivener Publishing, Hoboken. NJ, USA, pp 61–100
127. Kumar V (2020) Ultrasonic-assisted de-agglomeration and power draw characterization of silica nanoparticles. Ultrason Sonochem 65:105061
128. Kunaver M, Medved S, Čuk N, Jasiukaitytė E, Poljanšek I, Strnad T (2010) Application of liquefied wood as a new particle board adhesive system. BioresTechnol 101(4):1361–1368
129. Lago-Olveira S, Arias A, Rebolledo-Leiva R, Feijoo G, González-García S, Moreira MT (2024) Monitoring the bioeconomy: Value chains under the framework of life cycle assessment indicators. Clean Circ Bioecon 7:100072
130. Łebkowska M, Radziwiłł MZ, Tabernacka A (2017) Adhesives based on formaldehyde—Environmental problems. Biotechnologia 98:53–65
131. Lee SH, Md Tahir P, Lum WC, Tan LP, Bawon P, Park BD, Osman Al Edrus SS, Abdullah UH (2020) A review on citric acid as green modifying agent and binder for wood. Polymers 12(8):1692
132. Lee W, Liu C (2002) Preparation of liquefied bark based reso resin and its application to particle board. J Appl Polym Sci 87:1837–1841
133. Lei H, Du G, Pizzi A, Celzard A (2008) Influence of nanoclay on urea-formaldehyde resins for wood adhesives and its model. J Appl Polym Sci 109(4):2442–2451
134. Lengowski EC, Bonfatti Júnior EA, Nishidate Kumode MM, Carneiro ME, Satyanarayana KG (2019) Nanocellulose-reinforced adhesives for wood-based panels. In: Inamuddin, Thomas S, Kumar Mishra R, Asiri AM (eds) Sustainable polymer composites and nanocomposites, no 167. Springer, Berlin, pp 1001–1025
135. Liang W, Lv M, Yang X (2016) The effect of humidity on formaldehyde emission parameters of a medium-density fiberboard: Experimental observations and correlations. Buld Env 101:110–115
136. Lin Q, Yang G, Liu J, Rao J (2006) Property of nano-$SiO_2$/urea formaldehyde resin. Front Forest China 1:230–237
137. Lubke H, Ihnát V, Kuňa V, Balberčák J (2020) A multi-stage cascade use of wood composite boards. Wood Res 65:843–854
138. Lubis MAR, Hong MK, Park BD (2018) Hydrolytic removal of cured urea-formaldehyde resins in medium-density fiberboard for recycling. J Wood Chem Technol 38:1–14
139. Luo T, Hao Y, Wang C, Jiang W, Ji X, Yang G, Chen J, Janaswamy S, Lyu G (2022) Lignin nanoparticles and alginate gel beads: Preparation, characterization and removal of methylene blue. Nanomaterials 12(1):176
140. Łukawski D, Hochmańska-Kaniewska P, Bałęczny B, Martin A, Janiszewska-Latterini D, Lekawa-Raus A (2024) Phenol-formaldehyde resin enriched with graphene nanoplatelets as an electroconductive adhesive for wood composites. Int J Adhes Adhes 132:103678
141. Luo X, Shuai L (2020) Lignin-based adhesives. In: Green adhesives. Wiley, pp 25–56
142. Mantanis GI, Athanassiadou ET, Barbu MC, Wijnendaele K (2018) Adhesive systems used in the European particleboard, MDF and OSB industries. Wood Mater Sci Eng 13:104–116
143. Maulana MI, Lubis MAR, Febrianto F, Hua LS, Iswanto AH, Antov P, Kristak L, Mardawati E, Sari RK, Zaini LH, Hidayat W, Lo Guidice V, Todaro L (2022) Environmentally friendly starch-based adhesives for bonding high-performance wood composites: A review. Forests 13:1614

144. Migneault S, Koubaa A, Riedl B, Nadji H, Deng J, Zhang T (2011) Potential of pulp and paper sludge as a formaldehyde scavenger agent in MDF resins. Holzforschung 65:403–409
145. Mirski R, Banaszak A, Bekhta P (2021) Selected properties of formaldehyde-free polymer-straw boards made from different types of thermoplastics and different kinds of straw. Materials 14:1216
146. Miyashiro D, Hamano R, Umemura K (2020) A review of applications using mixed materials of cellulose. Nanocellulose Carbon Nanotubes Nanomater 10(2):186
147. Mondal S (2018) Review on nanocellulose polymer nanocomposites. Polym-Plast Technol Eng 57(13):1377–1391
148. Moslemi A, Behzad T, Pizzi A (2020) Addition of cellulose nanofibers extracted from rice straw to urea formaldehyde resin; effect on the adhesive characteristics and medium density fiberboard properties. Int J Adhes Adhes 99:102582
149. Muñoz F, Moya R (2018) Effect of urea formaldehyde resin modified with nano-clay on physical and mechanical properties of particleboards manufactured with wood from plantation species. J Biobased Mater Bioenergy 12(5):482–492
150. Muñoz F, Moya R, Muñoz F, Moya R (2018) Effect of nanoclay-treated UF resin on the physical and mechanical properties of plywood manufactured with wood from tropical fast growth plantations. Maderas Ciencia y Tecnología 20(1):11–24
151. Ndiwe B, Pizzi A, Tibi B, Danwe R, Konai N, Amirou S (2019) African tree bark exudate extracts as biohardeners of fullybiosourced thermoset tannin adhesives for wood panels. Ind Crop Prod 132:253–268
152. Neves RM, Ornaghi HL Jr, Zattera AJ, Amico SC (2021) Recent studies on modified cellulose/nanocellulose epoxy composites: A systematic review. Carbohyd Polym 255:117366
153. Neykov N, Sedliacikova M, Antov P, Potkány M, Kitchoukov E, Halalisan A-F, Poláková N (2024) Efficiency of micro and small wood-processing enterprises in the EU—Evidence from DEA and fractional regression analysis. Forests 15:58
154. Ng H-M, Sin LT, Bee S-T, Tee T-T, Rahmat AR (2017) Review of nanocellulose polymer composite characteristics and challenges. Polym-Plast Technol Eng 56(7):687–731
155. Ninikas K, Mitani A, Koutsianitis D, Ntalos G, Taghiyari HR, Papadopoulos AN (2021) Thermal and mechanical properties of green insulation composites made from Cannabis and bark residues. J Compos Sci 5:132
156. Nisha A, Bhardwaj R, Verma M, Acharya A (2024) Functionalized nanoclays in bioactive materials. Compos Interf 31(7):857–881
157. Nordström E, Demircan D, Fogelström L, Khabbaz, F, Malmström, E (2017) Green binders for wood adhesives. In: Applied adhesivebonding in science and technology. Intechopen Books, London, UK, pp 47–71
158. Norizan MN, Shazleen SS, Alias AH, Sabaruddin FA, Asyraf MRM, Zainudin ES, Abdullah N, Samsudin MS, Kamarudin SH, Norrrahim MNF (2022) Nanocellulose-based nanocomposites for sustainable applications: A review. Nanomaterials 12(19):3483
159. Oksman K, Bismarck A (2014) Handbook of green materials: processing technologies, properties and applications (in 4 volumes), vol 5. World Scientific, Hackensack, NJ
160. Oktay S, Pizzi A, Köken N, Bengü B (2024) Chemical modification techniques of corn starch for synthesis wood adhesive. Int J Adhes Adhes 128:103545
161. Oktay S, Kızılcan N, Bengü B (2021) Development of bio-based cornstarch-Mimosa tannin-sugar adhesive for interior particleboard production. Ind Crop Prod 170:113689
162. Papadopoulos AN, Bikiaris DN, Mitropoulos AC, Kyzas GZ (2019) Nanomaterials and chemical modifications for enhanced key wood properties: A review. Nanomaterials 9(4):607
163. Park BD, Causin V (2013) Crystallinity and domain size of cured urea–formaldehyde resin adhesives with different formaldehyde/urea mole ratios. Eur Polym J 49:532–537
164. Park BD, Jeong H-W (2011) Effects of acid hydrolysis on microstructure of cured urea-formaldehyde resins using atomic force microscopy. J Appl Polym Sci 122(5):3255–3262
165. Park BD, Kim J (2008) Dynamic mechanical analysis of urea-formaldehyde resin adhesives with different formaldehyde-to-urea molar ratios. J Appl Phys 108:2045–2051

166. Pędzik M, Auriga R, Kristak L, Antov P, Rogoziński T (2022) Physical and mechanical properties of particleboard produced with addition of walnut (*Juglans regia* L.) Wood Residues. Materials 15(4):1280
167. Pędzik M, Janiszewska D, Rogoziński T (2021) Alternative lignocellulosic raw materials in particleboard production: A review. Ind Crop Prod 174:114162
168. Phanthong P, Reubroycharoen P, Hao X, Xu G, Abudula A, Guan G (2018) Nanocellulose: Extraction and application. Carbon Resour Convers 1:32–43
169. Pichelin F, Nakatani M, Pizzi A, Wieland S, Despres A, Rigolet S (2006) Thick wood panels bonded industrially with formaldehyde free tannin adhesives. For Prod J 56(5):31–36
170. Pipíška T, Nociar M, Král P, Ráheľ J, Bekhta P, Réh R, Krišťák L, Jopek M, Pijáková B, Wimmer R, Šernek M (2024) Characterization of randomly oriented strand boards manufactured from juvenile wood of underutilized wood species. Eur J Wood Wood Prod. https://doi.org/10.1007/s00107-024-02080-4
171. Pizzi A (1994) Advanced wood adhesives technology. CRC Press
172. Pizzi A (2023) Little secrets for the successful industrial use of tannin adhesives: A review. J Renew Mater 11(9):3403–3415
173. Pizzi A (2024) Recent developments in advanced lignin-based adhesives and binders. In: Advanced lignin technologies. https://doi.org/10.5772/intechopen.1004627
174. Pizzi A (2016) Wood products and green chemistry. Ann For Sci 73(1):185–203
175. Pizzi A, Laborie MP, Candan Z (2024) A review on sources, extractions and analysis methods of a sustainable biomaterial: Tannins. J Renew Mater 12(3):397–425
176. Pizzi A, Papadopoulos A, Policardi F (2020) Wood composites and their polymer binders. Polymers 12:1115
177. Puttasukkha J, Khongtong S, Chaowana P (2015) Curing behavior and bonding performance of urea formaldehyde resin admixed with formaldehyde scavenger. Wood Res 60(4):645–654
178. Pylypchuk IV, Karlsson M, Lindén PA, Lindström ME, Elder T, Sevastyanova O, Lawoko M (2023) Molecular understanding of the morphology and properties of lignin nanoparticles: unravelling the potential for tailored applications. Green Chem 25(11):4415
179. Raji M, Mekhzoum MEM, Qaiss AEK, Bouhfid R (2016) Nanoclay modification and functionalization for nanocomposites development: Effect on the structural, morphological, mechanical and rheological properties. In: Jawaid M, Qaiss AEK, Bouhfid R (eds) Nanoclay reinforced polymer composites. Springer Singapore, Singapore, pp 1–34
180. Rashid AB, Hoque ME, Kabir N, Rifat FF, Ishrak H, Alqahtani A, Chowdhury MEH (2023) Synthesis, properties, applications, and future prospective of cellulose nanocrystals. Polymers 15(20):4070
181. Reh R, Kristak L, Sedliacik J, Bekhta P, Wronka A, Kowaluk G (2024) Molded plywood with proportions of beech bark in adhesive mixtures: production on an industrial scale. Polymers 16:966
182. Réh R, Krišťák Ľ, Sedliačik J, Bekhta P, Božiková M, Kunecová D, Vozárová V, Tudor EM, Antov P, Savov V (2021) Utilization of birch bark as an eco-friendly filler in urea-formaldehyde adhesives for plywood manufacturing. Polymers 3(4):511
183. Ridho MR, Lubis MAR, Nawawi DS, Fatriasari W (2024) Optimization of areca leaf sheath nanolignin synthesis by a mechanical method for in situ modification of ultra-low molar ratio urea-formaldehyde adhesives. Int J Biol Macromol:132614
184. Roffael E (2011) On the responsiveness of hardened UF-resins of different molar ratio towards ammonia fumigation. Eur J Wood Wood Prod 69(4):675–676
185. Röllig P, Tudor EM, Barbu MC, Direske M (2024) Effect of the particle geometry on lightweight particleboard from Paulownia using high-frequency pressing technology. Wood Mater Sci Eng:1–8. https://doi.org/10.1080/17480272.2023.2286625
186. Roumeli E, Papadopoulou E, Pavlidou E, Vourlias G, Bikiaris D, Paraskevopoulos KM, Chrissafis K (2012) Synthesis, characterization and thermal analysis of urea–formaldehyde/nano$SiO_2$ resins. Thermochim Acta 527:33–39
187. Saito K, Hirabayashi Y, Yamanaka S (2021) Reduction of formaldehyde emission from urea-formaldehyde resin with a small quantity of graphene oxide. RSC Adv 11:32830–32836

188. Salari A, Tabarsa T, Khazaeian A, Saraeian A (2012) Effect of nanoclay on some applied properties of oriented strand board (OSB) made from underutilized low quality paulownia (Paulownia fortunei) wood. J Wood Sci 58:513–524
189. Salari A, Tabarsa T, Khaezaian A, Saraeian A (2013) Improving some of applied properties of oriented strand board (OSB) made from underutilized low quality paulownia (*Paulownia fortunie*) wood employing nano-$SiO_2$. Ind Crop Prod 42:1–9
190. Santos J, Pereira J, Ferra J, Magalhães FD, Martins JM, Carvalho LH (2021) New cardoon (*Cynara cardunculus* L.) particleboards using cardoon leaf extract and citric acid as bioadhesive. Mater Circ Econ 3(1):1–7
191. Santos J, Pereira J, Escobar-Avello D, Ferreira I, Vieira C, Magalhães FD, Martins JM, Carvalho LH (2022) Grape canes (*Vitis vinifera* L.) applications on packaging and particleboard industry: New bioadhesive based on grape extracts and citric acid. Polymers 14(6):1137
192. Saražin J, Pizzi A, Amirou S, Schmiedl D, Šernek M (2021) Organosolv Lignin for Non-Isocyanate Based Polyurethanes (NIPU) as wood adhesive. J Renew Mater 9:881–907
193. Sarika PR, Nancarrow P, Khansaheb A, Ibrahim T (2020) Bio-based alternatives to phenol and formaldehyde for the production of resins. Polymers 12:2237
194. Saud AS, Maniam GP, Rahim MHA (2021) Introduction of eco-friendly adhesives: source, types, chemistry and characterization. In: Jawaid M, Khan TA, Nasir M, Asim M (eds) Eco-friendly adhesives for wood and natural fiber composites; composites science and technology. Springer, Singapore
195. Savov V, Antov P, Panchev C, Lubis MAR, Taghiyari HR, Lee SH, Krišťák Ľ, Todorova M (2023) The impact of hydrolysis regime on the physical and mechanical characteristics of medium-density fiberboards manufactured from recycled wood fibers. Fibers 11:103
196. Savov V, Antov P, Trichkov N (2021) Properties of high-density fiberboards bonded with urea-formaldehyde and phenol-formaldehyde resins. Innov Wood Ind Eng Des 2(20):17–26
197. Savov V, Valchev I, Antov P (2019) Processing factors for production of eco-friendly medium density fibreboards based on lignosulfonate adhesives. In: Proceedings of the 2nd International Congress of Biorefinery of Lignocellulosic Materials (IWBLCM 2019), Córdoba, Spain, pp 165–169
198. Sedelnikova M, Bakina O, Ugodchikova A, Tolkacheva T, Khimich M, Uvarkin P, Kashin A, Miller A, Egorkin V, Schmidt J, Sharkeev Y (2022) The role of microparticles of β-TCP and wollastonite in the creation of biocoatings on Mg0.8Ca Alloy. Metals 12(10):1647
199. Sedliačiková M, Stroková Z, Klementová J, Šatanová A, Moresová M (2020) Impacts of behavioral aspects on financial decision-making of owners of woodworking and furniture manufacturing and trading enterprises. Acta Fac Xylologiae Zvolen 62:165–176
200. Segovia F, Blanchet P, Essoua GGE (2021) Potential of the crude glycerol and citric acid mixture as a binder in medium-density fiberboard manufacturing. Eur J Wood Wood Prod 79(5):1141–1151
201. Selakjani PP, Dorieh A, Pizzi A, Shahavi MH, Hasankhah A, Shekarsaraee S, Ashouri M, Movahed SG, Abatari MN (2021) Reducing free formaldehyde emission, improvement of thickness swelling and increasing storage stability of novel medium density fiberboard by urea-formaldehyde adhesive modified by phenol derivatives. Int J Adhes Adhes 111:102962
202. Solt P, Konnerth J, Gindl-Altmutter W, Kantner W, Moser J, Mitter R, van Herwijnen HWG (2019) Technological performance of formaldehyde-free adhesive alternatives for particleboard industry. Int J Adhes Adhes 94:99–131
203. Song J, Chen S, Yi X, Zhao X, Zhang J, Liu X, Liu B (2021) Preparation and properties of the urea-formaldehyde res-in/reactive halloysite nanocomposites adhesive with low-formaldehyde emission and good water resistance. Polymers 13(14):2224
204. Sun J, Li L, Cheng H, Huang W (2018) Preparation, characterization and properties of an organic siloxane-modified cassava starch-based wood adhesive. J Adhes 94:78–293
205. Sutiawan J, Hermawan D, Massijaya MY, Kusumah SS, Lubis MAR, Marlina R, Purnomo D, Sulastiningsih IM (2021) Influence of different hot-pressing conditions on the performance of eco-friendly jabon plywood bonded with citric acid adhesive. Wood Mater Sci Eng. https://doi.org/10.1080/17480272.2021.1884898

206. Taghiyari HR, Mohammad-Panah B, Morrell JJ (2016) Effects of wollastonite on the properties of Medium-Density Fiberboard (MDF) made from wood fibers and camel-thorn. Maderas Ciencia y tecnologia 18(1):157–166
207. Taghiyari HR, Arbabi Ghamsari F, Salimifard E (2018) Effects of adding nano-wollastonite, date palm prunings and two types of resins on the physical and mechanical properties of medium-density fiberboard (MDF). Bois et Forets des Tropiques 335(1):49–57
208. Taghiyari HR, Hosseini SB, Ghahri S, Ghofrani M, Papadopoulos AN (2020a) Formaldehyde emission in micron-sized wollastonite-treated plywood bonded with soy flour and urea-formaldehyde resin. Appl Sci 10:6709
209. Taghiyari HR, Tajvidi M, Taghiyari R, Mantanis GI, Esmailpour A, Hosseinpourpia R (2020b) Nanotechnology for wood quality improvement and protection. In: Husen A, Jawaid M (eds) Nanomaterials for agriculture and forestry applications. Elsevier, Amsterdam, The Netherlands, pp 469–489
210. Taghiyari HR, Esmailpour A, Majidi R, Morrell JJ, Mallaki M, Militz H, Papadopoulos AN (2020c) Potential use of wollastonite as a filler in UF resin based medium-density fiberboard (MDF). Polymers 12:1435. https://doi.org/10.3390/polym12071435
211. Taghiyari HR, Militz H, Antov P, Papadopoulos AN (2021) Effects of wollastonite on fire properties of particleboard made from wood and chicken feather fibers. Coatings 11(5):518
212. Tajvidi M, Gardner DJ, Bousfield DW (2016) Cellulose nanomaterials as binders: Laminate and particulate systems. J Renew Mater 4:365–376
213. Tasooji M, Tabarsa T, Khazaeian A, Wool RP (2010) Acrylated epoxidised soy oil as an alternative to urea-formaldehyde in making wheat straw particleboard. J Adhes Sci Technol 24:1717–1727
214. Tayeb AH, Amini E, Ghasemi S, Tajvidi M (2018) Cellulose nanomaterials—binding properties and applications: A review. Molecules 23(10):2684
215. Tisserat B, Eller F, Mankowski M (2019) Properties of composite wood panels fabricated from Eastern red cedar employing various bio-based green adhesives. BioResources 14:6666–6685
216. Tobisch S, Dunky M, Hänsel A, Krug D, Wenderdel C (2023) Suvey of wood-based materials. In: Niemz P, Teischinger A, Sandberg D (eds) Springer handbook of wood science and technology. Springer Nature Switzerland AG
217. Tomak ED, Gonultas O (2018) The wood preservative potentials of Valonia, chestnut, tara and sulphited oak tannins. J Wood Chem Technol 38
218. Trache D, Hussin MH, Brosse N (2020) Editorial: Recent trends in preparation, characterization and applications of nanocellulose. Front Chem 8:594379
219. Umemura K, Inoue A, Kawai S (2003) Development of new natural polymer-based wood adhesive I: Dry bond strength and water resistance of Konjac glucomannan, chitosan and their composites. J Wood Sci 49:221–226
220. Valchev I, Savov V, Yordanov I, Petrin S, Antov P (2024) Effect of hot-pressing temperature on the properties of eco-friendly fiberboard panels bonded with hydrolysis lignin and phenol-formaldehyde resin. Polymers 16:1059
221. Veigel S, Rathke J, Weigl M, Gindl-Altmutter W (2012) Particle board and oriented strand board prepared with nanocellulose-reinforced adhesive. J Nanomater
222. Vineeth SK, Gadhave RV, Gadekar PT (2019a) Nanocellulose applications in wood adhesives—Review. Open J Polym Chem 9:63–75
223. Vineeth SK, Gadhave RV, Gadekar PT (2019b) Chemical modification of nanocellulose in wood adhesive: Review. Open J Polym Chem 9:86–99
224. Wang H, Chen G, Zhang S, Zhang W, Wu H, Li C, Xiao Z (2022) Camellia meal-based formaldehyde-free adhesive with self-crosslinking, and anti-mildew performance. Ind Crop Prod 176:114280
225. Wang X, Wang S, Xie X, Zhao L, Deng Y, Li Y (2017) Multi-scale evaluation of the effects of nanoclay on the mechanical properties of wood/phenol formaldehyde bondlines. Int J Adhes Adhes 74:92–99
226. Wedaïna AG, Pizzi A, Nzie W, Dawne R, Konai N, Amirou S, Segovia C, Kueny R (2021) Performance of unidirectional biocomposite developed with *Piptadeniastrum Africanum* tannin resin and *Urena Lobata* fibers as reinforcement. J Renew Mater 9:477–493

227. Wibowo ES, Lubis MAR, Park B-D (2021) Simultaneous improvement of formaldehyde emission and adhesion of medium-density fiberboard bonded with low-molar ratio urea-formaldehyde resins modified with nanoclay. J Korean Wood Sci Technol 49(5):453–461
228. Wibowo ES, Lubis MAR, Park BD, Kim JS, Causin V (2020) Converting crystalline thermosetting urea-formaldehyde resins to amorphous polymer using modified nanoclay. J Ind Eng Chem 87:78–89
229. Wibowo ES, Park B-D (2020) Cure kinetics of low-molar-ratio urea-formaldehyde resins reinforced with modified nanoclay using different kinetic analysis methods. Thermochim Acta 686:178552
230. Wohlgemuth R, Twardowski T, Aguilar A (2021) Bioeconomy moving forward step by step—A global journey. N Biotechnol 61:22–28
231. Wood based panels market size and share analysis—growth trends and forecasts (2024–2029). https://www.mordorintelligence.com/industry-reports/wood-based-panel-market (accessed on 5 April 2024)
232. Wronka A, Kowaluk G (2022) The influence of multiple mechanical recycling of particleboards on their selected mechanical and physical properties. Materials 15:8487
233. Wu L, Liu X, Lv G, Zhu R, Tian L, Liu M, Li Y, Rao W, Liu T, Liao L (2021) Study on the adsorption properties of methyl orange by natural one-dimensional nano-mineral materials with different structures. Sci Rep 11(1):10640
234. Xi X, Wu Z, Pizzi A, Gerardin C, Lei H, Du G (2020) Furfuryl alcohol-aldehyde plywood adhesive resins. J Adhes 96(9):814–838
235. Xiao J, Guo D, Xia C, Li T, Lian H (2022) Application of nano-$SiO_2$ reinforced urea-formaldehyde resin and molecular dynamics simulation study. Materials 15:8716
236. Yadav R (2021) Development of low formaldehyde emitting particle board by nano particle reinforcement. J Appl Nat Sci 13(4):1187–1197
237. Yadav SM, Lubis MAR, Park B-D (2022) Modification of nanoclay with different methods and its application in urea-formaldehyde bonded plywood panels. Wood Mat Sci Eng 17(6):734–743
238. Yadav SM, Lubis MAR, Wibowo ES, Park B-D (2021) Effects of nanoclay modification with transition metal ion on the performance of urea–formaldehyde resin adhesives. Polym Bull 78(5):2375–2388
239. Yang G, Gong Z, Luo X, Chen L, Shuai L (2023) Bonding wood with uncondensed lignins as adhesives. Nature 621:511–515
240. Yang Y, Zhang L, Ren Y, Huo H, Zhong Y, Huang K, Yue K, Zhang Z (2024) Glueless formaldehyde-free biocomposites with high strength based on the three-dimensional structure of wood fibres. Chem Eng J 488:150591
241. Younesi-Kordkheili H (2024) Reduction of formaldehyde emission from urea-formaldehyde resin by maleated nanolignin. Int J Adhes Adhes 132:103677
242. Younesi-Kordkheili H, Pizzi A (2022) Preparation and properties of a modified corn flour-lignin-glyoxal as a green wood adhesive. Int Wood Prod J. https://doi.org/10.1080/20426445.2022.2048338
243. Yue X, Huang L, Huang L, Luo X (2021) A sustainable strategy for Medium-Density Fiberboards preparation from waste hybrid Pennisetum Straws. Waste Biomass Valor 12(9):5161–5173
244. Zahedsheijani R, Faezipour M, Tarmian A, Layeghi M, Yousefi H (2012) The effect of Na+ montmorillonite (NaMMT) nanoclay on thermal properties of Medium Density Fiberboard (MDF). Eur J Wood Prod 70(5):565–571
245. Zahedsheijani R, Gholamiyan H, Tarmian A, Yousefi H (2011) Mass transfer in medium density fiberboard (MDF) modified by Na+ montmorillonite (Na+MMT) nanoclay. Maderas Ciencia y tecnología 13(2):163–172
246. Zanuttini R, Negro F (2021) Wood-based composites: Innovation towards a sustainable future. Forests 12:1717
247. Zare Y (2016) Study of nanoparticles aggregation/agglomeration in polymer particulate nanocomposites by mechanical properties. Compos A Appl Sci Manuf 84:158–164

248. Zeng QH, Yu AB, Lu GQ, Paul DR (2005) Clay-based polymer nanocomposites: Research and commercial development. J Nanosci Nanotechnol 5(10):1574–1592
249. Zhang H, She Y, Song SP, Pu JW (2013) Modified nanocrystalline cellulose used for improving formaldehyde emission and bonding strength of urea formaldehyde resin adhesive. In: Key engineering materials. Trans Tech Publ, pp 846–851
250. Zhang B, Zhang F, Wu L, Gao Z, Zhang L (2020) Assessment of soybean protein-based adhesive formulations, prepared by different liquefaction technologies for particleboard applications. Wood Sci Technol 55:33–48
251. Zhang H, Zhang J, Song S, Wu G, Pu J (2011) Modified nanocrystalline cellulose from two kinds of modifiers used for improving formaldehyde emission and bonding strength of urea-formaldehyde resin adhesive. BioResources 6(4):4430–4438
252. Zheng K, Chen T, Zhang J, Tian X, Ge H, Qiao T, Lei J, Li X, Duan T, Zhu W (2019) Nano-montmorillonite regulated crystallization of hierarchical strontium carbonate in a microbial mineralization system. Materials 12(9):1392
253. Zhou X, Du G (2019) Applications of tannin resin adhesives in the wood industry. In: Tannins—structural properties, biological properties and current knowledge. IntechOpen, London, United Kingdom
254. Zhou X, Pizzi A, Du G (2012) The effect of nanoclay on melamine-urea-formaldehyde wood adhesives. J Adhes Sci Technol 26(10–11):1341–1348

# Fire Retarding Treatments of Wood and Wood-Based Materials

**Linda Makovicka Osvaldova and Widya Fatriasari**

**Abstract** The flame-retardant treatment of wood and wood-based materials is a type of technology that can convert combustible wood into flame retardant materials. The main method of this kind of transformation is the addition of chemical substances to wood. Flame-retardant wood not only has fire-retardant performance but also retains the original excellent properties of wood. The ideal flame-retardant materials for wood and wood-based materials should have the following characteristics: high potency, lack of toxicity, no environmental effects during use, durability, no heat or light decomposition, no ease of hydrolysis or breaching, good dimensional stability of the wood after fire retardant treatment, no effect on the physical and mechanical properties of the wood, a low cost, a rich source and ease of use.

**Keywords** Fire retardant · Wood · Wood-based materials

## 1 Introduction

Patents for fire-retarding treatments of materials were first granted in the eighteenth century. Wilde (1735) was granted a patent for a fire retardant made of alum and borax, Gay Lucas (1781) for a solution of inorganic salts, while Fuchs (1820) recommended the use of "water glass" to prevent the spread of fire. In the nineteenth century, ammonium salts of phosphoric acid began to be used for fire protection [21].

As we can see, fire retardants have been used for centuries. However, fire retardants have never been used, although fire retardancy is still an unfamiliar concept for many people. Surprisingly, wood or other fibrous natural materials are classified as noncombustible materials with a certain low degree of flammability/fire reaction

L. M. Osvaldova (✉)
Department of Fire Engineering, Faculty of Security Engineering, University of Žilina, Žilina, Slovakia
e-mail: linda.makovicka@uniza.sk

W. Fatriasari
Research Center for Biomass and Bioproducts, National Research and Innovation Agency (BRIN), Jakarta, Indonesia

 
P. Antov et al. (eds.), *Nanomaterials Additives in Bioadhesives for Wood Composites*, Smart Nanomaterials Technology, https://doi.org/10.1007/978-981-96-3400-2_8

class. According to the National Fire Safety Survey, many people underestimate the danger of fire. Perhaps people's underestimation of fires stems from the fact that most building materials are now protected by technologies and materials that help to mitigate the danger of fires.

Fire retardants are chemical substances that, via chemical, physical or combined methods, prevent the rapid ignition and burning of wood [18]. In conjunction with other fire protection equipment, such as smoke detectors, alarms, and sprinklers, fire retardants offer one of the most effective means available to protect people and their property from the effects of fire. Statistical studies and extensive scientific research have repeatedly proven the importance of fire retardants.

We modified all flammable materials to suppress this negative property as much as possible. Such adjustment is usually economically demanding, as it must meet several conditions at once: the adjustment, while demanding efficacy, must not negatively affect other technical, hygienic, or aesthetic properties of the material [18]. The fire-retarding treatment, i.e., reduction of flammability, interferes with the combustion process primarily through its ability to remove heat, to degenerate, or even self-combust.

Methods for improving flammability properties from the point of view of fire protection can be specified as follows:

- stabilize the material's decomposition into flammable products,
- fire resistant or low flammability insulating coatings,
- additions of additives that melt under the influence of heat and form nonflammable coatings,
- additions of additives that have an antioxidant effect, i.e., their decomposition creates an inert coating that extinguishes flame,
- support for the formation of a carbon structure, a carbon residue that prevents the spread of combustion deeper into the material, and
- addition of additives that interrupt the mechanism of chain reactions by binding free radicals.

A fire retardant usually has the function of a catalyst, which can change the speed of reaction necessary to achieve chemical equilibrium but cannot change the intensity of the heat flow, the lack of oxygen, or an excess of oxygen, resulting in changes in the conditions and manifestations of combustion. However, it can change or affect the fuel formation process and flammability. In the later stages of the combustion process, especially at high heat flows, the possibility of fire retardation is already very limited. For effective fire retardancy, the reactions that occur at the beginning of the individual stages of combustion, i.e., initiation (contact of the wood—heat system), propagation (release of flammable gases—oxygen), or termination (formation of solid residues—oxygen), are decisive [4].

Flame retardation is a relatively complicated process and is mostly based on the use of retardation systems that complement and influence each other. A separate but related problem is the increased toxicity of the combustion products of retarded materials.

During surface treatment, a retardation system is applied using common impregnation procedures (soaking, spraying, etc.). A favorable reduction in flammability is thus achieved, especially for natural cellulose-based materials. The most effective method of retardation is the addition of a reactive retardant directly to the polymerization or polycondensation process, which is not possible with wood.

Fire retardants can be divided into four groups:

*The first group* consists of retardants that release nonflammable gases in the thermal range where flammable gases are also formed as products of wood decomposition. This results in the dilution of flammable gases, such that their concentration decreases and their ignition becomes more difficult. The most widespread compounds used to protect wood and cellulosic materials against fire are various inorganic salts. Their advantage is good solubility in water, and thus, different impregnation procedures can be used [17, 21].

*The second group* consists of retardants that absorb the heat of the heat source and thus "cool down" the source. These retardants are currently not popular, as they age quickly, decreasing their effectiveness.

*The third group* consists of intumescent flame retardants, whose mechanism involves the formation of an insulating protective layer as the retardant carbonizes and foams (there was no description given here, which means that the reader who does not recognize the term "intumescent" is totally lost for the moment, despite probably being familiar with the concept itself). I see it is explained later on but I think a very short description here would be helpful and put it at a similar level of explanation as the other groups listed here. Their effectiveness is the highest and is two-staged (physical and chemical); thus, their application is the widest.

*The fourth group* consists of mechanical retardants, such as foils and various facings made of noncombustible materials. The application of such retardants to wood is effective but entails certain risks.

In the case of intumescent coatings, one component of the retardant reacts in the first stage of heat loading, creating a foam of several centimeters from the thin film. In this way, a gap between the surface of the wood and the heat source is created. This is the first method of physical retardation [21]. The second mechanism of physical retardation lies in the fact that foam is a very poor conductor of heat and considerably slows the heating of the wood. The third method is chemical retardation, when chemical reactions occur during further heating and significantly slow the burning process. Thermofoaming paint, for example, is a mixture of various substances (urea condensates with formaldehyde, methyl-melamine condensates, phosphates, plasticizers, pigments, etc.).

We must not forget that flame retardants are also chemical substances that require a safety data sheet. The sheet contains a whole range of data that must be given (see below). In many cases, the parameters of the retardants found on these data sheets determine the choice of the retardant for a given situation.

## 2 Theory of Thermal Degradation in Relation to Flame Retardancy

During the thermolysis of cellulose, 1,6-anhydro-D-glucopyranose (levoglucosan) is formed as the main depolymerization product. It is assumed that its formation occurs through oxygen–carbon 1,4-biradical structures, which create an anhydride after isomerization. Levoglucosan is thermally and thermo-oxidatively labile and easily decomposes into many unsaturated, carbonyl, and other compounds that continue to react further (Fig. 1) [4].

This method clarified the main degradation mechanism involved in the burning of crystalline and high-purity cellulose. In amorphous structures, oxidative degradation to form peroxy radicals is observed even at low temperatures (150–200 °C). It was found that the hydroperoxide oxygen is bound to the first carbon atom of the D-glucopyranose unit. Simultaneously, with the binding of oxygen, a decrease in the degree of polymerization of the heated cellulose is observed.

The amorphous structures of cellulose are thermo-oxidatively degraded down to the monomer unit, while the crystalline structure remains essentially unchanged. By heating the cellulose in an inert atmosphere, the average degree of polymerization does not essentially decrease from the beginning. The first changes are observed at 200–280 °C when a slow weight loss occurs. According to spectral studies and analysis of volatile products, this process is dehydration, which also occurs mainly in amorphous structures.

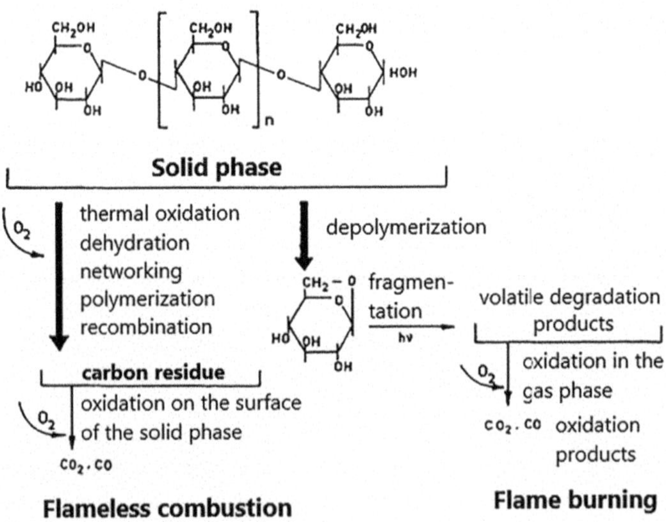


**Fig. 1** Thermolytic degradation of cellulose [4]

The main volatile products at the indicated temperatures were water and a small amount of carbonyl and furan compounds. In addition to intramolecular dehydration reactions, intermolecular water splitting and bond renewal are also possible. From the point of view of the propagation of the combustion process, the dehydration reactions of depolymerization associated with the formation of 1,6-anhydrous D-glucopyranose have very different meanings. Dehydration reactions lead to the formation of an oxygen-enriched residue and volatile products that dilute the fuel. Although the carbon-enriched residue reacts with oxygen at temperatures between 450 and 600 °C, the formation of carbon oxides occurs almost exclusively on the surface of the solid phase to a kinetically different degree.

## 3 Influence of External Conditions on the Flammability Parameters of Cellulosic Materials

Determination of the flammability parameters (flash point $T_{fig}$ and ignition temperature $T_{sig}$) is used as a basic characteristic of the thermal conditions for the initiation of the burning process of polymeric materials. The disadvantage of the standard methodology is that it does not provide information about the influence of the geometry and amount of material or the concentration of oxygen in the surrounding atmosphere on the flammability parameters.

The oxygen concentration significantly affects the ignition temperature, which has a linear relationship in the range of 10–80% vol. $O_2$. The ignition temperature of most cellulosic materials in an air atmosphere is 405–430 °C. The simultaneously determined ignition induction periods ($\tau_{sig}$), together with the ignition temperature, allow us to determine the activation energy of the ignition process depending on the oxygen concentration in the surrounding atmosphere. The induction period is expressed as the time interval from inserting the sample into the heated vertical reactor until ignition.

By increasing the concentration of oxygen, the apparent activation energy decreases, and in the interval of 10–99.6% $O_2$, it displays a linear character. The addition of flame retardants can significantly alter the thermal conditions of initiation. Flame retardants usually have a pro-degradative effect on cellulosic materials, and the temperature of active decomposition decreases compared to that of nonretarded materials. The active thermal decomposition of lignocellulosic materials can be influenced by flame retardants, and $(NH_4)_2HPO_4$ is among the most effective. Lignocellulosic materials treated in this way begin to decompose at a lower temperature (Fig. 2), but on the other hand, there is a significant shift to higher values of both flash and ignition temperatures compared to nonretarded materials [4].

A more noticeable effect of $(NH_4)_2HPO_4$ on the thermal conditions of the initiation of spruce wood occurs at the flash point. Although this retardant lowers the initial temperature of active thermal decomposition, dehydration reactions involving the formation of a significant amount of water are preferred. Water affects the overall

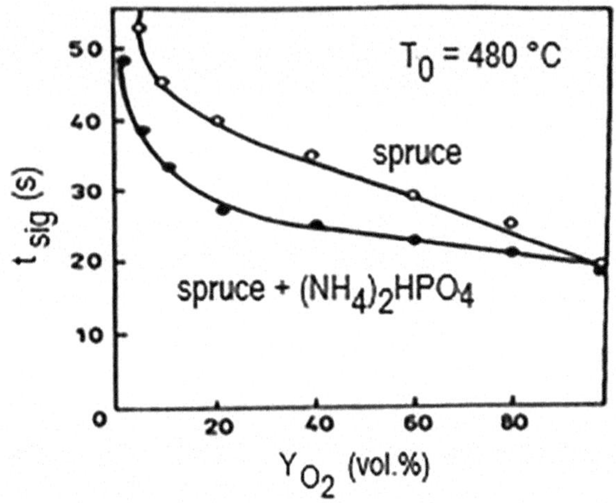


**Fig. 2** Dependence of the induction period for ignition of natural and fire-retarded spruce wood on the oxygen concentration [4]

flammability of the mixture of volatile products and the enthalpy of flame combustion, which is manifested by a significant increase in the flash point. For many flammable solid materials, there is another limiting concentration of oxygen above which the material spontaneously switches from flameless combustion to flame combustion (Fig. 3).

This value is different from the defined limiting oxygen number and is called the upper limit oxygen number of flameless combustions. Above this value, after stabilization of the burning of the material, if the flame is suppressed by the flow of $CO_2$, it extinguishes, and spontaneous ignition and flame burning occur immediately. Above the measured limit, subject to constant test conditions (sample ignition time, sample geometry, surface quality, and oxidation mixture flow rate), the material is subjected to flame burning only [4, 6].

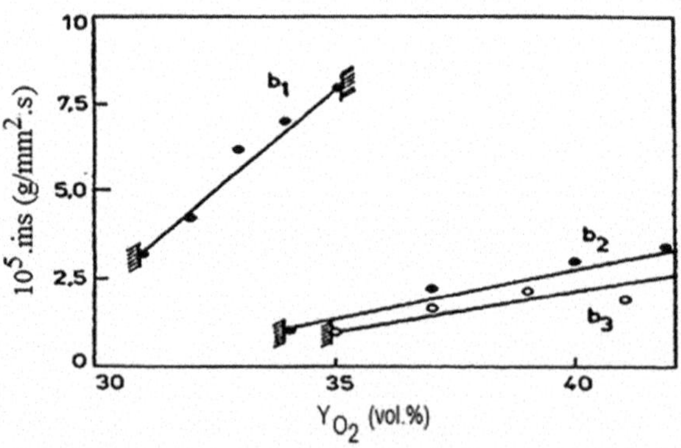


**Fig. 3** Effect of oxygen concentration on the burning rate of spruce wood (b—sample thickness, while b3 > b2 > b1) [4]

## 4 Limit Conditions of the Flame and Flameless Combustion Process

When we analyze in detail the results of determining the limited oxygen level, we conclude that cellulosic materials promote the burning process in a flame or flameless way. It can be influenced by a change in the external conditions (oxygen concentration, the temperature of the oxidation mixture, material geometry, additives, preliminary thermal stress, etc.). Using the high-temperature oxygen number method, both flame and flameless stationary countercurrent methods of combustion propagation can be modeled. Based on the established oxygen limit at 23 ± 2 °C, the materials can be classified into four groups (Table 1) [4].

For materials with a tendency to form a nodular residue, four qualitatively distinct zones of the combustion process can be determined, delimited by the lower limit oxygen concentration for flameless combustion ($YO_2/S1$), the lower limit oxygen concentration for flame combustion ($YO_2/F1$) and the upper limit oxygen concentration for flame burning ($YO_2/Su$). At an oxygen concentration greater than $YO_2/Su$, if the flame is suppressed by the $CO_2$ stream, it extinguishes and immediately reignites spontaneously. The method of propagation in individual zones bounded by limited concentrations of oxygen is shown in Table 2 and Fig. 4 [6].

Pure cellulose is not able to prevent flameless combustion. In the process of formation, depolymerization reactions prevail with a minimum yield of carbon residue, which is not sufficient to maintain flameless propagation. There are only two zones of the combustion process delimited by the lower limit for flame combustion.

**Table 1** Classification of the examined samples according to the method of combustion in an air atmosphere [4]

| Group | Method of burning | $Y_{O2}$ (% vol. $O_2$) | Material |
|---|---|---|---|
| 1 | Flameless | ≤ 21 | Cellulose + $KHCO_3$ |
| 2 | Flame | ≤ 21 | Cellulose, hard fiberboard |
| 3 | Flame and flameless | ≤ 21 | Soft fiberboard, linen, textile |
| 4 | Flame and flameless | ≤ 21 | Spruce wood, beech wood, fire-retarded wood |

**Table 2** Different methods of burning polymers that form carbon residues during thermolysis or thermo-oxidative degradation [4]

| Burning | Intermediate stages of the combustion process | | |
|---|---|---|---|
| | I | II | III |
| Flame and flameless | Gasification | Combustion | Oxidation of the carbon residue |
| Flame | Gasification | Combustion | |
| Flameless | Gasification | | Oxidation of the carbon residue |

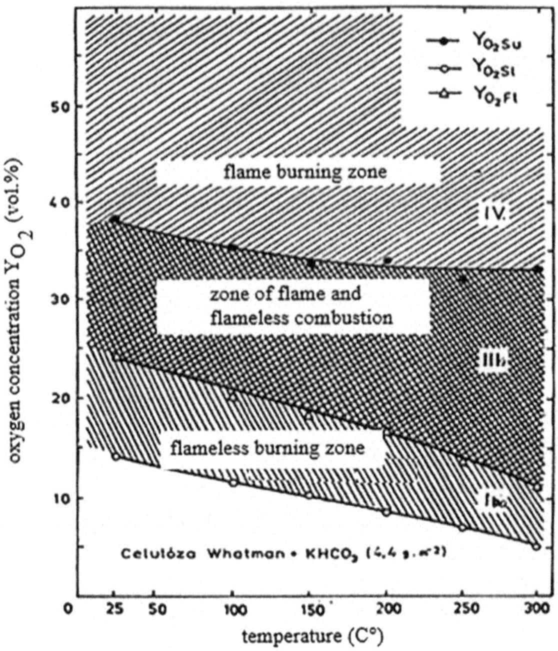


**Fig. 4** Temperature dependence of the burning method of Whatman cellulose impregnated with potassium bicarbonate [4]

The way the combustion process spreads in pure cellulose can be influenced by additives that catalyze dehydration reactions. In cellulose impregnated with aqueous solutions of KCl, NaCl, $NaHCO_3$, and $KHCO_3$, a strong tendency toward flameless combustion with a tendency toward thermal self-ignition was found. $KHCO_3$ is often used as a heterogeneous chemical inhibitor in firefighting. When applied to matter, as in the case of cellulose, it acts as an accelerator of flameless combustion at lower concentrations. A more noticeable retardation effect on the flameless combustion induced by it will be manifested only at concentrations above 25% of $KHCO_3$ [4].

The addition of $KHCO_3$ induces flameless combustion, and the lower limit concentration of oxygen for flameless propagation ($YO_2$/S1) at an oxidation temperature of 23 ± 2 °C is lower than that for flame propagation ($YO_2$/F1) of pure and modified cellulose. The presence of inorganic salts $Na^+$ and $K^+$ causes cellulose materials to have a strong tendency toward flameless combustion, and for this reason, we do not recommend applying retarding additives with the mentioned cations. During the thermal stress of materials, various exo- and endothermic processes occur, associated with a weight change, which is characteristic of the tested material. Various thermoanalytical methods can be used to monitor these changes. During thermal stress, gaseous decomposition products are created depending on the degradation temperature, which can be continuously sampled and determined.

Several additives, as well as impurities in lignocellulosic materials, can cause a change in the course of active thermal decomposition by catalyzing reactions occurring at lower temperatures (dehydration, thermal oxidation). To obtain an overview of the overall thermal behavior of the monitored substances for the entire temperature range, a thermogravimetric method with a defined heating rate can be used (a heating rate of 5 °C/min is often used). However, with this procedure, it is not possible to observe the processes of self-heating or the influence of external factors on this process or to determine the critical conditions for the combustion process. To obtain more detailed information about the progress of the combustion process, which is a cascade process, an electrically heated hot-air furnace can be used according to the standard STN ISO 871. In this procedure, for better monitoring of the influence of airflow, a steel sieve is used as a sample holder and is connected to the scales using a glass rod located under the stove. The sampling probe of the gas analyzer can be placed above the opening of the hot air furnace cover. With such a configuration of test equipment, it is possible to find the lowest temperature at which it is still possible to observe the decomposition of the samples depending on the airflow rate in the furnace [6].

Figure 5 shows the course of thermal decomposition of cellulose impregnated with a 5% aqueous solution of $(NH_4)_2HPO_4$. The thermal decomposition of cellulose treated in this way occurs in four stages. The second stage started at 170.3 °C, during which the highest weight loss of 60.25% occurred, and the maximum weight loss rate reached a maximum of 0.01133 g/s at 256.2 °C. During the third stage, a weight loss of 9.8% occurred and occurred in the temperature range of 331.4 °C to 400 °C; the maximum weight loss rate reached 0.00147 g/s at a temperature of 383.5 °C. In the fourth stage, the oxidation of the carbonaceous residue occurred, which was associated with a slow weight loss down to the value of the resistant residue of 6.1% at a temperature of 608.5 °C.

The curves of the temperature dependence of CO and $CO_2$ during decomposition (Fig. 6) show two extremes, which correspond to the maximum rate of decomposition of the retarded cellulose. During the second stage of decomposition, the $CO/CO_2$ ratio reached 0.73; during the third stage, it reached 0.43; and during the fourth stage, it reached 0.12 [6].

The data on the released gases CO and $CO_2$ (Fig. 6) also provide significant information about the initiation phase of the self-heating and thermal degradation process. When monitoring the effect of cellulose modification with an aqueous solution of $(NH_4)_2HPO_4$, it was found that the decrease in the temperature of the main decomposition and the increase in the resistant residue were dependent on the increasing concentration of $(NH_4)_2HPO_4$ (Fig. 7).

When using sodium or potassium salts as flame retardants, it is necessary to apply an effective retardant against flameless burning simultaneously. Compared to pure cellulose, lignin, hemicelluloses, and associated components were shown to affect the limiting oxygen concentrations for flame combustion of wood and wood-based materials.

Based on the intermediate stages of the combustion process, which include the formation of fuel, the burning of generated flammable gases and steam, and the oxidation of carbonaceous residue, the combustion process methods can be classified into three groups (Table 2) [4].

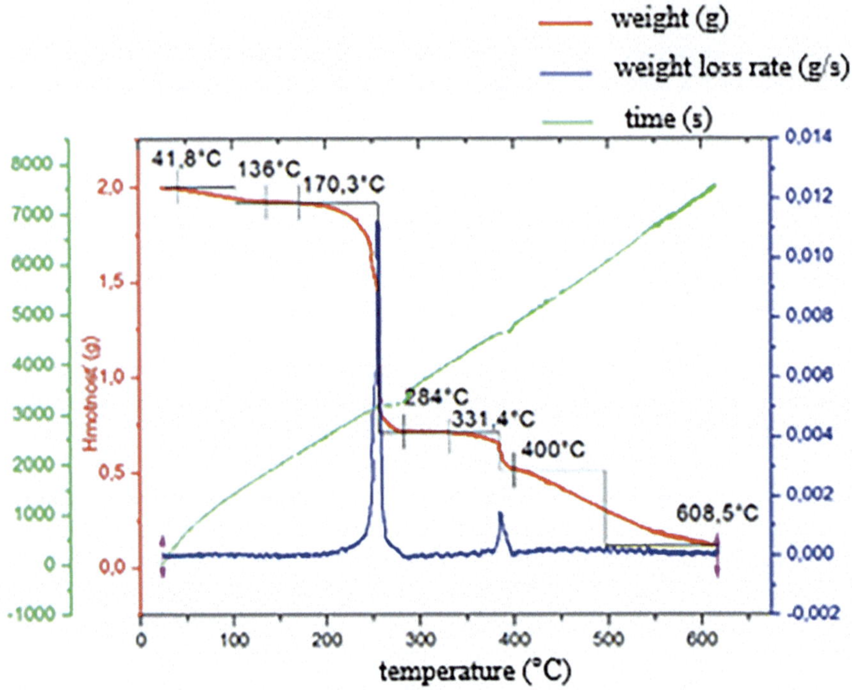


**Fig. 5** TG/DTG recording of the thermal decomposition of cellulose impregnated with a 5% $(NH_4)_2HPO_4$ solution monitored in a modified hot air furnace [6]

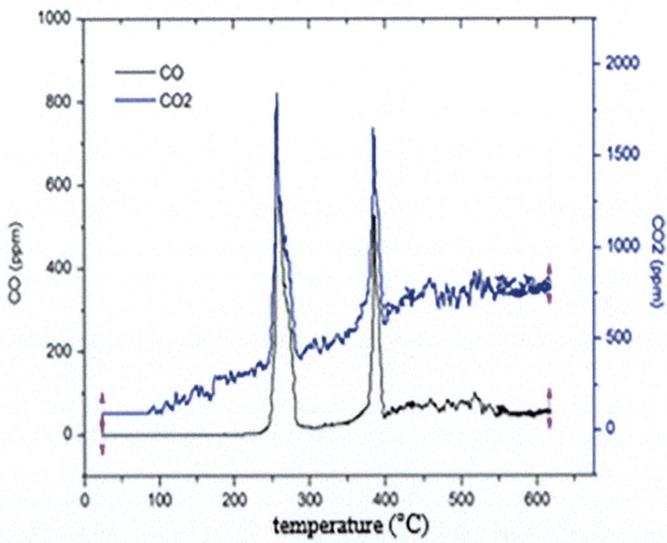


**Fig. 6** Record of released gases during thermal decomposition from a cellulose sample impregnated with a 5% aqueous solution $(NH_4)_2HPO_4$ in a modified hot air furnace [6]

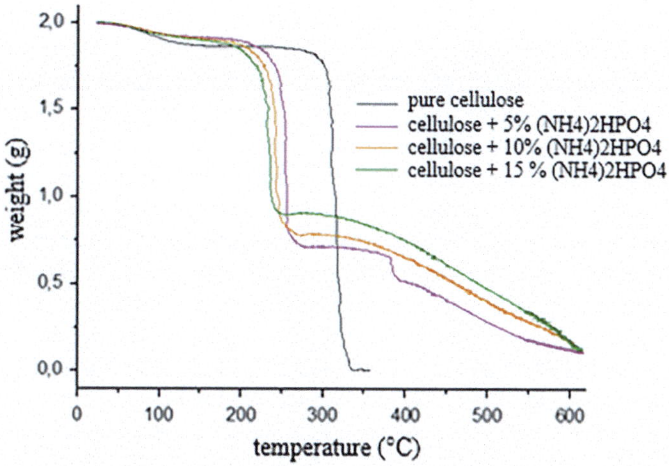


**Fig. 7** Effect of $(NH_4)_2HPO_4$ concentration on the weight loss of fire-retardant and pure cellulose as a function of degradation temperature [6]

## 5 Possibilities of Applying Retardants to Natural Wood

The method of application is determined by the type of retardant; for example, intumescent retardants can only be applied by coating. Retardants of the first group—water-soluble-based solutions of inorganic salt—can be applied by coating, dipping, or impregnation. The results of the experiment in which the effectiveness of the retardant was compared depending on its application are shown in Fig. 8.

It can be seen from the given figure that the influence of the retardant application technology is very significant and affects the Q value in the flammability test according to STN 73 0862 [21] (a method no longer used today). This experiment evaluated a low-quality retardant produced from waste products for the production of artificial fertilizers. The classification of the modified material in the evaluation of

**Fig. 8** Retardant effectiveness depending on its application (N—untreated material, 1—one-sided coating, 2—double-sided coating, I—impregnation, Q—flammability class according to STN 73 0862)

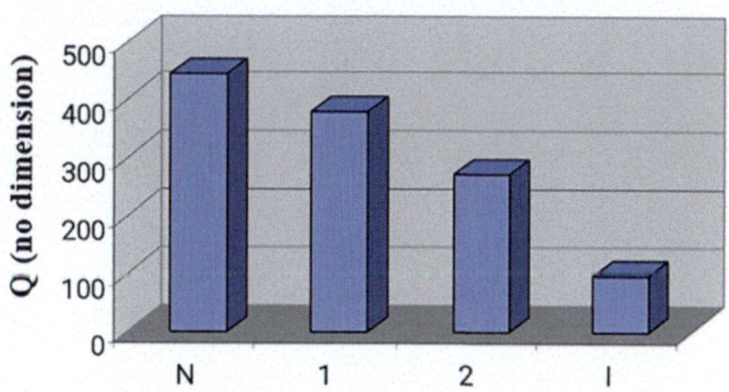

the "degree of flammability" differed significantly according to the individual modifications. If the untreated material is classified as flammable **C2**, the same material treated with either a one-sided coating or a double-sided coating is classified as flammable **C1**. Impregnated material is flammable at level **B**, of course, when using the same retardant [21]. The application of retardants by impregnation cannot be recommended everywhere. However, there are places, for example, in the mass production of buildings, where this procedure can be recommended.

With the many structural elements of wooden buildings, we encounter the need to apply two protective substances, a fungicide insecticide and a fire retardant. Logically, fungicide insecticides must be applied earlier, and practice has shown that the retardant should be applied later, even with a time gap, so that the first protective substance can be fixed onto the wood and mature. Sometimes the prevailing opinion is that painting the wood with, say, a fungicidal-insecticidal substance will fill the pores, making it difficult for the wood to catch fire, so any protective substance has a certain degree of retardation. However, [22] carried out interesting experiments to investigate this claim, and her measurements refuted this opinion. The fungicidal-insecticidal substance did not show any retarding ability. When the fire-retardant treatment was evaluated using the same method, the superiority of the retardant was clearly demonstrated.

Osvaldová also evaluated the relative burning speed of 5 different treatment methods: untreated wood (N), treatment with a fungicidal-insecticidal substance (KA), a flame retardant (OH), a mixture of a fungicidal-insecticidal substance and a retardant (50/50%) (KAOH), and wood treated first with a fungicidal-insecticidal substance and 6 weeks later with a flame retardant (KAOHM). Figure 9 shows that the burn-out rate between unprotected wood and wood treated with a fungicide-insecticidal substance shifts slightly in favor of the fungicide-insecticidal substance at approximately 200 s, but the peak of both compared materials is reached at the same time and in the same size. The effect of the retardant is then noticeable in the other 3 curves.

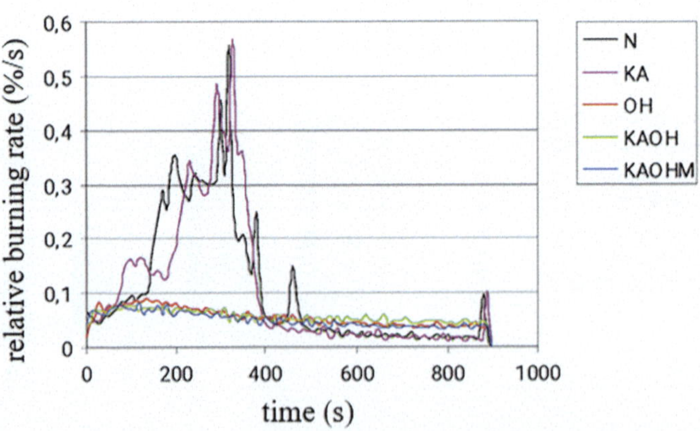


**Fig. 9** Course of the relative burn-out rate for the first evaluated group of modifications

## 6 Retardation of Some Large-Scale Materials

In addition to the retardation treatment of natural wood, it is also necessary to pay attention to wood-based materials, especially large-scale materials, which are quite popular in wooden buildings. In general, several methods for the retardation treatment of large-scale materials exist. The simplest modification is a coating, in which the retardant is applied onto the surface of the material. A high-quality intumescent coating can guarantee good retardation, but in many cases, the long-term effect of this modification cannot be guaranteed. Although the manufacturer ensures the durability of the retardant and the repeatability of the coating, mechanical damage to the coating, such as abrasion or the action of chemicals, reduces the retarding effect. This is why we often turn to a more complex solution to this problem, which is the retardation treatment of the entire large-scale material—treatment inside the matter.

Most retardants, however, affect the physical and mechanical properties of large-scale materials. This statement applies to all large-scale materials. In such a case, it must be ensured that the mechanical properties of the fire-treated materials do not decrease (see Fig. 10, curve 1). If curve *n* represents the untreated material, the retardation treatment must ensure, during further heating, the course according to curve 2. It would be ideal for the course of change in tensile properties to depend on the time of fire development according to curve 3 (see Fig. 1) [17].

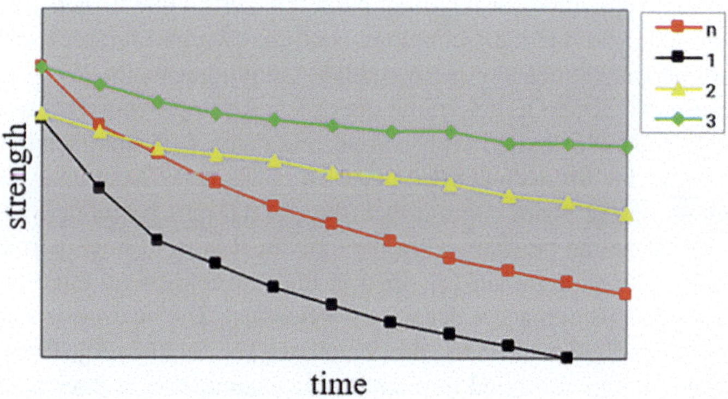


**Fig. 10** Possible effect of retardation treatment on the mass of the material on the change in mechanical properties during fire exposure

## 6.1 Retardation of Chipboards

The fire-technical properties of chipboards are mainly characterized by the input materials, i.e., wood, glue, or other additives used. In addition to the basic composition of chipboards, their fire-technical properties are influenced byproduction technology. Surface-pressed chipboards have better fire-technical properties than roller-pressed chipboards. When comparing boards with the same composition of wood and glue, as well as density and thickness, there is a diametric difference in fire-technical properties between the two technologies (surface vs roller pressing) [5]. It is possible to claim that roller-pressed chipboards cause large-scale fires in wooden buildings that are difficult to extinguish, which is why they are no longer used. Another limiting criterion for the fire-technical properties of chipboards is their density. Here, it is possible to claim, to some degree, that there is a directly proportional relationship between density and fire-technical properties. Chipboards with a higher density show positive fire-technical properties. In addition to thickness, density is also a limiting factor in value standards for the selection of particle boards and OSB for wooden buildings.

Particle boards (DTDs) can be fire-retarded in several ways. The first method is the impregnation of chips with a flame retardant. This procedure results in relatively good retardation, but it is economically very demanding and practically unfeasible. Adding a certain percentage of powder retardant appears to be an economically advantageous and effective method for retarding particle boards. Glue is then applied with the retardant. DTD production technology continues without further technological requirements and adjustments or increased labor requirements [5, 25].

The experiments aimed to determine the reason for improving the fire-technical properties of chipboard treated with a solid retardant. With this technology, it is possible to adjust the fire-technical properties of DTD at different levels according to the retardant on the board. With a three-layer DTD, the percentage representation of the retardant can be changed in the individual layers (in the peripheral and central layers). The experiment described in the text concerns a three-layer DTD with a thickness of 18 mm and a density of 600 $kg/m^3$. The board was subjected to thermal stress according to the then-valid standard STN 73 0862 [20]. This thermally degraded material was subjected to microscopic examination at three positions, as shown in Fig. 11. Figure 12 is a shot from position 1. Figure 13 is from position 2 shown in Fig. 11 (from the center of the board), and Fig. 14a and b are shots from position 3 according to Fig. 11. Figure 12 shows a sample taken 2 mm below the charred layer. Position 1 is the side that has been thermally loaded with both a flame and a radiant source according to the mentioned methodology. Figure 12 shows the "moon landscape", which resulted from the melting of the powder retardant; the melt covered the surface of the chips. This is the first success of the retardation, i.e., consuming heat to change the state from solid to liquid, which solidifies again after cooling.

An interesting effect occurred during microscopic observation. The samples under the microscope started to form some kind of bubbles. This phenomenon is shown

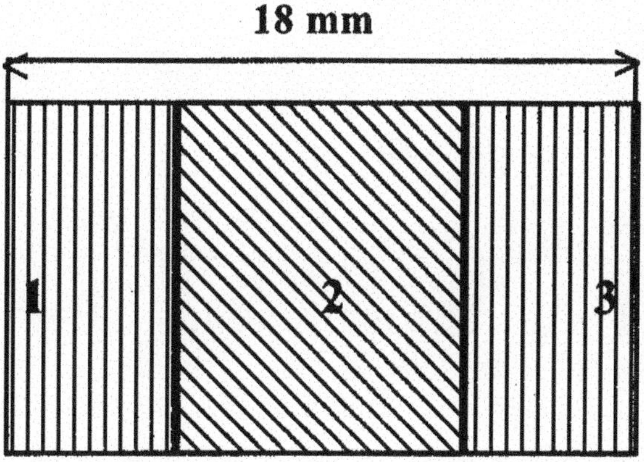


**Fig. 11** Scheme of positions for microscopic observation of the retardation treatment of DTD with powder retardant

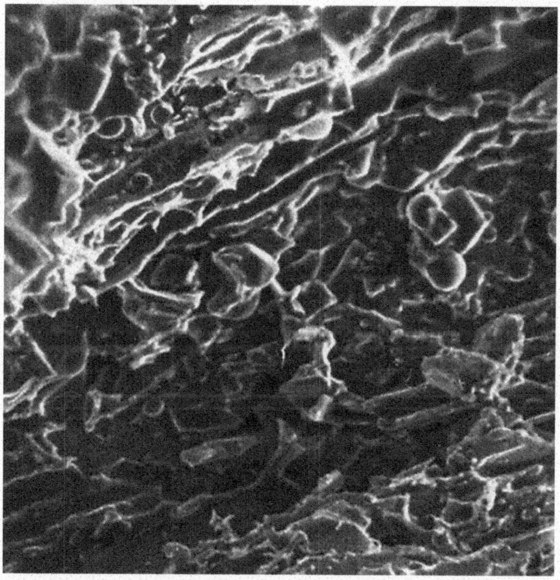

**Fig. 12** Formation of a melt on the sample from position 1

in Fig. 13. Figure 13a shows a bubble beginning to form, and Fig. 13b shows an enlarged bubble just before bursting. These bubbles were formed by the heat of the electron microscope beam. Therefore, we can assume that the retardant still had a preserved function and reacted to thermal influences. Figure 14a is a photo from the center of the DTD, i.e., 9 mm from the fire source. The retardant crumbled but still retained its powder form. This is created either by thermal degradation of the crystals

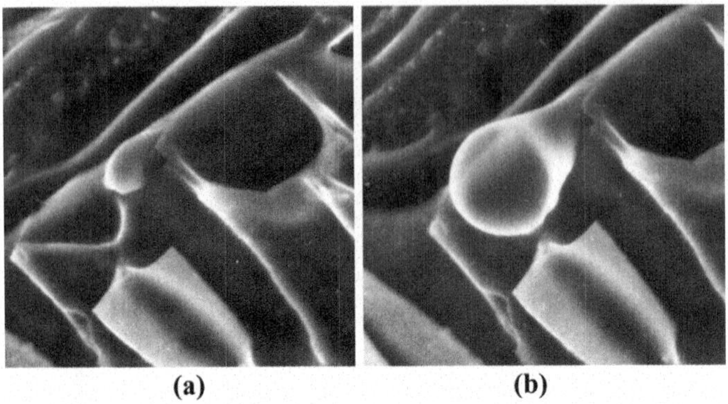

Fig. 13 Observation of the retardant melt from position 1: **a** The formation of bubbles, which are caused by the heat of the electron microscope beam; **b** Enlargement of the bubble during a longer observation period

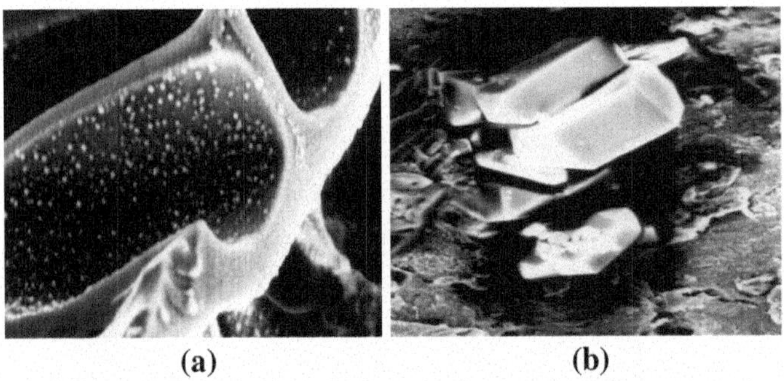

Fig. 14 **a** Microscopy image from the center of DTD treated with powder retardant in position 3 after thermal load **b** Presence of crystalline retardant in position 3 after thermal test

or by recondensation of retardant vapors, which are more strongly concentrated and condensed into a crystalline form after cooling.

Figure 14b represents the position of the sample approximately 16 mm from the fire. In this picture, it is visible that the retardant retained its crystalline form and that it did not change. It can be chemically retarded by changing the state. The DTD modified in this way was classified as flammability class **A** according to the regulations of the time.

## 6.2 Retardation of Plywood

From a fire engineering point of view, plywood is a completely different large-scale material than chipboard. The basic building element is veneer, a relatively more compact wood than chipboard. Because of this characteristic, the character of wood remains more preserved than that of chipboard. The density of plywood can be adjusted only by changing its thickness or the number of veneers used in the plywood. In addition to the type of wood, the fire-technical properties are influenced by the glue and the quality of the gluing technology. Fire retardation of plywood is possible in the following ways. A very effective but economically demanding method is to impregnate veneers with a flame retardant. The impregnation of already dried and sorted veneers, which must be dried after impregnation, requires another technique. This treatment procedure requires additional technological equipment, increased energy consumption, and increased labor [14, 15]. As will be demonstrated by the results of the experiment described below, the choice of a suitable retardant is very important. Not every retardant meets the conditions of use for impregnation of veneers and their subsequent pressing.

Three types of retardants were used in the experiment. These were retardants of the types FR101, Ohňostop retardant (marked as K according to the manufacturer's initials), and FR A. All were used at a concentration of 20%. The retardant was applied in the amounts prescribed by the manufacturers of the given retardant. A retardant was applied to each group of veneers. These were first coated 3 times and then soaked for 2, 6, and 24 h. We used the same veneer treatment technology for the veneers, which were subsequently heated in the press at the given pressing temperature and for the prescribed pressing time. The main evaluation criterion was weight loss.

Finally, we divided all the retardants into four groups according to the treatment technology. In the first group, we compared the burning characteristics immediately after coating. Ohňostop had the best results in both cases (before and after pressing, Fig. 15a). The second group consisted of veneers soaked for 2 h. The results were similar to those of the first group (see Fig. 15b). The third group consisted of veneers after 6 h of soaking (see Fig. 15c). The fourth group consisted of veneers after 24 h of soaking. In general, except for FR A, the retardants achieved good results. The weight loss was minimal for FR101 and the Ohňostop retardant (Fig. 15d).

At the end of the experiment, the pressed veneer achieved worse results than the unpressed veneer. The experiments confirmed the difference between pressed veneers (veneers heated in the press) and those not heated in the press. The retardant exposed to the press temperature begins to act while still in the press: it degrades, and its future effectiveness decreases. Therefore, for the production of retardant-treated plywood, we recommend paying attention to the selection of a suitable retardant.

Another method is to modify the glue used to glue the plywood veneers together. Of course, this requires "cooperating" chemicals (glue and retardant) to avoid reactions with each other such that neither the function of the glue nor the retardant is reduced. Such cooperating chemicals were found in the phenolic resin F 20 and the retarding element Retar 131 [13]. Plywood glued with such an adhesive mixture demonstrated improved fire-technical properties. We must not forget the flame propagation criterion. In this case, the contact top veneer is not treated and may show signs of flame spread, reducing the overall retardation effect. A hot–cold bath is recommended for spraying the retardant on the top veneer of the finished plywood

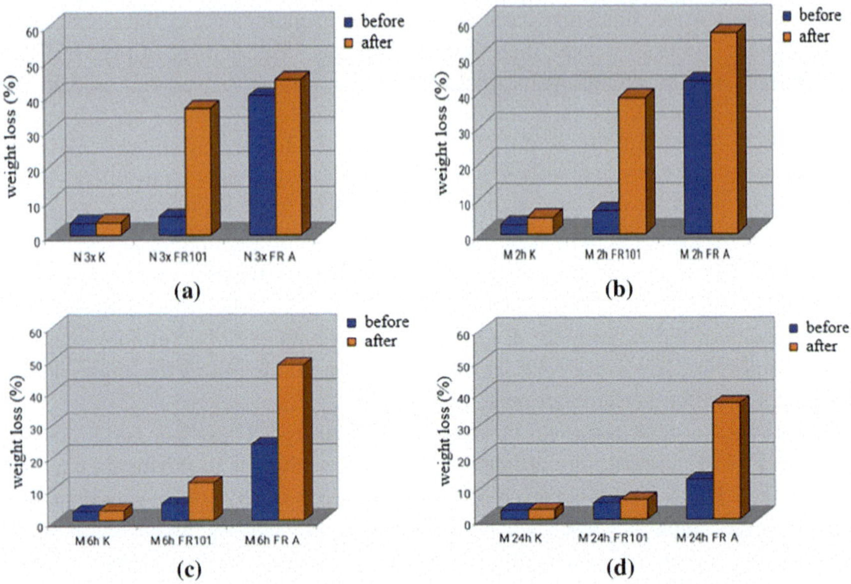


**Fig. 15** **a** Evaluation of the retardant effect-coating 3x **b** Evaluation of retardant effect—soaking for 2 h **c** Evaluation of retardant effect—soaking for 6 h **d** Evaluation of retardant effect—soaking for 24 h

immediately after pressing; when the hot surface cools down, the air in the intercellular spaces cools down, and the retardant is somehow absorbed into the wood of the veneer, which leads to better preimpregnation.

The combination of both procedures (impregnation of veneers and glue modification) creates a very high-quality plywood material in all aspects, both technical and fire-related. This material is used for special purposes. Among the possible modifications, we must not forget the coating, the application of a retardant on the surface of the material, or retardation foils [19].

The fire-technical properties of fiberboards can clearly be classified among the worst of all large-scale materials since it is difficult to consider the nature of the wood. Another limiting factor is the size of the wooden element from which the material is made, i.e., the wood fiber. The density of fiberboards affects some fire-technical properties but not to such an extent that it might have a significant effect on, e.g., the degree of flammability.

## 7 Biopolymeric Nanomaterials for the Fire-Retardant Treatment of Wood and Wood-Based Materials

Wood and wood composite products are manufactured with fire-retardant additives, after which wood or wood products are impregnated, painted, or surface coatings are applied in situ after the wood products are installed in the building [10]. Surface treatments, such as fire-retardant coatings, are also commonly employed to reduce material flammability. There are several advantages to these coating modification methods. However, most raw materials used in modified coatings are not environmentally sound. The flame retardancy of materials, particularly wood and wood-based products, can be improved by adding fillers, additives, or treatments based on organic and inorganic methods, including halogenated-based fire retardants, nitrogen-based fire retardants, and nanofillers [28]. The development of fire-retardant impregnations and coatings using nanotechnologies can assure that wooden constructions and materials have such distinctive qualities. Additionally, impregnation prevents ignition of the primary component of wood by altering the pyrolysis process, inhibiting the combustion chain reaction, and diluting pyrolysis gases by the byproducts of flame-retardant degradation. According to Vakhitoca [31], the fire resistance of materials with relatively low levels of nanocomponents (3–5 wt.%) often improves significantly with the addition of nanofillers [31].

Most flame-retardant filler materials are synthetic, nonbiodegradable, and environmentally unfriendly [23]. The use of biopolymers in daily life has increased; however, biopolymers pose a fire risk since they burn quickly and melt when heated, releasing a large amount of hazardous smoke [2, 28]. Consequently, biopolymers can be advantageous in terms of the environment and biodegradability. One of the issues in the future will be the research of biomass-based materials for fire retardants. Biomass replaces traditional carbon sources with high carbon content, such as chitosan, cellulose, lignin, starch, DNA, and phytic acid, and is often applied as a char-forming agent [7, 26, 29].

Biomass-derived nanomaterials and nanostructured materials have garnered increasing interest recently due to their superior thermal and mechanical properties [9]. Nanomaterials are a fascinating class of materials that include a wide variety of samples with at least one dimension between 1 and 100 nm. Through the thoughtful design of nanomaterials, exceptionally high surface areas can be attained [3]. Nanomaterials are preferred for the creation of composites due to their small size, low degree of aggregation, homogenous size distribution, and high dispersion tendency. There are a variety of processes used to make nanoparticles, including biological, chemical, and mechanical processes [1, 12]. Nanomaterials are preferred for the creation of composites due to their small size, low degree of aggregation, homogenous size distribution, and high dispersion tendency. There are a variety of processes used to make nanoparticles, including biological, chemical, and mechanical processes [9]. Nanoparticles from biomass, such as nanolignin, nanochitosan, and nanocellulose, are considered to be more environmentally benign for wood and wood product treatment (Fig. 15); however, they need to be modified for integration, resulting in effective performance as flame retardant materials.

Chemical modifications can be made to the reactive sites in lignin, such as hydroxyl, methoxyl, carboxyl, and carbonyl groups, which can be chemically modified. According to Yang et al. [34], phosphorus-nitrogen-modified lignin with metal ions, phosphorus or nitrogen-modified lignin, and lignin-and-flame-retardant combinations containing silica and phosphorus-nitrogen-modified lignin are the four main types of chemically modified lignin used as fire retardants [26]. Because of how similar the chemical structures of lignin and nanolignin are, their functions are very comparable. Because of its small size, nanolignin is more promising than lignin. To increase the usefulness of nanolignin, it may be dispersed in polymers at the nanoscale [29]. Nanolignin could serve as a filler to increase the thermal conductivity of composites by plugging the pores in the polymer mixtures [33].

The possibility of using chitosan for the construction of flame-retardant layer-by-layer coatings on various types of fabrics is justified by the protonation of the amino groups positioned at the C-2 position of the D-glucosamine repeat unit. The chemical structure of chitosan suggests its potential for use as a carbon and nitrogen source in the design of low environmental impact flame retardant formulations [16]. Chitosan is widely used as a charring agent since it includes high nitrogen (NH2) and hydroxyl (OH) concentrations [32]. In addition, such reactive groups can be successfully utilized to chemically modify biomacromolecules to introduce phosphorus-containing functionalities [16]. Ammonia ($NH_3$) is created when the amino groups in chitosan are heated to high temperatures. The $NH_3$ that is produced lowers the oxygen content of the system. The ring-opening reaction of chitosan is also accelerated by combustion, which creates an aromatic ring structure and a dense carbon layer in the condensed phase [32]. Chitosan could make polymer carbonization easier, resulting in more carbon residue and condensed phase flame retardancy.

Chitosan is easily chemically altered, primarily by adding phosphorus-based groups that, when exposed to a flame or an irradiation heat source, can interact synergistically with the carbon source (Fig. 16). The successfully obtained chitosan derivatives have also demonstrated excellent potential for the development of efficient flame-retardant systems [16]. The wood treated by chitosan layering has a higher limiting oxygen index (LOI) value. An increased value is also found in the treatment of wood using chitosan sodium phytate, chitosan/sodium phytate/nano-$TiO_2$, and chitosan/sodium phytate/nano-$TiO_2$-ZnO [35].

The other biopolymer is cellulose, which is a linear chain of glucose units that are highly polymerized and joined together by bonds made of oxygen, carbon, and hydrogen (H). Compared to the branching structure of hemicellulose, it has better thermal stability [11]. To obtain nanocellulose, top-down and bottom-up methods have been reported previously. Most commonly, chemical processes, including esterification, silanization, oxidation, and polymer grafting, are used to modify cellulose to produce acceptable fire-retardant materials (Fig. 15). Nanocellulose improves the fire-retarding properties of biobased polymers such as PLA [30]. The flame-retardant properties of nanocellulose cause the pHRR to decrease more than those of microcrystalline cellulose. The nanoscale dimensions of nanocellulose exacerbated

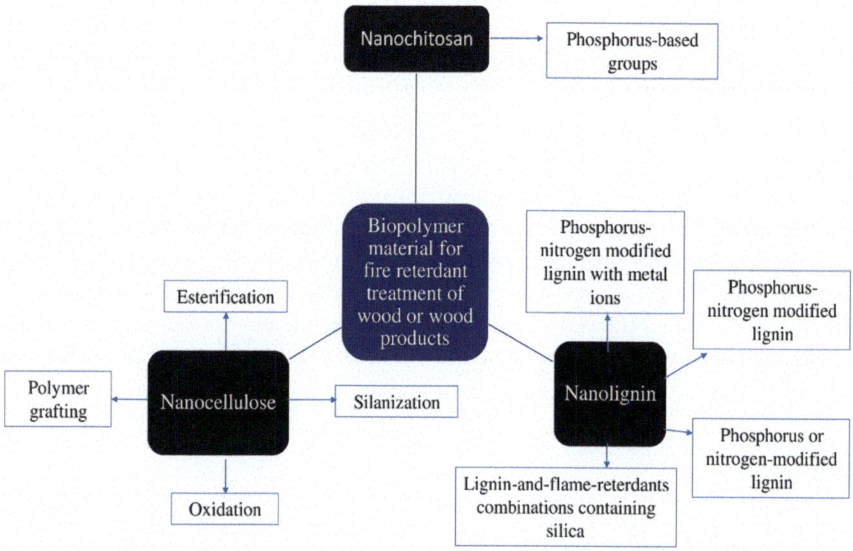


**Fig. 16** Biopolymer nanomaterial as a fire-retardant treatment for wood and wood products

the charring effect due to the increased interfacial contact between the cellulose nanoparticles and the PLA matrix [8].

## 8 Conclusion

Fire-retarding treatments for wood and wood-based materials are critical for enhancing fire safety in various applications. These treatments involve applying chemical substances to wood to convert it into flame-retardant materials. The primary objective is to reduce the combustibility of wood and slow the spread of fire. There are several methods for treating wood to make it fire-retardant:

1. Chemical impregnation. This method involves treating wood with fire-retardant chemicals that penetrate the wood fibers, decreasing the susceptibility to ignition and combustion.
2. Surface coatings: Fire-retardant coatings can be applied to the surface of wood to form a protective barrier that inhibits the ignition and spread of flames.
3. Pressure treatment: In this method, fire-retardant chemicals are forced into the wood under pressure, ensuring deep penetration and long-lasting fire protection.
4. Intumescent coatings: These coatings expand when exposed to high temperatures, forming a protective char layer that insulates the wood and slows the progress of the fire.

5. Inorganic treatment: Certain inorganic compounds, such as borates and phosphates, can be used as fire-retardant treatments for wood by disrupting the combustion process.

The choice of fire-retardant treatment depends on various factors, including the type of wood, the intended application, and regulatory requirements. It is essential to select treatments that not only provide effective fire protection but also meet safety, environmental, and durability standards. Overall, fire-retardant treatments for wood and wood-based materials play a crucial role in improving fire safety in buildings, furniture, transportation, and other industries, helping to protect lives and property from the devastating effects of fires. Continued research and innovation in this field are essential for developing safer and more sustainable fire-retardant solutions.

## References


1. Abe K, Yano H (2010) Comparison of the characteristics of cellulose microfibril aggregates isolated from fiber and parenchyma cells of Moso bamboo (*Phyllostachys pubescens*). Cellulose 17(2):271–277
2. Altarawneh M, Saeed A, Al-Harahsheh M, Dlugogorski BZ (2019) Thermal decomposition of brominated flame retardants (BFRs): Products and mechanisms. Prog Energ Combust Sci 70:212–259
3. Baig N, Kammakakam I, Falath W (2021) Nanomaterials: a review of synthesis methods, properties, recent progress, and challenges. Mater Adv 2(6):1821–1871
4. Balog K (1999) Spontaneous ignition (Samovznietenie). SPBI, Ostrava. ISBN 80–86111–43–1
5. Bucko P, Osvald A, Reh R (1990) Properties of fire-retarded waterproof chipboards. (Vlastnosti vodovzdorných drevotrieskových dosák retardovaných proti ohňu). Drevo, 45(2):38–39. ISSN 0012–6144
6. Chrebet T, Balog K (2010) Monitoring the effect of ammonium hydrogen phosphate on cellulose pyrolysis in a modified electrically heated hot air furnace. (Sledovanie vplyvu hydrogénfosforeč—nanu amónneho na pyrolýzy celulózy v modifikovanej elektricky vyhrievanej teplovzdušnej peci). In: Fire protection 2010: Collection of lectures XIX. Yearbook of the international conference, Ostrava, VSB-TU, 8–9 September 2010. University of Mining and Technology in Ostrava, Ostrava, pp 92–96. ISBN 978–80–7385–087–6
7. Costes L, Laoutid F, Brohez S, Dubois P (2017) Bio based flame retardants: When nature meets fire protection. Mater Sci Eng: R: Reports 117:1–25
8. Costes L, Laoutid F, Khelifa F, Rose G, Brohez S, Delvosalle C, Dubois P (2016) Cellulose/phosphorus combinations for sustainable fire retarded polylactide. Eur Polym J 74:218–228
9. Jawaid M, Kian LK, Alamery S, Saba N, Fouad H, Alothman OY, Sain M (2024) Development and characterization of fire retardant nanofiller from date palm biomass. Biomass Conv Bioref 14:9109–9119
10. Jones D, Brischke C (2017) Performance of bio-based building materials, Woodhead Publishing Series in Civil and Structural Engineering, 1st edn. ISBN: 0081009828
11. Karimah A, Ridho MR, Munawar SS, Ismadi Y, Amin RD, Lubis MAR, Wulandari AP, Nurindah IAH, Fudholi A, Asrofi M, Saedah E, Sari NH, Pratama BR, Fatriasari W, Nawawi DS, Rangappa SM, Siengchin S (2021) A comprehensive review on natural fibers: Technological and socio-economical aspects. Polymers 13(24):4280
12. Liu H, Liu D, Yao F, Wu Q (2010) Fabrication and properties of transparent polymethylmethacrylate/cellulose nanocrystals composites. Biores Technol 101(14):5685–5692
13. Mahut J, Horsky D, Osvald A (1983) Retardation treatment of plywood. (Retardačná úprava preglejok.) Drevo 38(8):219–221

14. Mahut J, Osvald A, Koboriová E, Kulhavy L (1983) Treatment of phenol-formaldehyde resin F—20 with antioxidant 49. (Úprava fenolformaldehydovej živice F—20 Antioxidantom 49) In: Collection of lectures—VI seminar "Glues in the woodworking industry". P CSVTS VSLD, Elected, pp 240–249
15. Mahut J, Osvald A, Reinprecht L, Krakovsky A (1985) Fire retard treatment of phenoplast with Triarylphosphate. In: Flammability of polymers. Polymer Institute Slovak Academy of Science, Smolenice, pp 44–53
16. Malucelli G (2020) Flame-retardant systems based on chitosan and its derivatives: State of the art and perspectives. Molecules 25(18):4046
17. Osvald A (1997) Assessment of fire safety of wood and wood-based materials and products. (Scripts), (Hodnotenie požiarnej bezpečnosti materiálov a výrobkov z dreva a na báze dreva). Technical University of Zvolen, Zvolen, p 104. ISBN 80–228–0595–5
18. Osvald A (1995) Evaluation of the flammability of thermal, acoustic and fireproof insulating materials. (Hodnotenie horľavosti tepelných, zvukových a protipožiarnych izolačných materiálov). In: Collection of lectures from the international professional forum "Application of thermal insulation materials in the solution of fire safety of buildings". Izomat, Nová Bana, pp 31–40
19. Osvald A, Mahút J, Sedliačik M (1982) Surface treatment of waterproof plywood against fire. (Povrchová úprava vodovzdorných preglejok proti ohňu). In: Syntema ´82—chemicalization of wood industry. House of Technology CSVTS, Brno, pp 56–60. ISBN 60/751/82
20. Osvald A, Réh R, Bešenda M (2001) DTD in terms of fire protection in wooden buildings (DTD z hľadiska požiarnej ochrany v drevostavbách). In.: Proceedings "Drevostavby" volyne: Higher vocational school and secondary industrial school volyne
21. Osvald A, Osvaldová L (2003) Fire retardation of spruce wood. Scientific studies. (Retardácia horenia smrekového dreva). Technical University Zvolen, Zvolen, p 61
22. Osvaldová L (2005) Flame retardants (Retardéry horenia). Arpos 18–19:18–21
23. Rahman MR, Huibin Hamdan S JLC (2018) Introduction and reinforcing potential of silica and various clay dispersed nanocomposites. Elsevier, Silica and Clay Dispersed Polymer Nanocomposites, pp 1–24
24. Réh R, Vargicová K, Osvald A, Rychlý J (1990) Waterproof particleboard against fire. In: Degradation, stabilization and combustion of polymers. Bratislava: SAV, 1990, (poster)
25. Reh R, Osvald A, Bucko P (1991) Reducing the degree of flammability of waterproof chipboard (Zníženie stupňa horľavosti vodovzdornej drevotrieskovej dosky). Drevársky výskum 128. Alfa, Bratislava, pp 43–56. ISSN 1336–4561
26. Solihat NN, Hidayat AF, Taib MNAM, Hussin MH, Lee SH, Ghani MAA, Edrus SSOA, Vahabi H, Fatriasari W (2022) Recent developments in flame-retardant lignin-based biocomposite: manufacturing, and characterization. J Polym Env 30(11):4517–4537
27. TaibMNAM AP, Savov V, Fatriasari W, Madyaratri EW, Wirawan R, Osvaldová LM, Hua LS, Ghani MAA, Edrus SSAOA, Chen LW, Trache D, Hussin MH (2022) Current progress of biopolymer-based flame retardant. Polym Degrad Stab 205:110153
28. Taib MNAM, Hamidon TS, Garba ZN, Trache D, Uyama H, Hussin MH (2022) Recent progress in cellulose-based composites toward flame retardancy applications. Polymer 244:124677
29. Vahabi H, Brosse N, Latif NHA, Fatriasari W, Solihat NN, Hashim R, Hussin MH, Laoutid F, Saeb MR (2021) 24—Nanolignin in materials science and technology— does flame retardancy matter? Biopolymeric Nanomaterials. S. Kanwar, A. Kumar, T. A. Nguyen, S. Sharma and Y. Slimani, Elsevier: 515–559.
30. Vahabi H, Shabanian M, Aryanasab F, Mangin R, Laoutid F, Saeb MR (2018) Inclusion of modified lignocellulose and nanohydroxyapatite in development of new bio based adjuvant flame retardant for poly(lactic acid). Thermochim Acta 666:51–59
31. Vakhitova LN (2019)16—Fire retardant nanocoating for wood protection. Nanotechnology in eco-efficient construction, 2nd edn. In: Pacheco-Torgal F, Diamanti MV, Nazari A et al., Woodhead Publishing, pp 361–391
32. Wang M, Yin GZ, Yang Y, Fu W, Díaz Palencia JL, Zhao J, Wang N, Jiang Y, Wang DY (2023) Bio based flame retardants to polymers: A review. Adv Ind Eng Polym Res 6(2):132–155

33. Wang X, Ji SL, Wang XQ, Bian HY, Lin LR, Dai HQ, Xiao H (2019) Thermally conductive, super flexible and flame-retardant BN-OH/PVA composite film reinforced by lignin nanoparticles. J Mater Chem C 7(45):14159–14169
34. Yang H, Yu B, Xu X, Bourbigot S, Wang H, Song P (2020) Lignin-derived biobased flame retardants toward high-performance sustainable polymeric materials. Green Chem 22(7):2129–2161
35. Zhou L, Fu Y (2020) Flame-retardant wood composites based on immobilizing with Chitosan/Sodium Phytate/Nano-$TiO_2$-ZnO coatings via layer-by-layer self-assembly. Coatings 10(3):296

# Wood Protection: Nanomaterials to Improve Durability in Wood & Wood-Based Composites

**Hamid R. Taghiyari, Mahdi Arabi, Dorina Camelia Ilies, Elham Nadali, Zharas Berdenov, and Vasile Grama**

**Abstract** As a renewable biological material that has a myriad of applications, wood has played major roles in human life and history, acting as a shelter, fuel (heating and cooking), food, urban design, heritage objects, and even control on the surrounding environment and atmosphere. Natural forests cannot satisfy the growing needs for industrial timber. Therefore, cultivation of low-density fast-growing tree species has become popular. However, they have drawbacks, like susceptibility to biological degradation. Though these trees can be used for production of wood-based composite panels, the products have themselves weaknesses, such as thickness swelling, and consumption of large quantities of synthetic adhesives. Over the past decades, researchers carried out studies to find modern ways to overcome the above-mentioned drawbacks and weaknesses using nanomaterials and nanotechnology. The present Chapter takes a brief look at the studies focused on protection of wood and wood-based composites against organisms (mainly fungi and insects) and improvement in their biological durability, and limitation of water absorption and thickness swelling.

---

H. R. Taghiyari (✉)
Department of Wood Science and Technology, Faculty of Materials Engineering and Interdisciplinary Sciences, Shahid Rajaee Teacher Training University (SRTTU), Tehran, Iran
e-mail: htaghiyari@sru.ac.ir

M. Arabi
Department of Surveying and GIS Engineering, Faculty of Civil Engineering, Shahid Rajaee Teacher Training University (SRTTU), Tehran, Iran

D. C. Ilies · V. Grama
Department of Geography, Tourism and Territorial Planning, Faculty of Geography, Tourism and Sport, University of Oradea, Oradea, Romania

E. Nadali
Ph.D. Graduate at Department of Wood & Paper Science and Technology, Faculty of Natural Resources, University of Tehran, Karaj, Iran

Z. Berdenov
Department of Physical and Economic Geography, Faculty of Science, L.N. Gumilyov, Eurasian National University, Astana, Kazakhstan

 
P. Antov et al. (eds.), *Nanomaterials Additives in Bioadhesives for Wood Composites*, Smart Nanomaterials Technology, https://doi.org/10.1007/978-981-96-3400-2_9

**Keywords** Biodegradation · Modification · Nanomaterials · Natural durability · Preservation · Protection

## 1 Introduction

Nanotechnology generally refers to the production and application of materials with nanoscale dimensions (ranging 1–100 nm). The main criterion to identify nanomaterials is that at least one of their dimensions has to range lower than 100 nm, categorizing them into nanoparticles (as zero-dimensional nanomaterials), nanorods and nanotubes (as one-dimensional nanomaterials), and nanosurfaces or nanosheets (as two-dimensional nanomaterials, like graphene) [49, 86]. There is also another group of nanomaterials, having three dimensions called nanoprisms or nanoflowers [23]. At this infinitesimally small dimension, properties (physical, chemical, thermal, electrical, and even biological) of the nanoscaled material may fundamentally change. For instance, conductivity can be changed or improved in some conductors and semiconductors [54].

The terms "Nanomaterial" and "Nanotechnology" were first coined about five decades ago by Norio Taniguchi [113]. While the cited author originally defined these terms to describe semiconductors, they were later broadened to cover a wide range of different simple and complex materials and composites including molecular nanotechnology, carbon nanotubes, biomimetic nanotechnology, and a variety of experimental advancements on nanoindentation, nanolithography, atomic force microscopy (AFM), and scanning tunneling microscopy (STM).

Nanotechnology is considered a rather new branch in modern science, providing various capabilities to develop new materials with distinguished and sometimes customized properties. However, the nature has long been using nanomaterials synthesized through natural processes [117]. Some visionary scientists consider nanotechnology as the basis to initiate a new Industrial Revolution to enter a new civilization era, which can even be more influential when compared to the Stone, Bronze, and Iron Ages. In different industrial sectors including the forestry sector, nanotechnology can provide a host of opportunities to develop a variety of unprecedented wood and wood-based products with improved and unique properties to be used in daily life of people [20, 32, 38, 57].

Wood is a renewable material in myriad of types and patterns with unique and versatile properties that has been used for different applications and has played a major role in the history of civilization, human development, cultural heritage, and air and climate sustainability [15, 26, 42, 53, 68, 101, 119]. The degradability and renewability aspects of wood make it favorable in terms of environmental concerns. However, some of its inherent drawbacks limit its applications. Over the past few decades, researchers around the world have used nanomaterials to address these drawbacks, including dimensional instability towards water and vapor, susceptibility to biodegradation especially in fast-growing tree species (mainly different deteriorating fungi, insects, and termites), vulnerability to fire and high temperature, and even UV radiation [15, 29].

Nanotechnology not only can improve different properties in wood itself by mitigating the above mentioned drawbacks, but it also provides new potential and options to upgrade engineered wood products through the development of new bio-resins that increase the ability to use small diameter logs and other lignocellulosic materials [4, 5, 6, 7, 8, 9, 10, 19, 39, 55, 56, 65, 44, 112, 120, 121, 32, 62, 78, 79, 81, 80, 84 87, 88, 95, 96]. In this connection, different nanomaterials can be added to improve the properties of coatings, biocides, resins, polymer composites, and reinforced wood-based materials [93]. Nanomaterials can also be derived from the cellulose, lignin and extractive components in wood allowing us to create new materials from the existing resources.

The present chapter intends to review some of the main and recent applications of nanomaterials in wood and wood-based composite panels to improve their biological durability against different species of wood-decay fungi, insects (woodboring beetles and termites), to increase resistance to fire, and to decrease susceptibility to UV radiation. Furthermore, and with an aim to decrease production costs, some of the successful nanomaterials were chosen to check their effectiveness at micro-scale.

## 2 Degradability of Wood

Wood is mainly comprised of three main natural polymers, including cellulose, hemicellulose(s), and lignin [58, 59, 60, 89]. These natural polymers can be degraded as a result of attack by different living organisms, fire, UV radiation, and chemicals. Wood can also be easily damaged physically, like scratch and breakage.

In terms of biodegradability, wood is considered the main source of food for a variety of different living organisms, including wood-decay fungi, various insects and termites, marine borers, and even microorganisms. Different wood species may also have extractives that can in some cases be considered nutritious for the above-mentioned organisms. These extractives function as an initial attraction and therefore, they are the basis for a consequent severe attack, though they comprise only a small portion in wood species (Fig. 1). This organic nature of wood necessitates utilization of effective preservation and protection methods and materials so that the service life of wooden structures is prolonged to a great extent.

There are four basic requirements for most of the wood degrading organisms, namely a source of food, water, suitable temperature, and oxygen (Fig. 2). Practically speaking, it would be difficult to control oxygen level and temperature, when medium to large scale wood structures are considered as they are usually located in open air where the level of oxygen and temperature cannot be easily controlled, or the process of controlling would not be acceptable at an industrial scale because of cost preferences. Therefore, most of the industrial wood protection strategies are designed either to prevent water and moisture getting into wood texture, or to chemically

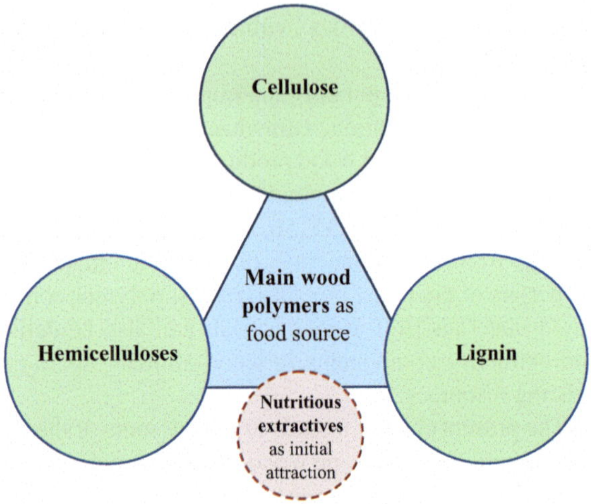


**Fig. 1** The main wood polymers (cellulose, hemicelluloses, and lignin), acting as the main source of food for different wood degrading organisms, plus nutritious extractives (like starch, and some carbohydrates, …) acting as an initial attraction to the organisms

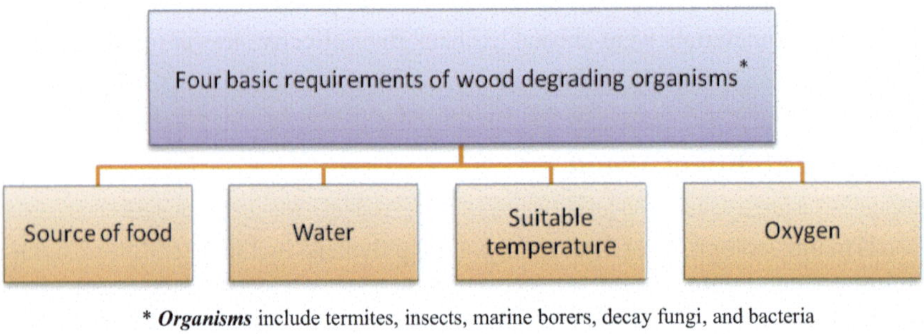


**Fig. 2** The four basic requirements of all wood degrading organisms

modify wood polymers in a way that can no longer be digested by the organisms' enzymes.

Burning and degradation of the main wood polymers by fire is one of the biggest risks for wood and wood-based composites. Although the risk decreases in large timbers because the rate of burning is predictable and can therefore be taken into account in design [36, 123], the yearly damages are still considerable even in forests and standing trees. One of the favorable properties of wood is that it tends to fail slowly as it burns while other constructional materials such as steel fail at a very high rate once they become ductile. Wood substrates are sometimes covered with non-combustible materials, like gypsum boards. However, many people would rather have the beauty of wood with different natural patterns and textures. These applications of wood generally require using either a kind of fire retardant coatings or impregnation with fire retardant chemicals.

In terms of damages caused by ultra-violet (UV) radiation, it is believed that UV energy is rather high and damages the surfaces exposed to this light. It creates some free radicals that decompose lignin structure, ultimately leading to a fast and irreversible depolymerization of lignin. While the depth of penetration is rather shallow, the process is fast and decomposition of the exposed surfaces begins within the first hours, sharply reducing the quality of the surface appearance as a vital factor in all materials, including garden furniture and outdoor wooden structures.

## 3 Wood Protection

Engineers generally design timber in buildings and structures in a way to avoid direct and close contact with soil, away from water and moisture, and they usually apply an initial protection front like a coating or covering, thereby preventing the biodegrading organisms and degrading agents to easily attack wood and composite texture. However, there exist many applications in which exposure to insects, fungi spores, water and moisture is inevitable. For cases such as these, a variety of protection strategies have been developed, implementing a variety of different preservation techniques and materials [11, 52, 64, 71, 74, 82, 97–99]. Surface protection of timber and composites (like brushing and dipping of solid wood in preservatives and toxic chemicals, and synthetic veneers for composite panels) is considered one of the simplest methods that have long been used. More effective preservation methods require that the wood be impregnated (using combinations of vacuum and pressure) with different preservatives so that the chemicals are driven deep into the heart of timber. In composite panels, preservatives can be either applied on wood particles and fibers, or they can be mixed with resins and sprayed on wood-based furnish to form the mat.

Another effective method is to alter the chemical structure of the main polymers in wood so that the wood enzymes secreted by the organisms cannot easily degrade them [92, 94, 102]. Thermal modification is considered the most commercially used technique to chemically alter the wood basic polymers [36]. As usually no chemical is used in this technique, it is considered to have little or no hazard to both human and wildlife, making it favorable for a wide range of non-soil contact applications in Europe and many other parts of the world [21, 28, 36, 94, 115, 116]. The above mentioned nanometals and nanominerals may have double functions. Firstly, they can enhance thermal conductivity, by which thermal modification of wood can be carried out more effectively [118]. Secondly, they may even have inherent antifungal property, adding to the efficiency of the modification, as is the case of nanowollastonite and nanometals (like nano-zinc oxide and nanocopper) [3, 14].

Enhancement of heat transfer by nanomaterials (mostly metals and some minerals) during production of wood-based composite panels reduces pressing times as a bottle-neck in composite production process. Some of these nanomaterials (including nanocopper, nanosilver, and nanowollastonite) have fungicidal property. Apart from enhancement in resin curing and fungicidal effect, wollastonite acts as a reinforcing agent in the resin, improving physical and mechanical properties in lignocellulosic composite panels [58, 65, 104, 105, 106].

Many scientists and researchers reported effectiveness of nano-copper to be used for wood preservation and protection [24, 34, 37, 43, 63, 85, 114]. Chemical compounds containing nano-copper can usually be dissolved in water, which is a cheap carrier and can easily be found everywhere. Copper compounds can have both basic and acidic formulations. One of the most widely used systems that have been used for a long time is chromate copper arsenate, known as CCA. It uses chromic acid for solubilizing the copper compounds in the solvent. However, applications of CCA in any forms and formulations have encountered severe restrictions in many countries around the world. Therefore, the industry soon turned to systems containing ammonia or amine. Still, the above mentioned systems had two major disadvantages, namely high costs and corrosiveness to metal fastenings. Therefore, copper carbonate was ground to be micronized and suspended as very fine powder in water. This eventually reduced the need for costly co-solvents. Development of nano-sized copper particles provided wood preservation industry with the potential for a deeper and more effective impregnation. Moreover, the overall water solubility is lower which in turn reduces the potential of copper contamination in the surrounding soil and environment [31, 47], though more studies should be carried out on this unexplored attribute. Nano-particles have even a slight advantage compared with conventional solubilized preservatives in terms of their ability to move into wood through the cell pits. However, it is to be noted that there is a strong limitation for penetration of the preservatives, vis. the size of cell wall pores. Therefore, it would be unlikely to expect that nano-metals can overcome the problem of complete fluid flow in the heartwood of some wood species. However, they can more easily carry active ingredients deep into the inner parts and cavities [25, 70].

The size of particles plays an important role in the effectiveness of delivering nanaomaterials into the texture of wood. Therefore, copper particles at a nano-scale can more easily penetrate into wood cell walls, in comparison to other copper systems [45]. In this connection and based on previous research studies, nano-zinc oxide successfully protected Paulownia against an attack by *T. versicolor* [2], moreover, nano-copper proved to be highly effective in hindering termite attack [3]. Still, more studies should be carried out to make a firm conclusion if the enhanced mobility has a direct positive correlation with an increased efficacy against the attack by organisms.

It is unclear if the enhanced mobility can result into higher efficacy, although nano-zinc oxide was reported to effectively protect Paulownia against *T. versicolor* [2] and nano-copper showed more effectiveness against termites [3].

It is to be considered that impregnation of solid wood specimens with nanomaterials have some limitations, mainly the need to use special devices with materials that are not chemically reactive with the nanomaterials, and the delivery processes that are to be used for the impregnation. In terms of wood-based composite panels, there are easy options to choose as the delivery system; the wood particles can be pretreated; nano-particles can also be included in the composite furnish before it is subjected to cold press and even before the furnish is taken to hot press. In some recent studies, aqueous nano-particles were mixed with the resins (in these examples, urea formaldehyde resin, and polyvinyl acetate resin). The mixtures were then sprayed in a rotary drum on the wood fibers and wood chips at a commercial scale

(Gazvin Industrial Complex, Gazvin, Iran) [77, 104]. The decay resistance of the produced panels were reported to be significantly improved. Nano-materials were also reported to protect from termite attack [18, 65]. In a recent study, a common formulation of zinc was used to successfully hinder termite attack [65]. However, the cited authors indicated that copper compounds demonstrated only a promising result. Though the effectiveness of some nano-metals is repeatedly proved, there is a real concern about the production costs. Therefore, nano-minerals are economically more accepted. Wollastonite ($CaSiO_3$ is a good substitute to metals. As an inosilicate mineral, its thermal conductivity coefficient is much higher than wood; moreover, it has significant fungicidal property that makes it beneficial to be used in solid wood and wood-based composites [1, 46, 108, 109]. It is proved that wollastonite is not toxic to the organisms that are not primarily the target [40, 41, 100], still, it has demonstrated significant improved decay resistance in solid wood and wood-based composites against different fungi species [46, 105, 106]. It is also widely available in many countries around the world. Wollastonite (at both nano- and micro-scales) has been proved to have significant reinforcing effect in two wood adhesives, viz. urea–formaldehyde and polyvinyl acetate. The effect resulted in significant improvements in many properties, both in solid wood species, and wood-based composite panels as well [108, 109, 110]. In addition to wollastonite, other mineral nanomaterials like nanoclay were reported to reduce water uptake while limiting fungal attack in wood-based composites [17, 108].

Metal and mineral nano-particles have also improved biological resistance in solid wood species, and wood-based composites. Polymeric nano-carriers like encapsulated materials [72] were applied in solid wood specimens to transfer active ingredients deep into wood bodies [50, 69].

The results indicate that metal and mineral nanomaterials have the promising potentials to improve biological resistance to different wood-deteriorating organisms (including different wood-decay fungi, insects, and termites), physical and mechanical properties, and to reduce hot-pressing and the production costs thereof.

### *3.1 Surface Protection of Wood*

Surface coatings and finishes are primarily applied on wood to satisfy two initial purposes, firstly for protection, and secondly for keeping or adding to the natural beauty of the wood substrate. Wood surfaces can be easily exposed to water, fire, insects, fungal spores, UV radiation from the sunlight, physical damages like scratch, and even degrading chemicals. A surface coating can, to a great extent, protect the substrate and the underneath layers from potential damages caused by the above-mentioned factors.

In terms of water exposure, wood is generally known as a hydrophilic material which in turn results in swelling and shrinkage of wood. This is considered one of its main draw-backs of all wood species. Therefore, a coating with which the hydrophobicity of the surface is improved would be a more desirable attribute. The contact angle of a water droplet on the surface of piece of wood is a good criterion to assess the hydrophilicity. Hydrophilic surfaces generally have a contact angle smaller than 90º. Hydrophobic surfaces have contact angles larger than 90º. Coatings with contact angles more than 150º are very hydrophobic. There have been some achievements to improve the hydrophobicity in wood coatings through studying lotus and ramie leaves. These leaves demonstrate a property that is known among scientists as superhydrophobicity, resulting in an extreme high resistance against penetration of water droplets [33, 90, 125–128]. A successful attempt was made to electrostatically deposit a fibrous membrane, composed of cellulose acetate at nano-scale, to create a superhydrophobic surface [76].

In terms of damages caused by ultra-violet (UV) light, there have been some successful achievements, using nano-coatings. Nano-materials can improve the protection time of both the coatings and the wood substrate, decreasing the intervals of renewing the protective coatings. A suitable UV protection of rubberwood substrate (*Hevea brasilensis*) was provided by nano-zinc oxide amended maleic anhydride grafted polypropylene (MAPP) based coatings, reported by Salla et al. [91]. In other studies, titanium dioxide and zinc oxide nanoparticles were successfully used to improve nano-coatings by Zuccheri et al. [129] and Nosál' and Reinprecht [75], respectively. While low levels of zinc or titanium nano-particles are not practically effective in the protection of the wood substrate (*Pinus radiata*) against UV radiation, iron-oxide nano-particles result in higher degrees of protection [73].

Protection against wood-decay fungi are reported based on the average weight loss in the specimens, after different exposure periods to specific fungi (ASTM D-2017 (1981)). Findlay [30] provided a Table based on which different wood species are graded according to the average weight losses (Table 1). In this connection, a new research studies are in progress, using nanominerals (mainly wollastonite and sepiolite) to improve the protective property of a water-based acrylic-latex paints against wood-decay fungi, the results of which are not yet published (Fig. 3). However, based on the visual observation it was obvious that the paint containing a kind of nanomaterial maintained its integrity after being exposed to *Trametes versicolor* fungus, while the paint with no nanomaterial (as control) lost its integrity and was almost removed from the surface of wood specimens (Fig. 3).

While suitable nano-coatings with improved water repellency, biological resistance, and UV protection properties remain elusive, a potential market welcomes all

**Table 1** Wood grading according to the weight loss (%) caused by Xylophagous Fungi and its strength grade [30]

| Weight loss (in %) | Durability grade (resistance category of wood) |
|---|---|
| Lower than 5 | Very resistant |
| 5–10 | Resistant |
| 10–20 | Moderately resistant |
| 20–30 | Not resistant |
| Higher than 30 | Perishable or without resistance |

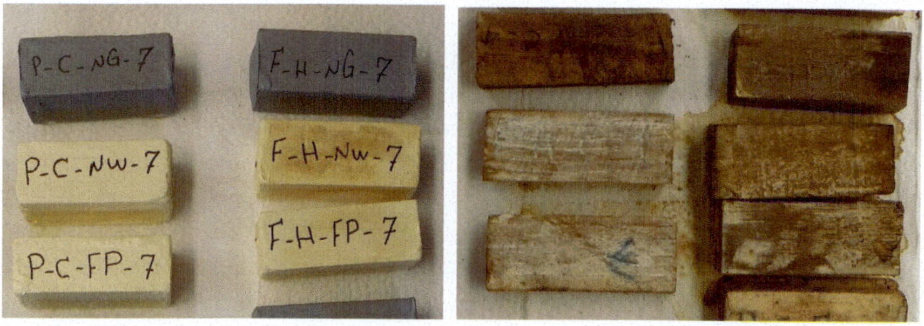


**Fig. 3** Specimens with an acrylic-latex paint containing with and without nanomaterial exposed to *Trametes versicolor* for five months

successful attempts as they can extend the service life of exterior surfaces of wood structures.

While wood coatings and finishes can add to the beauty of the substrate and protect it from biological, physical, and chemical damages, they tend to collect debris and dust over time. The removal of the dirt on the surface of coatings requires costly cleaning, and it shortens the service life of the coating as well. Some recent studies have focused on the development of coatings with self-cleaning and anti-fouling properties to decrease the maintenance costs, to increase the service life, and even to prevent microbial or algal growth on cellulosic-based packaging and biomedical implants [16]. Functionalization of surfaces is a practical approach to increase the resistance of surfaces to adhesion of dirt, and to minimize the damages caused by fouling. In this connection, Poly(ethylene glycol) and oligo(ethylene glycol) are considered two popular compounds for surface functionalization [22].

In addition to the materials, techniques, and processes mentioned above, some biocides and anti-corrosion additives are sometimes added to coatings and finishes to protect them both during storage in cans, and to limit microbial penetration to some extent once applied on surfaces. Nanomaterials can enhance the effectiveness of coatings in this regard. For instance, an enhanced resistance property to corrosion was achieved, using a POSS-NH2 (polyhedraloligomeric silsesquioxane) nano-coating [51]. They reported a rather high impedance value of $109\ \Omega\ \text{cm}^2$ for specimens that were tested for 30 days.

## 3.2 *Protection in Depth*

When the protecting materials (chemicals, preservatives, fire-retardants, and nanomaterials) are forced deep into the voids and empty spaces in a piece of wood (vessels and cell cavities), or dispersed over the micro surfaces of chips and fibers in wood-based composites, an in depth protection against different degrading agents is achieved. In wood-based composites, preservatives and nanomaterials can either be directly sprayed on wood particles and fibers prior to resin application, or they can first be mixed with the resin and be sprayed on the furnish if pH and other chemical interactions are avoided to prevent pre-curing of the resin before the hot press.

Over the past century, many pressure methods were developed to force preservatives into the texture of solid wood species. Depending on whether the preservatives and fire-retardants need to only cover the surfaces of cellular voids and spaces, or they need to fill them, a variety of vacuum and pressure regimes were developed. Among the most popular preservation methods are full-cell and empty-cell processes. Full-cell process was invented by John Bethell in the 19th Century. There is an initial vacuum in the pressure vessel to remove air within cell cavities. Once the vessel is filled with the preservative under pressure, cell cavities are filled with it and the compressed air does not force out the preservative once the pressure is lifted. While this method is considered effective, there are environmental concerns because of rather high leaching and chemicals that remain inside the impregnated wood. Empty-cell process was presented by Max Rüping at the turn of the twentieth century. In this process, there is no initial vacuum and therefore, the compressed air inside cell cavities drives out the preservatives once the pressure is lifted. Moreover, there is a last vacuum step which facilitates driving out of the preservative out of wood much faster and more efficiently [71].

In terms of using nanomaterials as preservatives against decay fungi, application of metal- and mineral-based nanoparticles was reported to be effective. These nanomaterials may react in separate or combined mechanisms. The first mechanism is the toxicity of the nanomaterial used as a preservative. In this connection, copper compounds (as a metal-based nanomaterial) and wollastonite (as a mineral-based nanomaterial) were reported to be significantly effective in preventing decay fungi to grow [40, 41].

In a study, poplar specimens were impregnated with wollastonite nanoparticles (NW) to be exposed to *Trametes versicolor* fungus [46]. Though the average NW retention was reported to be only 0.032 g/cm$^3$ (±0.004), the results demonstrated significant decrease of more than 40% in the mass loss in the NW-impregnated specimens compared to the control ones (Fig. 4). The average mass loss was 47% (1.2 standard deviation) in the control specimens (sixteen weeks of exposure to the decay fungus), while the mean amount of mass loss in the NW-impregnated specimens was only 3.6% (0.7 standard deviation). Findlay scale (ASTM D-2017, 1981) indicates that wood with a weight loss of more than 30% is classified as perishable or without resistance against biological decay by fungi; therefore, the control poplar specimens were classified as perishable. However, impregnation with NW made the specimens be classified in the very resistant category.

Visual observations also illustrated distinct difference in the growth of growth of *T. versicolor* mycelium over control and NW-impregnated specimens, based on Willeitner scale. Fungal colonization was 100% on the control poplar wood specimens (Fig. 5a), while in the NW-impregnated specimens, growth of the fungus mycelium was inhibited to a significant extent (Fig. 5b). The limitation of growth was not only obvious on the wood specimens, but it was also evident on the malt-agar in Petri dishes.

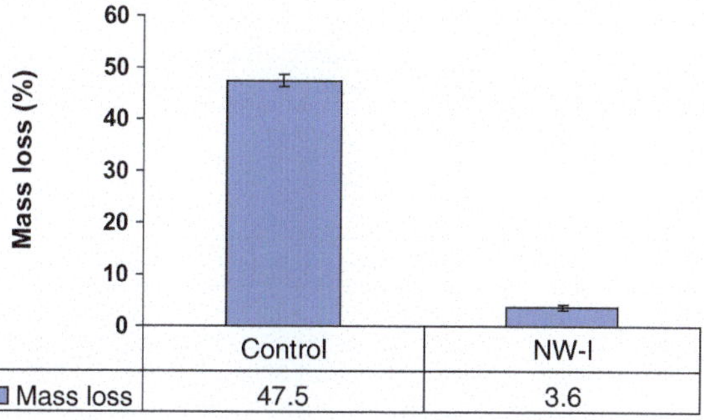


**Fig. 4** Mass loss (%) in the control and NW-impregnated specimens (NW = nano-wollastonite; I = impregnated) [46]

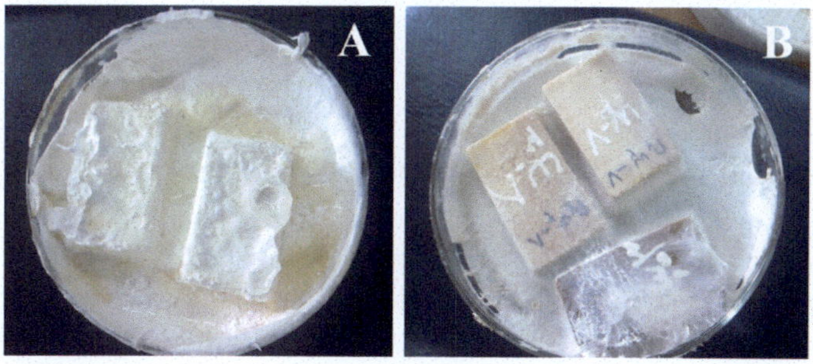


**Fig. 5** Mycelium growth of *Trametes versicolor* fungus on control poplar specimens (**a**) and NW-impregnated poplar specimens (**b**) [46]

Karimi et al. [46] further carried out a leaching test. The results revealed that only 4% (based on the dry weight) of the deposited NW leached out. This low leaching out of wollastonite can be attributed to the adsorption energy and low contact distance of wollastonite on wood cell-wall polymers, resulting in a rather powerful bond between wollastonite nanoparticles and cell-wall. Using density functional theory (DFT) analysis, it was calculated that wollastonite has sufficient adsorption energy to interact with cellulose and hemicelluloses [62, 83, 107]. The optimal adsorption distance of 1.7 Å, and adsorption energy of –6.6 eV provided enough boding strength for wollastonite nanoparticles to be fixed in wood so that they could not be easily leached out. However, it is to be noted that fixation of wollastonite on wood cell-wall polymers has a weak point when encountered with a large number of water molecules. In this connection, MDF panels produced with only wollastonite as the binder (and having no resin) had good mechanical properties but performed poorly in terms of water absorption and thickness swell [10]. The poor performance was attributed to the ability of abundant water molecules (with hydroxyl groups) to break the wollastonite-cellulose bonds and to replace them.

In terms of protection of wood-based composites, nano-wollastonite (NW) was added to UF resin at 10% (based on the dry weight of wood particles and fibers) to produce MDF and particleboard. Specimens with and without NW were exposed

**Table 2** Mass losses (%) in medium-density fiberboard (MDF) and particleboard (PB) specimens once being exposed to *Antrodia vaillantii* for four months (NW = nanowollastonite content; CF = chicken feather content) [105, 106]

| Type of wood-based composite | Weight loss (%) | | | | | |
|---|---|---|---|---|---|---|
| | Wood 100% | | Wood (95%) + CF (5%) | | Wood (90%) + CF (10%) | |
| | NW (0%) | NW (10%) | NW (0%) | NW (10%) | NW (0%) | NW (10%) |
| MDF | >26 | >2.5 | >14.7 | >4.5 | >25 | >3 |
| PB | >37 | >2.5 | >22.5 | >4.5 | >21 | >2.5 |

to *Antrodia vailantii*. Results demonstrated significant improved resistance in both MDF and particleboard composites [105, 106] (Table 2).

## 3.3 Modification of Chemical Structure of Wood Polymers

Wood is comprised of three main polymers, namely cellulose, hemicellulose, and lignin. Tow of the main costly and troublesome drawbacks in wood are directly connected to the chemical structure of these polymers. The first is the vulnerability of wood to decay fungi, including white-rot, brown-rot, and soft-rot fungi. The second is the dimensional instability of wood towards water and vapor as a result of adsorption of water molecules to hydroxyl groups that are abundant in cellulose and hemicellulose. Both these two main drawbacks can be substantially improved by alteration in the chemical structure of wood cell-wall polymers. The main purpose of the chemical alteration is to change the structure in a way that either the fungi do not recognize and their enzymes cannot digest them, or the hydroxyl groups are inter-bonded with themselves or other chemicals so that they can not make bonds with water molecules.

At a commercial scale, alteration in the chemical structure of wood polymers can be achieved by two main modification methods, either by thermal modification (which involves no chemicals in the process), or by addition of chemicals to wood (like acetylation and furfulysation). Thermal modification is the most commercially used modification technique that has a vital advantage over other techniques; it is considered an environmentally friendly method to alter the chemical structure of wood because no chemical is used in the process.

Nanomaterials that can alter specific wood attributes or functionalize wood can be considered another approach to improve wood properties. Functionalization is a challenging process since it requires uniform and complete penetration of the material into the wood cell-wall [32]. Failure to accomplish deep penetration can leave large amounts of wood cell wall polymer in the original untreated state which can still be easily capable of interacting with water molecules and fungi enzymes. In a study, inorganic clay nanoplatelets were applied to balsa wood specimens to create a nanostructured wood hybrid. There was an initial delignification process needed, with peracetic acid to create pathways for easy particle movement throughout wood specimens [32]. Energy dispersive x-ray analysis illustrated concentrations of nano-clay deep inside the cell walls. Thermogravimetric analyses (TGA) showed that modified wood had more residual when heated to 800 °C. This implied that nano-clay resided in the cell wall improved char formation, potentially acting as a thermal barrier and insulator. Wollastonite and sepiolite were also reported to similarly act as an insulating layer, which further prevented penetration of piloted fire to deeper into the wood.

## 4 Protection Against Fire Hazards

Wood and wood-based products are vulnerable to fire and high temperatures. Human beings were aware of this from the beginning of creation as it has natural causes too which damages forests and jungles. Fire is hazardous to wildlife and produces large amount of carbon dioxide to the atmosphere. However, occurrence of fire in urban areas can be considered from two other perspectives, too. Firstly, it is hazardous for human life, making it a vital factor for urban planners; and secondly, it demolishes wood products and structures in the blink of an eye; structures that were initially produced to last for decades.

Fire retardants have long been introduced to wood industry to protect timber from fire. Older houses were sometimes coated with local clays as a surface protection; or the surface of timbers was charred to limit oxygen access to the woody parts beneath to create a protective layer to prevent the spread of flame to the inner parts. There are two most common methods for applying fire retardants on exposed timber. The first is surface coating of the timbers. The second which is more effective is in-depth protection by impregnation. Treatment with fire retardants include combinations of a number of different chemicals [12, 61, 66, 67, 123], as well as some of the organic compound systems. The organic systems usually contain urea formulations that are not very acidic. The fire retardants that are currently used have formulations and chemical ingredients that hinder the spread of fire to other parts, and they control emissions of smoke at the same time. Previous formulations had lower pH numbers, therefore reacting with wood and metal fastenings, ultimately resulting in the weakening of the structure. There are still some more negative aspects of these reactions, including damage to the appearance (usually by darkening woody members), and increasing water absorption and dimensional instability. Impregnating wood lumbers with the above mentioned compounds and formulations has an extra negative effect on gluing and coating properties, making the treated timber difficult to be used in furniture industry [103, 124]

Performance of different metal and mineral nanoparticles as fire retardants have been studied, using an invented apparatus using piloted flame (Fig. 6) [103, 123]. The nanomaterials that have been investigated so far were chosen not to be toxic to human and wildlife. Though the selected metal and mineral compounds were proved to be effective as fire retardants in wood and wood-based products, their functions may be different [103]. Nanosilver (as a metal nanomaterial) improved fire properties mainly by improving thermal conductivity in wood [103, 118]. Wollastonite, clay, and sepiolite (as mineral nanomaterials) additionally provided an incombustible insulating and protective layer, making it harder for the flame to access the inner parts. It is to be noted that no material and chemical has so far been found to fully prevent woody materials from combustion. However, the ability of a material to delay the process of catching fire in a piece of wood has a vital importance since it provides building occupants with longer time to evacuate, and causes less damage until the fire fighters can put out fire.

The improved thermal conductivity caused by the addition of metal and mineral nanoparticles (nanosilver and nanowollastonite) facilitated transfer of heat to the inner parts of wood and wood-based composites [104, 118]. More rapid dissipation

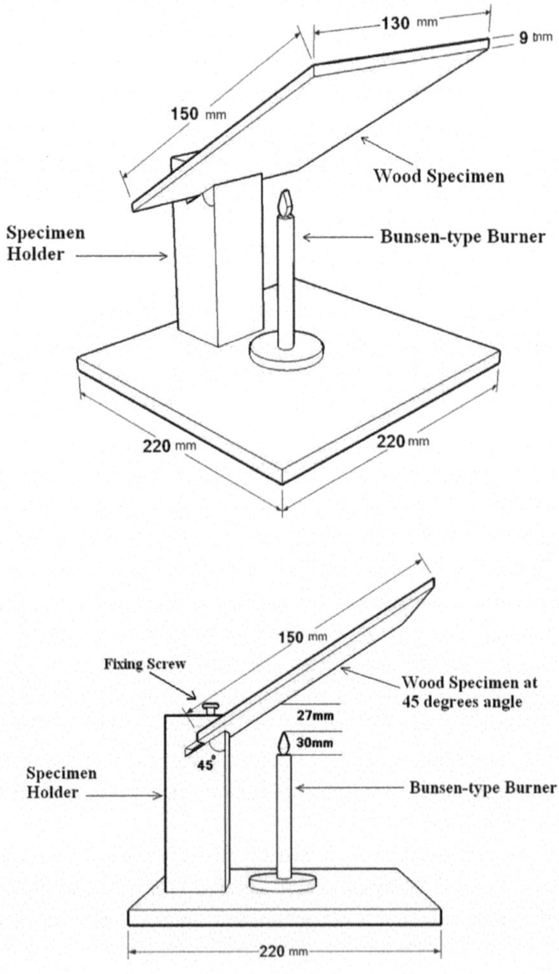


**Fig. 6** Schematic diagram of an invented fire testing apparatus (Fixed fire test apparatus; Iranian Patent No. 67232; license No. 3407 certified and issued by Iranian Research Organization for Science and Technology; USPTO Pub. No.: US 2019/0212283 A1; Appl. No. 15/866,752) [103]

of heat delayed the time for a given portion of the wood to reach the combustion temperature. Wollastonite is a mineral material with rather high thermal conductivity coefficient in comparison to wood. In addition to dissipation of heat, nano-wollastonite formed an insulating and protective layer on the surface, delaying the wood beneath to catch fire.

Graphene is another nanomaterial that was reported to successfully improve different fire properties ignited by piloted fire in both solid wood species and wood-based panels [27]. The thickness dimension of graphene is comprised of only one layer of carbon atoms that are arranged hexagonally to form a sheet (Fig. 7). Ultimately, these hexagonally-arranged carbon sheets pack together densely in an arrangement that look like a honeycomb lattice. This material has special structure with exceptional properties which made it ideal to be used in many emerging materials that can be used in different electronic parts and utensils, in devices that can save energy, and in a myriad of composite materials [35, 48]. In a study, graphene was added in an aqueous paint (acrylic-latex-based) [27]. The cited authors concluded that addition of graphene demonstrated statistically significant improvements in many of the fire properties that were studied, including times to onset of ignition, and glowing as well (Fig. 8).

There is another type of fire retardants that expands when exposed to high temperatures, called intumescent coatings. These coatings create a barrier to oxygen to reach the inner wood layers, thereby preventing from further spread of fire. Different nanomaterials and nano-fillers were found to incorporate phosphorous, boron, and nitrogen compounds in intumescent coatings [13, 120]. A study on silicon dioxide ($SiO_2$ at nano-scale reported two significant improvements in the coating that was used; the first improvement was in the fire performance; and the second one was in reducing corrosion effects on metals. The nanomaterial was dispersed in ammonium polyphosphate-pentaerythriotol-melamine. Sepiolite was another mineral material that was reported to be effective in improving fire properties. Nano-sepiolite was dispersed in a paint (viz. an aqueous acrylic-latex-based paint and coated on fir wood specimens. The fir specimens were thermally-modified to compare the results with the unheated ones. Sepiolite demonstrated significant performance and improvement in different fire properties, including times to onset of ignition and glowing. Moreover, the time that the piloted fire penetrated into the wood specimens and burnt it through to reach the back side was also improved significantly (Fig. 9). Wollastonite and sepiolite have both proved to be promising to be used at industrial scales to improve different fire properties and retardancy when mixed with acrylic-latex paints. However, there is little or no study that elaborated on these two mineral materials when mixed with and added to intumescent paints and coatings. This can be the topic of research studies for young researchers for further exploration and investigation.

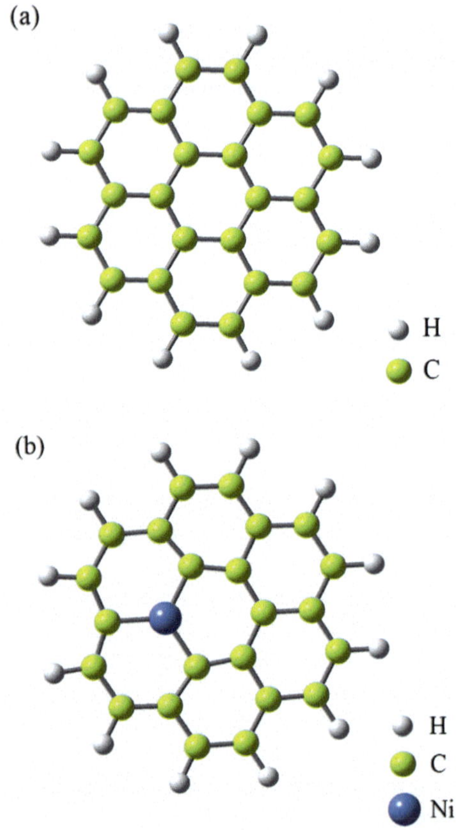


**Fig. 7** Atomic structure of graphene flakes (**a**: pure; and **b**: Ni-doped) [27]

## 5 Conclusion

Wood is a biological material that can either be used as it is found in nature or be engineered and used as a kind of composite material. In either case, they have some drawbacks that may limit their applications. These drawbacks can be categorized in biological susceptibility to some organisms (including different fungi, insects, termites, and marine borers), dimensional instability to water and vapor, fire and high temperature, and physical damages. Nanotechnology has tremendous potential in protecting wood against the above mentioned. Nanomaterials can be used as surface or in-depth protection. Though many promising studies have already been carried out, there are many boundaries that are remained unexplored for young researchers.

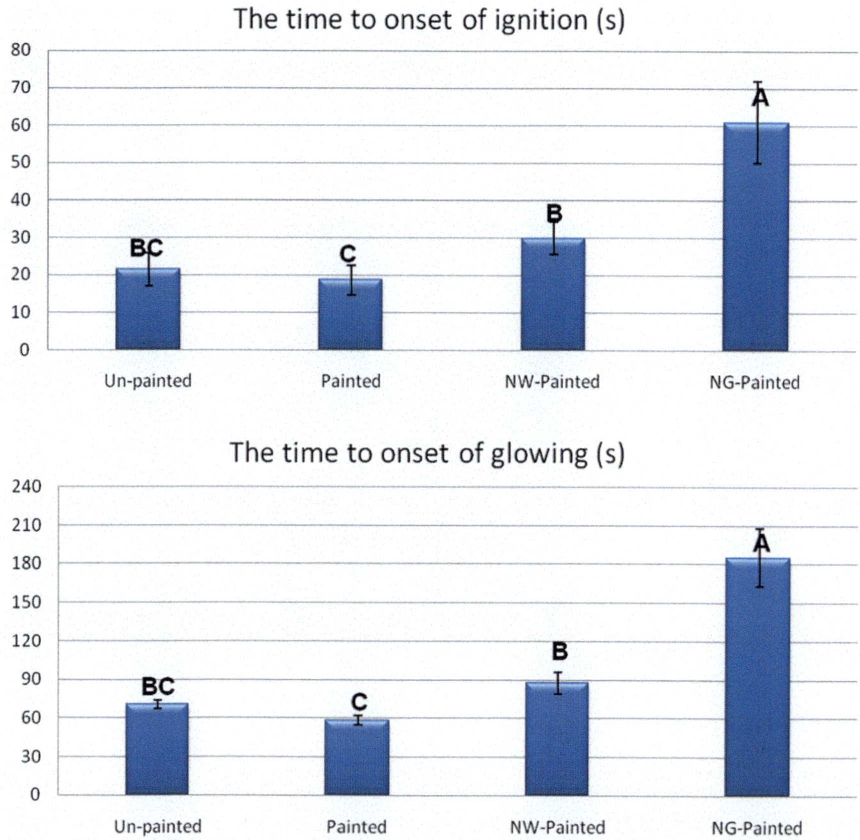


**Fig. 8** Time (in seconds) to onset of ignition and glowing, in the four treatments of beech specimens (NW = nanowollastonite; NG = nano-graphene) (Letters on each column represent Duncan groupings at 95% level of confidence) [27]

**Fig. 9** SEM image showing sepiolite nanostrands [111]

# References


1. Aitken E (2010) Analyses of the effect of silicon on Fusarium wilt on banana. Horticulture Australia, Sydney, NSW, Australia
2. Akhtari M, and Ganjipour M. (2013) Effect of nano-silver and nano-copper and nano- zinc oxide on Paulownia wood exposed to white-rot fungus—IRG/WP 13–30635. In: 44th annual meeting of the international research group on wood protection, Çesme, Turkey, pp 1–8
3. Akhtari M, Nicholas D (2013) Evaluation of particulate zinc and copper as wood preservatives for termite control. Eur J Wood Wood Prod 71(3):395–396. https://doi.org/10.1007/s00107-013-0690-7
4. Antov P, Savov V, Neykov N (2020a) Sustainable bio-based adhesives for eco-friendly wood composites—a review. Wood Res 65:51–62
5. Antov P, Mantanis GI, Savov V (2020b) Development of wood composites from recycled fibres bonded with magnesium lignosulfonate. Forests 11:613
6. Antov P, Jivkov, V, Savov V, Simeonova R, Yavorov N (2020c) Structural application of eco-friendly composites from recycled wood fibres bonded with magnesium lignosulfonate. Appl Sci 10:7526
7. Antov P, Savov V, Mantanis GI, Neykov N (2020d) Medium-density fibreboards bonded with phenol-formaldehyde resin and calcium lignosulfonate as an eco-friendly additive. Wood Mater Sci Eng 2020:1751279. https://doi.org/10.1080/17480272.2020.1751279
8. Antov P, Savov V, Krišťák Ľ, Réh R, Mantanis GI (2021a) Eco-friendly, high-density fiberboards bonded with urea-formaldehyde and ammonium lignosulfonate. Polymers 13:220. https://doi.org/10.3390/polym13020220
9. Antov P, Krišťák Ľ, Réh R, Savov V, Papadopoulos AN (2021b) Eco-friendly fiberboard panels from recycled fibres bonded with calcium ignosulfonate. Polymers 13(4):639. https://doi.org/10.3390/polym13040639

10. Aristri MA, Lubis MAR, Yadav SM, Antov P, Papadopoulos AN, Pizzi A, Fatriasari W, Ismayati M (2021) Iswanto AH Recent developments in lignin and tannin-based non-isocyanate polyurethane resins for wood adhesives—a review. Appl Sci 11:4242. https://doi.org/10.3390/app11094242
11. Auffan M, Masion A, Labille J, Diot MA, Liu W, Olivi L, Proux O, Ziarelli F, Chaurand P, Geantet C, Bottero JY, Rose J (2014) Long-term aging of a $CeO_2$ based nanocomposite used for wood protection. Environ Pollut 188(2014):1–7. https://doi.org/10.1016/j.envpol.2014.01.016
12. Ayrilmis N, Candan Z, White R (2007) Physical, mechanical, and fire properties of oriented strandboard with fire retardant treated veneers. Holz Roh Werkst 65:449–458
13. Aziz H, Ahmad F, Zia-ul-Mustafa M (2014) Nano filler reinforced intumescent fire retardant coating for protection of structural steel. In book: Proceedings of the international civil and infrastructure engineering conference 2014. Publisher: Springer, InCIEC, pp 831–844. ISBN: 978–981–287–289–0
14. Bak M, Yimmou BM, Csupor K, Németh R, Csóka L (2012) Enhancing the durability of wood against wood destroying fungi using nano-zinc. In: International scientific conference on sustainable development & ecological footprint, Sopron, Hungary, pp 1–6
15. Bal BC (2016) Some technological properties of laminated veneer lumber produced with fast-growing *Poplar* and *Eucalyptus*. Maderas-Cienc Tecnol 18:413–424
16. Banerjee I, Pangule RC, Kane RS (2010) Design of surfaces that prevent fouling by proteins, bacteria, and marine organisms. Adv Mat 23:690–718
17. Bari E, Taghiyari HR, Schmidt O, Ghorbani A, Aghababaei H (2015) Effects of nano-clay on biological resistance of wood-plastic composite against five wood-deteriorating fungi. Maderas-Cienc Tecnol 17(1):205–212
18. Bayatkashkoli A, Taghiyari HR, Kameshki B, Ravan S, Shamsian M (2016) Effects of zinc and copper salicylate on biological resistance of particleboard against *Anacanthotermes vegans* termite. Int Biodet Biodeg 115:26–30
19. Bekhta P, Noshchenko G, Réh R, Kristak L, Sedliačik J, Antov P, Mirski R, Savov V (2021) Properties of eco-friendly particleboards bonded with lignosulfonate-urea-formaldehyde adhesives and pMDI as a crosslinker. Materials 14:4875. https://doi.org/10.3390/ma14174875
20. Bhushan B (2017) Introduction to nanotechnology. In: Springer handbook of nanotechnology. Springer, pp 1–19
21. Borrega M, Karenlampi PP (2010) Hygroscopicity of heat-treated Norway spruce (*Picea abies*) wood. Eur J Wood Prod 68:233–235
22. Chapman J, Regan F (2011) Nanofunctionalized superhydrophobic antifouling coatings for environmental sensor applications. Advancing deployment with answers from nature. Adv Eng Mater 14:175-B184
23. Chouhan RS, Jerman I, Heath D, Bohm S, Gandhi S, Sadhu V, Baker S, Horvat M (2021) Emerging tri-s-triazine-based graphitic carbon nitride: A potential signal-transducing nanostructured material for sensor applications. Nano Select 2:712–743
24. Civardi C, Schlagenhauf L, Kaiser JP, Hirsch C, Mucchino C, Wichser A, Wick P, Schwarze FWMR (2016) Release of copper-amended particles from micronized copper-pressure-treated wood during mechanical abrasion. J Nanobiotechnology 14(77):1–10. https://doi.org/10.1186/s12951-016-0232-7
25. Clausen CA (2012) Enhancing durability of wood-based composites with nanotechnology. In: General Technical Report FPL–GTR–218, Forest Products Laboratory, Department of Agriculture, Madison, WI, USA. pp 1–3
26. Daly-Hassan H, Kasraoui M, Karra C (2014) Industrial timber production in Tunisia: Despite reforestation, dependence on imports is increasing. Bois For Trop 324:29–37
27. Esmailpour A, Majidi R, Taghiyari HR, Ganjkhani M, Mohseni Armaki SM, Papadopoulos AN (2020) Improving fire retardancy of beech wood by graphene. Polymers 12:303. https://doi.org/10.3390/polym12020303
28. Esteves B, Graca J, Pereira H (2008) Extractive composition and summative chemical analysis of thermally treated eucalypt wood. Holzforschung 62:344–351

29. Figueroa M, Bustos C, Dechent P, Reyes L, Cloutier A, Giuliano M (2012) Analysis of rheological and thermo-hygro-mechanical behaviour of stress-laminated timber bridge deck in variable environmental conditions Maderas. Cienc Tecnol 14:303–319
30. Findlay WPK (1951) The value of laboratory test on wood preservatives. (S1): Convention, British Wood Preserving Association
31. Freeman MH, McIntyre CR (2013) Micronized copper wood preservatives: Strong indications of the reservoir effect—IRG/WP 13–30609. 44th annual meeting of the international research group on wood preservation. Stockholm, Sweden, pp 1–16
32. Fu Q (2018) Wood nanotechnologies for transparency, fire retardancy and liquid separation. Doctoral Thesis supervised by Prof. Lars A. Berglund, Department of Fiber and Polymer Technology, KTH royal Institute of Technology, Stockholm, Sweden
33. Ganesh VA, Raut HK, Nair AS, Ramakrishna S (2011) A review on self-cleaning coatings. J Mater Chem 21:16304–16322
34. Geetha Devi P, SakthiVelu A (2016) Structural, optical and photoluminescence properties of copper and iron doped nanoparticles prepared by co-precipitation method,". J Mater Sci: Mater Electron 27(10):10833–10840. https://doi.org/10.1007/s10854-016-5190-1
35. Geim AK, Novoselov KS (2007) The rise of graphene. Nat Mater 6:183–191
36. Hill C (2006) Wood modification chemical, thermal and other processes. John Wiley & Sons, Ltd., p 239. ISBN: 0–470–02172–1
37. Hong X, Wen J, Xiong X, Hu Y (2016) Shape effect on the antibacterial activity of silver nanoparticles synthesized *via* a microwave-assisted method. Environ Sci Pollut Res 23(5):4489–4497. https://doi.org/10.1007/s11356-015-5668-z
38. Höglund M, Garemark J, Nero M, Willhammar T, Popov S, Berglund LA (2021) Facile processing of transparent wood nanocomposites with structural color from plasmonic nanoparticles. Chem Mater 33:3736–3745
39. Hubbe MA, Rojas OJ, Lucia, LA (2015) Green modification of surface characteristics of cellulosic materials at the molecular or nano scale: A review. BioResources 10(3):6095–6206. https://doi.org/10.15376/biores.10.3.hubbe
40. Huuskonen MS, Jarvisalo J, Koskinen H, Nickels J, Rasanen J, Asp S (1983) Preliminary results from a cohort of workers exposed to wollastonite in a Finnish limestone quarry. Scand J Work Env Hea 9:169–175
41. Huuskonen MS, Tossavainen A, Koskinen H, Zitting A, Korhonen O, Nickels J, Korhonen K, Vaaranen V (1983) Wollastonite exposure and lung fibrosis. Env Res 30:291–304
42. Ilies DC, Herman GV, Safarov B, Ilies A, Blaga L, Caciora T, Peres AC, Grama V, Bambang SW, Brou T, Taglioni F, Hassan TH, Hossain MA (2023) Indoor air quality perception in built cultural heritage in times of climate change. Sustainability 15:8284. https://doi.org/10.3390/su15108284
43. Jeevanandam J, Chan YS, Danquah MK (2016) Biosynthesis of metal and metal oxide nanoparticles. ChemBioEng Reviews 3(2):55–67. https://doi.org/10.1002/cben.201500018
44. Jiang W, Lyu G, Wu S, Lucia LA, Yang G, Liu Y (2016) Supercritical water-induced lignin decomposition reactions: A structural and quantitative study. BioResources 11:5660–5675
45. Jin L, Walcheski P, Preston A (2008) Laboratory studies on copper availability in wood treated with soluble amine copper and micronized copper systems—IRG/WP 08–30489. In: 39th annual meeting of the international research group on wood protection, Guanacaste, Costa Rica, pp 1–10
46. Karimi A, Taghiyari HR, Fattahi A, Karimi S, Ebrahimi GH, Tarmian A (2013) Effects of wollastonite nanofibers on biological durability of poplar wood (*Populus nigra*) against *Trametes versicolor*. BioResources 8:4134–4141
47. Kartal SN, Terzi E, Woodward B, Clausen CA, Lebow ST (2014) Chemical remediation of wood treated with micronised, nano or soluble copper preservatives. Holzforschung 68(7):831–837. https://doi.org/10.1515/hf-2013-0212
48. Katsnelson M, Iosifovich, M (2012) Graphene: Carbon in two dimensions, 1 edn. Cambridge University Press, Cambridge, NewYork

49. Kawalerczyk J, Dziurka D, Mirski R, Siuda J, Babicka M (2021) Possibility of use of NCC-reinforced melamine-urea-formaldehyde adhesive in plywood manufacturing. Drvna Ind 72 (3)
50. Khoee S, Hashemi A, Molavipordanjani S (2018) Synthesis and characterization of IUdR loaded PEG/PCL/PEG polymersome in mixed DCM/DMF solvent: Experimental and molecular dynamics insights into the role of solvent composition and star architecture in drug dispersion and diffusion. Eur J Pharm Sci 114:1–12. https://doi.org/10.1016/j.ejps.2017.11.019
51. Kumar SA, Sasikumar A (2010) Studies on novel silicone/phosphorus/sulphur containing nano-hybrid epoxy anticorrosive and antifouling coatings. Org Coat 68:189–200
52. Laks PE, Richter DL, Larkin GL (2000) Biological deterioration of wood-based composite panels. Wood Design Focus 11:7–14
53. Lekha MA, Motaung TE, Adnrew JE, Sitohle BB, Gibril M (2016) Effect of mechanical treatment on properties of cellulose nanofibrils produced from bleached hardwood and softwood pulps. Maderas-Cienc Tecnol 18:457–466
54. Li D (2012) Nanostructuring materials towards conventionally unachievable combination of desired properties. J Nanomater & Mol Nanotechnol 1(1). https://doi.org/10.4172/12324-8777.1000e102
55. Li Y, Fu Q, Yu S, Yan M, Berglund L (2016) Optically transparent wood from a nanoporous cellulosic template: Combining functional and structural performance. Biomacromol 17:1358–1364
56. Li Y, Fu Q, Rojas R, Yan M, Lawoko M, Berglund LA (2017) Lignin-retaining transparent wood. ChemSusChem, Chem-Sustain-Energy-Mater 10(17):3445–3451
57. Lidén G (2011) The European commission tries to define nanomaterials. Ann Occup Hyg 55:1–5
58. Lykidis C, Mantanis G, Adamopoulos S, Kalafata K, Arabatzis I (2013) Effects of nano-sized zinc oxide and zinc borate impregnation on brown-rot resistance of black pine (*Pinus nigra* L.) wood. Wood Mat Sci Eng 8:242–244
59. Lykidis C, Bak M, Mantanis G, Németh R (2016) Biological resistance of pine wood treated with nano-sized zinc oxide and zinc borate against brown-rot fungi. Eur J Wood Prod 74:909–911
60. Lykidis C, de Troya T, Conde M, Galvan J, Mantanis G (2018) Termite resistance of beech wood treated with zinc oxide and zinc borate nanocompounds. Wood Mat Sci Eng 13(1):45–49
61. Madyaratri EW, Ridho MR, Aristri MA, Lubis MAR, Iswanto AH, Nawawi DS, Antov P, Kristak L, Majlingova A, Fatriasari W (2022) Recent advances in the development of fire-resistant biocomposites—a review. Polymers 14(3):362
62. Majidi R, Taghiyari HR, Abdolmaleki D (2019) Molecular dynamics simulation evaluating the hydrophilicity of nanowollastonite on cellulose. J Struct Chem 6:1562–1569
63. Malviya KD, Chattopadhyay K (2015) High quality oxide-free metallic nanoparticles: A strategy for synthesis through laser ablation in aqueous medium. J Mater Sci 50(2):980–989. https://doi.org/10.1007/s10853-014-8658-5
64. Mankowski M, Morrell JJ (2000) Patterns of fungal attack in wood-plastic composites following exposure in a soil block test. Wood Fiber Sci 32:340–345
65. Mantanis G, Terzi E, Kartal SN, Papadopoulos A (2014) Evaluation of mold, decay and termite resistance of pine wood treated with zinc- and copper- based nanocompounds. Int Biodet Biodeg 90:140–144
66. Mantanis G, Athanassiadou E, Barbu M, Wijnendaele K (2018) Adhesive systems used in the European particleboard, MDF and OSB industries. Wood Mater Sci Eng 13:104–116
67. Mantanis G, Martinka J, Lykidis C, Ševčík L (2019) Technological properties and fire performance of medium density fibreboard (MDF) treated with selected polyphosphate-based fire retardants. Wood Mater Sci Eng; doi.org/https://doi.org/10.1080/17480272.2019.1596159.
68. Marcu F, Ilieş DC, Wendt AJ, Liliana Indrie, Ilies A, Burta L, Caciora T, Herman GV, Todoran A, Baias S, Albu A, Gozner M (2020). Investigations regarding the biodegradation of cultural heritage. Case study of traditional embroidered peasant shirt (Maramures, Romania), Rom Biotechnol Lett 2/2020, 25(2):1362–1368. https://doi.org/10.25083/rbl/25.2/1362.1368

69. Martínez Rivas CJ, Tarhini M, Badri W, Miladi K, Greige-Gerges H, Nazari QA, Galindo Rodríguez SA, Román RÁ, Fessi H, Elaissari A (2017) Nanoprecipitation process: From encapsulation to drug delivery. Int J Pharm 532(1):66–81. https://doi.org/10.1016/j.ijpharm.2017.08.064
70. Mattos BD, Tardy BL, Magalhães WLE, Rojas O (2017) Controlled release for crop and wood protection: Recent progress toward sustainable and safe nanostructured biocidal systems. J Control Release 262:139–150
71. Milton FT (1995) The preservation of wood, a self study manual for wood treaters. Minnesota Extension Service, University of Minnesota, College of Natural Resources, BU-6413-S, Revised 1995, p 103
72. Mora-Huertas CE, Fessi H, Elaissari A (2010) Polymer-based nanocapsules for drug delivery. Int J Pharm 385(1–2):113–142. https://doi.org/10.1016/j.ijpharm.2009.10.018
73. Morrell JJ, Stark NM, Pendleton DE, McDonald AG (2006) Durability of wood-plastic composites. Wood Des Focus 16:7–10
74. Nair S, Pandey KK, Giridhar BN, Vijayalakshmi G (2017) Decay resistance of rubberwood (*Hevea brasiliensis*) impregnated with ZnO and CuO nanoparticles dispersed in propylene glycol. Int Biodeterior Biodegradation 122:100–106. https://doi.org/10.1016/j.ibiod.2017.05.008
75. Nosáľ E, Reinprecht L (2017) Anti-bacterial and anti-mold efficiency of ZnO nanoparticles present in melamine-laminated surfaces of particleboards. BioResources 12(4):7255–7267. https://doi.org/10.15376/biores.12.4.7255-7267
76. Ogawa T Ding B, Sone Y, Shiratori S (2007) Super-hydrophobic surfaces of layer-by-layer structures film-coated electrospun nanofibrous membranes. Nanotechnol 18:165607. http://iopscience.iop.org/0957-4484/18/16/165607
77. Palanti S, Feci E, Predieri G, Francesca V (2012) Copper complexes grafted to amino-functionalized silica gel as wood preservatives against fungal decay: Mini-blocks and standard test. BioResources 7:5611–5621
78. Papadopoulos AN, Hill CAS (2002) The biological effectiveness of wood modified with linear chain carboxylic acid anhydrides against *Coniophora puteana*. Holz als Roh-und Werkstoff 60(5):329–332
79. Papadopoulos E (2009) Adhesives from renewable resources for binding wood-based panels. J Environ Prot Ecol 10:1128–1136
80. Papadopoulos AN, Foti D, Kyzas GZ (2020) Sorption behavior of water vapor of wood treated by chitosan polymer. Eur J Wood Wood Prod 78(3):483–491
81. Papadopoulos AN (2021) Advances in wood composites III. Polymers 13:163
82. Pendleton DE, Hoffard TA, Adcock T, Woodward B, Wolcott MP (2002) Durability of an extruded HDPE/wood composite. Forest Prod J 52:21–27
83. Peymanirad F, Majidi R, Vishkayi SI, Soleimani HR (2024) Exploring the potential of grapheme oxide frameworks as anode materials for Na-ion batteries applications: A density functional theory study. Appl Phys A Mater Sci Process 130(5):318
84. Pizzi A, Papadopoulos AN, Policardi F (2020) Wood composites and their polymer binders. Polymers 12:1115
85. Poletti Papi MA, Caetano FR, Bergamini MF, Marcolino-Junior LH (2017) Facile synthesis of a silver nanoparticles/polypyrrole nanocomposite for non-enzymatic glucose determination. Mater Sci Eng, C 75:88–94. https://doi.org/10.1016/j.msec.2017.02.026
86. Poole CP Jr, Owens FJ (2003) Introduction to nanotechnology. John Wiley & Sons
87. Réh R, Igaz R, Krišt'ák Ľ, Ružiak I, Gajtanska M, Božíková M, Kučerka M (2019) Functionality of beech bark in adhesive mixtures used in plywood and its effect on the stability associated with material systems. Materials 12:1298
88. Réh R, Kristak L, Sedliacik J, Bekhta P, Bozikova M, Kunecova D, Vozarova V, Tudor EM, Antov P, Savov V (2021) Utilization of birch bark as an eco-friendly filler in urea-formaldehyde adhesives for plywood manufacturing. Polymers 13(4):511
89. Rowell RM (2012) Handbook of wood chemistry and wood composites

90. Salca EA, Krystofiak T, Lis B, Hiziroglu S (2021) Glossiness evaluation of coated wood surfaces as function of varnish type and exposure to different conditions. Coatings 11:558. https://doi.org/10.3390/coatings11050558
91. Salla J, Pandey KK, Srinivas K (2012) Improvement of UV resistance of wood surfaces by using ZnO nanoparticles. Polym Degrad Stab 97:592–596. https://doi.org/10.1016/j.polymdegradstab.2012.01.013
92. Sandak A, Sandak J (2021) Wood modification: Characterization, modeling and applications. Coatings 11:869. https://doi.org/10.3390/coatings11070869
93. Sandberg D (2016) Additives in wood products—Today and future development. Wood Sci Eng. https://doi.org/10.1007/978-981-10-0655-5_4
94. Sandberg D, Kutnar A, Mantanis G (2017) Wood modification technologies—a review. Iforest-Biogeosciences For 10(6):895–908. https://doi.org/10.3832/ifor2380-010
95. Savov V, Valchev I, Antov P (2019) Processing factors for production of eco-friendly medium density fibreboards based on lignosulfonate adhesives. In: Proceedings of the 2nd International Congress of Biorefinery of Lignocellulosic Materials (IWBLCM 2019), ISBN: 978–84–940063–7–1, 4–7 June 2019, Córdoba, Spain, pp 165–169
96. Savov V, Antov P (2020) Engineering the properties of eco-friendly medium density fiberboard bonded with lignosulfonate adhesive. Drvna Ind 71(2):157–162
97. Schmidt O (2006) Wood and tree fungi: Biology, damage, protection, and use. Springer-Verlag Berlin Heidelberg, p 334
98. Schmidt O (2007) Indoor wood-decay basidiomycetes: damage, causal fungi, physiology, identification and characterization, prevention and control. Mycol Prog 6:261–279
99. Schmidt O, Gaiser O, Dujesiefken D (2012) Molecular identification of decay fungi in the wood of urban trees. Eur J Forest Res 131:885–891
100. Seaton A, Tran L, Aitken R, Donaldson K (2010) Nanoparticles, human health hazard and regulation. J R Soc Interface 7(Suppl 1):119–129. https://doi.org/10.1098/rsif.2009.0252.focus
101. Souza RGD, Dan ML, Dias-Guimaraes MA, Guimaraes LAOP, Braga JMA (2018) Fruits of the Brazilian Atlantic Forest: allying biodiversity conservation and food security. An Acad Bras Cienc 90:3583–3595
102. Spear MJ, Curling SF, Dimitriou A, Ormondroyd GA (2021) Review of functional treatments for modified wood. Coatings 11:327. https://doi.org/10.3390/coatings11030327
103. Taghiyari HR (2012) Fire retarding properties of nano-silver in solid woods. Springer: Wood Sci Technol 46(5):939–952. https://doi.org/10.1007/s00226-011-0455-6
104. Taghiyari HR, Moradi Malek B, Ghorbani Kookandeh M, Farajpour Bibalan O (2014) Effects of silver and copper nanoparticles in particleboard to control *Trametes versicolor* fungus. Int Biodet Biodeg 94:69–72
105. Taghiyari HR, Bari E, Schmidt O, Tajick Ghanbary MA, Karimi A, Tahir PMD (2014b) Effects of nanowollastonite on biological resistance of particleboard made from wood chips and chicken-feather against *Antrodia vaillantii*. Int Biodet Biodeg 90:93–98
106. Taghiyari HR, Bari E, Schmidt O (2014c) Effects of nanowollastonite on biological resistance of medium-density fiberboard against *Antrodia vaillantii*. Eur J Wood Prod 72:399–406
107. Taghiyari HR, Majidi R, Jahangiri A (2016) Adsorption of nano-wollastonite on cellulose surface: Effects on physical and mechanical properties of medium-density fiberboard (MDF). Cerne 22:215–222
108. Taghiyari HR, Bari E, Sistani A, Najafian M, Tajick Ghanbary MA, Ohno KM (2020a) Biological resistance of nanoclay-treated plastic composites with different bamboo contents to three types of fungi. J Thermoplast Compos Mater 33(8): 1048–1060. https://doi.org/10.1177/0892705718817339
109. Taghiyari HR, Esmailpour A, Majidi R, Morrell JJ, Mallaki M, Militz H, Papadopoulos AN (2020b) Potential use of wollastonite as a filler in UF resin based medium-density fiberboard (MDF). Polymers 12:1435. https://doi.org/10.3390/polym12071435
110. Taghiyari HR, Esmailpour A, Adamopoulos S, Zereshki K, Hosseinpourpia R (2020) Shear strength of heat-treated solid wood bonded with polyvinyl-acetate reinforced by nanowollastonite. Wood Res 65(2):183–194

111. Taghiyari HR, Soltani A, Esmailpour A, Hassani V, Gholipour H, Papadopoulos AN (2020d) Improving thermal conductivity coefficient in oriented strand lumber (OSL) using sepiolite. Nanomaterials 10, 599; https://doi.org/10.3390/nano10040599.
112. Tajvidi M, Gardner DJ, Bousfield DW (2016) Cellulose nanomaterials as binders: laminate and particulate systems. J Renewable Mat. https://doi.org/10.7569/JRM.2016.634103
113. Taniguchi N, Arakawa C, Kobayashi T (1974) On the basic concept of nanotechnology. In: Proceedings of the international conference on production engineering, Tokyo, Japan
114. Thandavan TMK, Gani SMA, San Wong C, Md. Nor R, (2015) Enhanced photoluminescence and raman properties of Al-doped ZnO nanostructures prepared using thermal chemical vapor deposition of methanol assisted with heated brass. PLoS ONE 10(5):1–18. https://doi.org/10.1371/journal.pone.0121756
115. Tjeerdsma BF, Boonstra M, Pizzi A, Tekely P, Militz H (1998) Characterization of thermal modified wood: Molecular reasons for wood performance improvement. CPMAS 13 CNMR characterization of thermally modified wood. Holz Roh und Werkstoff 56:149–153
116. Tjeerdsma BF, Militz H (2005) Chemical changes in hydrothermal treated wood: FTIR analysis of combined hydrothermal and dry heat-treated wood. Holz Roh Werkst 63:102–111
117. Uddin I, Poddar P, Kumar U, Phogat N (2013) A novel microbial bio-milling technique for the size reduction of micron sized Gd2O3 particles into nanosized particles. J Green Sci Technol 1:48–53
118. Vay O, De Borst K, Hansmann C, Teischinger A, Muller U (2015) Thermal conductivity of wood at angles to the principal anatomical directions. Wood Sci Technol 49:577–589
119. Verhaegen D, Randrianjafy H, Rakotondraoelina Andriatsitohaina H, Trendelenburg Rakotonirina MC, Andriamampianina N, Montagne P, Rasamindisa A, Chaix G, Bouillet JP, Bouvet JM (2014) *Eucalyptus robusta* for sustainable fuelwood production in Madagascar: Review of knowledge and future prospects. Bois Foret Trop 320:15–30
120. Wang Z, Enhou H, Liu F, Ke W (2009) Fire and corrosion resistances of intumescent nano-coating containing nano-$SiO_2$ in salt spray condition. J Mat Sci Technol 26:75–81
121. Wang Y, Runnerstrom EL, Milliron DJ (2016) Switchable materials for smart windows. Annu Rev Chem Biomol 7:283–304
122. Wang Z, Yang X, Sun B, Chai Y, Liu J, Cao J (2016) Effect of vacuum heat treatment on the chemical composition of larch wood. BioResources 11:5743–5750
123. White RH, Dietenberger MA (1999) Wood handbook, chapter 17: Fire safety; forest products laboratory, Gen. Tech. Rep. FPL—GTR—13, Madison, WI: U.S. Department of Agriculture, Forest Service, Forest Products Laboratory, p 463
124. Winandy JE, Lebow PK, Murphy JF (2002) Predicting current serviceability and residual service life of plywood roof sheathing using kinetics-based models. In: The 9th durability of building materials and components conference, Brisbane, Australia, p 7
125. Wulf M, Wehling A, Reis O (2002) Coatings with self-cleaning properties. Macromol Symposia 187:459–468
126. Zhang X-T, Sato O, Taguchi M, Einaga Y, Murakami T, Fujishima A (2005) Self-cleaning particle coating with antireflection properties. Chem Mater 17:696–700
127. Zhu J, Zhang H (2014) Self-cleaning coatings. Final Report, Department of Chemical and Biochemical Engineering, DRDC-RDDC-2014-C87. The University of Western Ontario, London, Ontario, Canada N6A 5B9
128. Zhu M, Song J, Li T, Gong A, Wang Y, Dai J, Yao Y, Wei L, Henderson D, Hu L (2016) Highly anisotropic, highly transparent wood composites. Adv Mater. https://doi.org/10.1002/adma.201600427
129. Zuccheri T, Colonna M, Stefanini I, Santini C, Gioia DD (2013) Bactericidal activity of aqueous acrylic paint dispersion for wooden substrates based on $TiO_2$ nanoparticles activated by fluorescent light. Materials 6(2013):3270–3283. https://doi.org/10.3390/ma6083270

# Life-Cycle Assessment of Nanoparticle in Wood Adhesive and Coating: The State-of-The Art

**Ria Aniza, Anelie Petrissans, and Mathieu Petrissans**

**Abstract** The life cycle assessment (LCA) of wood augmented with nanoparticles looks into the entire range of health and environmental effects related to this cutting-edge substance. By adding nanoparticles to wood, its mechanical qualities—such as strength, durability, and resistance to rot and pests—are greatly improved. This can increase the product's lifespan and decrease the need for replacements. By reducing the need for chemical treatments and preserving resources, these advancements may have a positive impact on the environment. However, there are worries over the environmental impact of these methods because producing nanoparticles requires a lot of energy and may release harmful compounds. Furthermore, the improved performance of the wood during the product's use phase may help to reduce total environmental consequences. On the other hand, there are significant obstacles in the disposal process of wood that has been infused with nanoparticles, especially when it comes to the release of nanoparticles into the environment during decomposition or burning. Because of the threats these particles bring to human health and ecosystems, proper management and disposal procedures are necessary. The Life Cycle Assessment (LCA) highlights that although the incorporation of nanoparticles into wood products presents encouraging advantages, it is imperative to implement sustainable production methods, carry out thorough evaluations of health and environmental risks, and create efficient regulatory structures. In order to balance innovation with sustainability and guarantee that technological improvements do not jeopardize ecological integrity or public health, a comprehensive approach is necessary.

R. Aniza (✉) · A. Petrissans · M. Petrissans
Université de Lorraine, INRAE, LERMAB, F88000, Epinal, France
e-mail: ria.aniza@univ-lorraine.fr

R. Aniza
International Doctoral Degree Program in Energy Engineering, National Cheng Kung University, 701, Tainan, Taiwan

Department of Environmental Engineering, National Cheng Kung University, 701, Tainan, Taiwan

Department of Environmental Engineering, Institut Teknologi Sepuluh Nopember, East Java 60111, Surabaya, Indonesia



P. Antov et al. (eds.), *Nanomaterials Additives in Bioadhesives for Wood Composites*, Smart Nanomaterials Technology, https://doi.org/10.1007/978-981-96-3400-2_10

**Keywords** LCA · Nanoparticle wood composite · Functional unit · System boundaries · Impact categories

## 1 Introduction

The proper disposal of waste produced prior to, during, and after manufacturing has become one of the major issues facing many industry sectors in recent times, and the reconstituted wood panel business is no exception. As an alternative to solid wood, wood panels are goods created from wood in various forms, such as veneers, battens, particles, and fibers that are subjected to pressure and temperature in a variety of stages [24, 43]. In the wood preservative stages, adhesives and coatings are then applied to the wood to create new products [44, 45]. With a focus on the furniture and civil construction sectors, wood-based panels are finding more and more uses across other industrial sectors. The development of a material with a more sustainable manufacturing that satisfies the demands of the industrial segment while maintaining durability and quality in comparison to other goods is necessary as a result of this increase.

Particles as small as 1–100 nm (nm) are called nanoparticles. Due to their high surface area-to-volume ratio and quantum effects, these particles' distinct physical and chemical properties set them apart from bulk materials [28]. The typical characteristics and properties of nanoparticles are related to size and surface area, quantum effects, and diverse compositions. Since nanoparticles are small compared to their volume, they have a huge surface area, which increases their reactivity and ability to interact with other substances. Quantum mechanical effects become substantial at the nanoscale, impacting a material's optical, electrical, and magnetic properties [8, 19]. Furthermore, a wide range of materials can be included in nanoparticle compositions, including metals (like gold and silver), metal oxides (like zinc oxide and titanium dioxide), semiconductors (like quantum dots), polymers, and lipids [28, 34]. These advantageous qualities have been used to enhance the quality of wood products.

In the present day, nanoparticles reach their utmost potential in a wide range of applications in most industrial sectors such as medical treatment (drug delivery systems, cancer treatment, imaging, and diagnostic) [29, 41], electronics (semiconductors devices) [7, 36], energy (energy batteries in solar cell and supercapacitors) [7, 16], environment (water and wastewater treatments as well as environment sensing) [22], pharmaceuticals (cosmetics, and material science (advanced textiles, composites, coatings) [6, 37]. Even with their enormous potential, nanoparticles present some challenges such as scalability, regulation-standardization, and toxicity as well as environmental impact.

The global market is becoming more and more competitive, making it vital to look for systems that provide everything a customer could possibly want while still being socially and environmentally conscious. By providing enhanced features like better durability [12], increased resistance to UV radiation [39], and antibacterial effects [34], the incorporation of nanoparticles into wood adhesives and coatings marks a significant achievement in materials research. These improvements are especially helpful in prolonging life and preserving the structural integrity and aesthetic appeal of wood products, which are extensively utilized in furniture, building, and other applications [25, 40]. Even though the practical advantages of wood adhesives and coatings boosted by nanoparticles are widely established, it is crucial to thoroughly evaluate the environmental effects of these materials over the course of their whole lifecycle.

An all-encompassing analysis of a product's environmental effects, from raw material extraction and production to use and disposal, is provided by a life cycle assessment (LCA) [5, 38]. Regarding wood adhesives and coatings improved by nanoparticles, life-cycle assessment (LCA) entails examining every phase of the product's existence to detect possible ecological drawbacks and sustainability issues. This entails investigating the extraction and synthesis of nanoparticles, the creation and use of adhesives and coatings, how well they function during use, and how to recycle or dispose of them after their useful lives. LCA comprises four stages: (1) purpose and scope; (2) functional unit and system boundaries; (3) impact categories; and (4) assumptions.

The current study's goal is to use life cycle assessment (LCA) to direct the development and application of circular business models (CBM) in the wood panel industry in order to produce cleaner products. Additionally, LCA offers a thorough assessment of the environmental effects of wood-based nanoparticles over the course of their whole lifecycle, from extraction and manufacture of raw materials to use and disposal or recycling. This aids in locating and estimating possible environmental advantages and disadvantages.

## 2 Bibliographic Portfolio

In order to construct the bibliographic portfolio, a list of keywords pertaining to the study's themes was established. The systematic literature review was conducted using the preferred reporting items for systematic reviews and meta-analysis (PRISMA) method, as shown in Fig. 1. Preferred Reporting Items for Systematic Reviews and Meta-Analyses, or PRISMA for short, is a popular methodology that aims to raise the caliber and transparency of reporting in these types of studies. It gives researchers a methodical way to make sure their reviews are thorough, organized, and transparent. Additionally, the Web of Science (WoS) is utilized to mine data regarding the bibliographic portfolio of nanoparticle application in wood research and industry.

The finding of the supporting material for the LCA of nanoparticle application for wood is defined in two ways including nanoparticles for wood adhesive and wood coating. The result of the keywords "Nanoparticle" AND "Wood" shows there are about 879 published documents. The search is further continued to add the adhesive into "Nanoparticle" AND "Wood" and "Adhesive". The results show for wood adhesive about 31 published documents. Additionally, the search moves to the wood coating, "Nanoparticle" AND "Wood" AND "Coating". The results show about 156

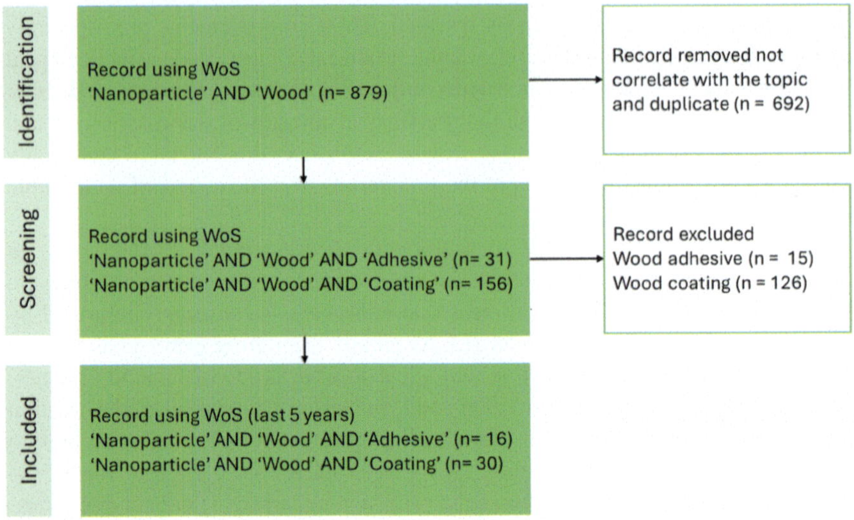


**Fig. 1** Bibliographic portfolio of nanoparticle application in wood adhesive and wood coating

published documents related to nanoparticle wood coating. In order to provide up-to-date information, the study focuses on a summary of the assessment of the recent technology and development of nanoparticle applications in wood for the past 5 years (2019–2024). Accordingly, the study related to nanoparticle wood adhesive and wood coating has identified about 16 and 60 published documents, respectively.

## 3 Purpose and Scope

This life cycle assessment's goal is to compare the environmental effects of traditional wood manufacture with creative uses of nanoparticles in wood. This analysis's objective is to assist the product development process in selecting a new product configuration and related procedures by evaluating each one's performance from an environmental perspective. The typical wood manufacture from raw or virgin wood of the forestry system into wood products in the timber industry is illustrated in the flow diagram in Fig. 2. The processes include two main parts which may take the matter from the forestry subsystem or timber industry subsystem. The major objective of this discussion is to identify the environmental impact of traditional and nanoparticle wood adhesive and coating in the wood products produced by the timber industry system production.

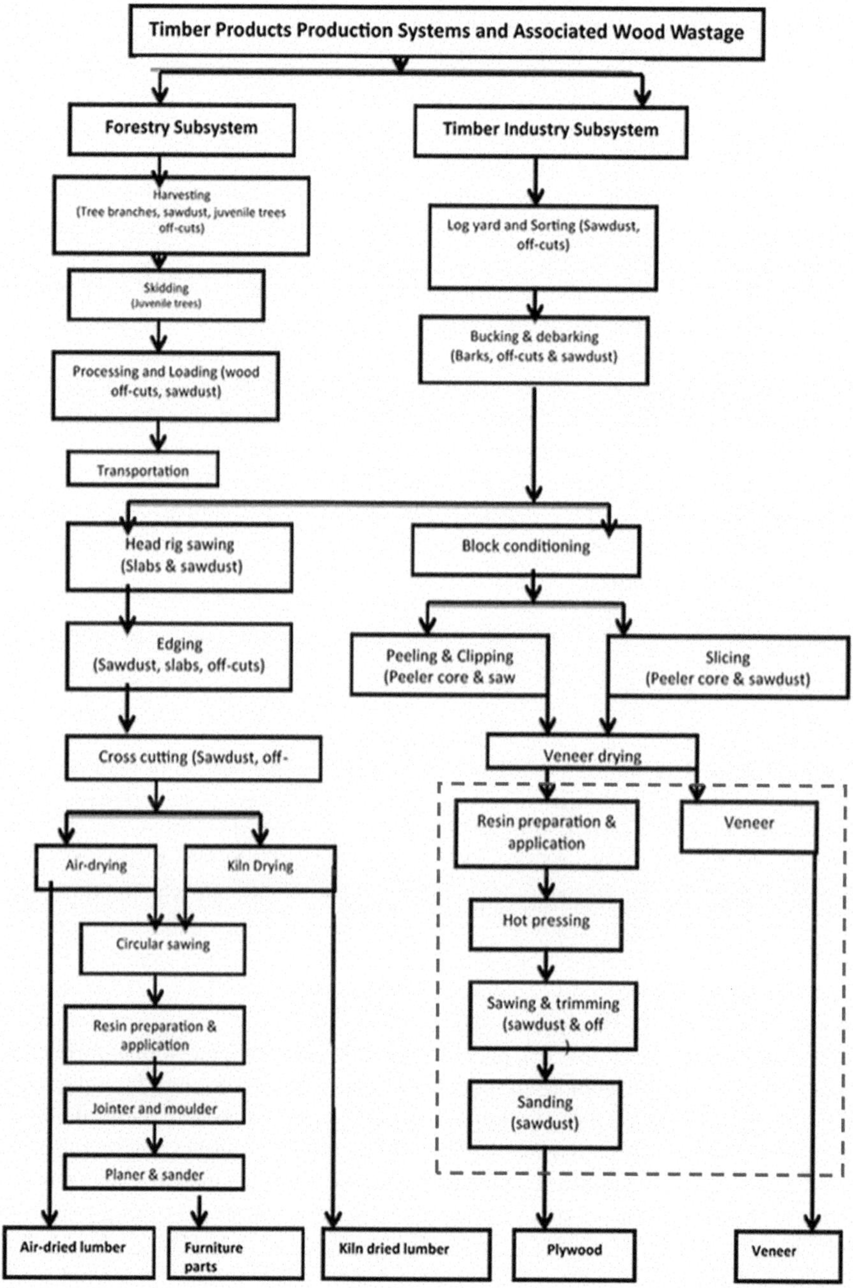


**Fig. 2** Production system, which is divided into the forestry and timber industry subsystems [1]

# 4 Functional Unit and System Boundaries

Among all processes, wood adhesive and coating are identified in the manufacturing process. The manufacturing process may involve sawing, drying, machining, jointing, gluing, and coating, which can be correlated to health, safety, and environmental concerns. In this regard, wood adhesive and wood coating are identified in the gluing and coating processes of wood manufacture which involve several chemical substances.

Several kinds of chemicals are employed in the production of wood, particularly in the coating, adhesive application, and preservation processes. Notwithstanding the fact that these chemicals have extended the life of timber products, their poisonous contents may have adverse effects on the environment. Another significant environmental issue is the disposal of wood from demolition sites that still contain high quantities of preservatives. Numerous nations have implemented regulations that forbid the application of hazardous chemicals. This discussion focuses on the LCA study with the cradle-to-grave system boundaries. This method chosen provides a better understanding of how the material is being harvested or collected, treated, and delivered to the consumers.

## *4.1 Wood Adhesive*

In the process of combining materials, an adhesive is positioned in between two adhering surfaces and solidifies to form a bond. This is known as adhesive bonding. Adhesives can offer a variety of flexibility and create bindings with extremely high strength and durability. Generally, the outcomes are highly reliable. Aesthetics, environmental and heat resistance requirements, adhesive consistency with the substrate, and needs for flexibility and stiffness must all be taken into account when choosing an adhesive. There are numerous kinds of adhesives available such as epoxies, cyanoacrylates, silicones, polyurethanes, phenolics, polyester, vinyl, and UV-curable adhesives [20, 30]. The traditional adhesives for wood usage are petroleum-based such as urea–formaldehyde (UF), phenol–formaldehyde (PF), melamine–formaldehyde (MF), phenol-resorcinol–formaldehyde (PRF), and resorcinol–formaldehyde (RF) [21, 43]. However, these traditional adhesives also come with harmful chemicals for the environment and human health. For instance, certain curative agents, like cycloaliphatic and aliphatic amines, may irritate or harm the liver, lungs, eyes, and skin. As a result, the usage of ecologically friendly adhesives is becoming more popular.

The ecologically friendly wood adhesive material is biologically derived from animals or plants. In this regard, the adhesive is commonly well-known as bio-adhesive. According to the data from Research and Markets, the demand for bio-based adhesive compound annual growth rate (CAGR) is projected to significantly increase by about 4.85% in 2027 [14]. The usage of green adhesive is required in a variety of applications due to rules and standards imposed by regulatory authorities like the US Green Building Council, California Air Resources Board (CARB), and the European Union. These regulations and standards include Registration, Evaluation, Authorization, and Restriction of Chemicals (REACH); Classification, Labelling, and Packaging (CLP); and the circular economy initiative. These rules fall into the categories of product category and application. The segmentation of the feedstock for the bio-adhesive is classified into 5 categories (Fig. 3) including sugar, glycerol, oil and fat, biogas, and others. According to the report bio-based adhesive market—industry outlook and forecast (2022–2027) by Research and Markets, the sugar-based feedstock (includes starch, cane, and beet) is expected to be the largest feedstock demand which will hold the highest global market share—forecasted about $3.87 billion by 2027. The second largest feedstock demand is projected to be glycerol bio-based adhesive, with a revenue share of about 28.83%.

The application of the adhesive has been classified into 6 major applications, including paper and packaging, construction, woodworking, personal care, medical equipment, and others [5, 13]. Among all the applications, paper and packaging holds the biggest share in the industry recorded in 2021. These phenomena occur due to the significant increase of the demand for ready-to-eat food, bonding pouches, bags, boards, permeable surface materials, and so on [20, 33]. According to the data, in 2021 North America (41%) has been recorded as the largest consumer of bio-adhesives products followed by Europe (27%).

## 4.2 *Wood Coating*

Wood products are typically shielded by wood coatings against environmental factors like moisture radiation, abrasion, chemical exposure, and biological degradation. On the other hand, they may release volatile organic compounds (VOCs) and contain liquid that is either composed of water or an organic solvent [19, 23]. VOCs, such as those containing chlorofluorocarbon, are regarded as a serious environmental issue from the standpoint of air pollution and public health and safety. A coating material for wood is a substance that can be liquid, paste, or powder that, when applied to a substrate, creates a layer with decorative, protective, or other particular qualities. The two main categories of synthetic resin chemistry are step growth (also called condensation), in which a tiny molecule like water is usually removed, and chain growth (sometimes called addition), in which repeating units are added to a developing chain.

The naturally derived film formers include drying oils and modified drying oils, natural resins and modified natural resins, and cellulose derivatives; the synthetically produced film formers are alkyds, polyurethanes, amino resins, polyester, acrylics, vinyls, and epoxies. Due to their opaque nature, many coatings need pigments for coloring and opacification, which in turn need wetting and stabilizing chemicals. Additives are materials that are incorporated into coating materials in small amounts (usually less than 1 percent) in order to alter or enhance one or more of their qualities.

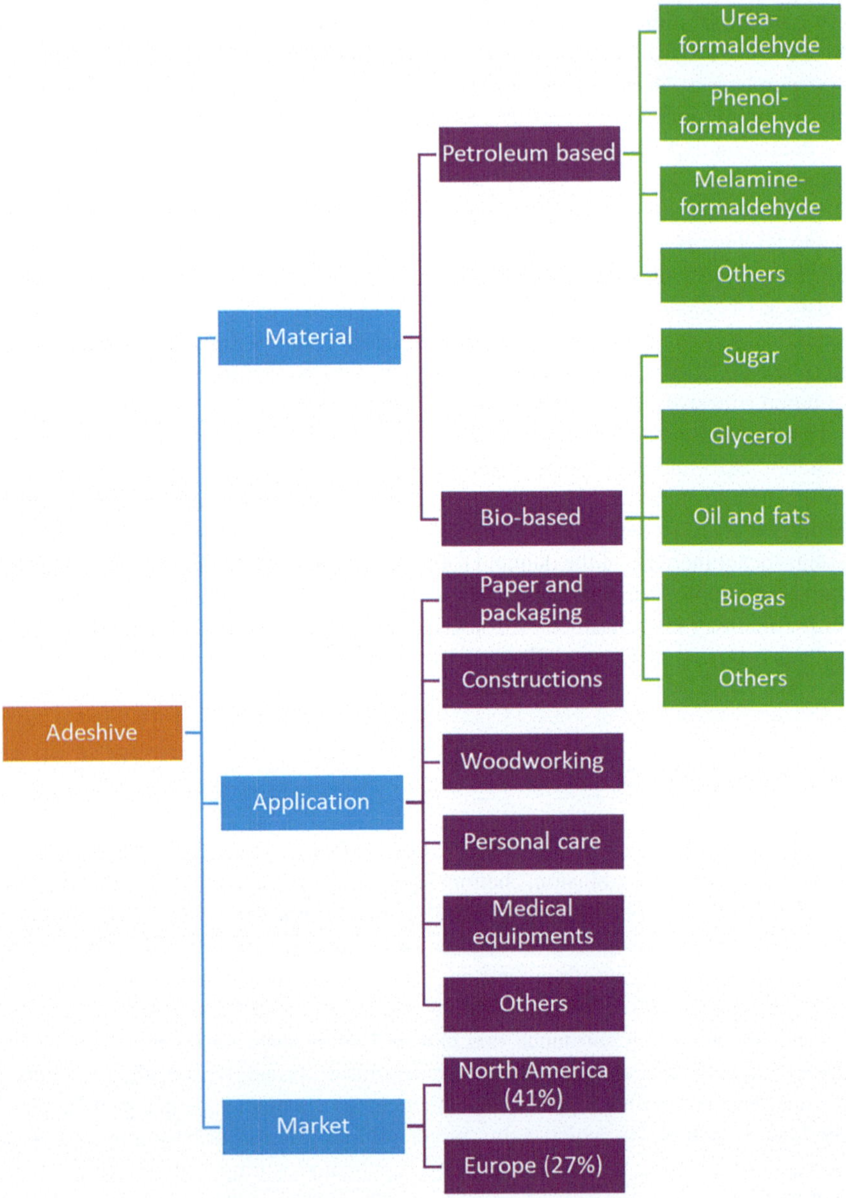


**Fig. 3** Global demand and application of adhesive

They can be classified as having an effect on liquid properties, qualities that convert to a film, and properties that are dry films based on their chemistry or functional role. They can also be split pragmatically into those that are required to actualize intrinsic properties and those that bestow additional desirable features.

The market for wood coatings is expected to reach a valuation of USD 10.90 billion globally in 2022 and expand at a CAGR of 4.9% between 2023 and 2030 [17]. The market is expected to increase due to a number of important factors, such as the increasing demand for home furnishings and the growing focus placed by consumers on aesthetics. The need for numerous furniture products, including beds, chairs, shelves, tables, and cabinets, has increased due to factors including the growth of the worldwide population and the housing market's recovery. Further fueling market expansion is the growing need for furniture due to the rising demand for interior design and décor, notably in hotels, corporate offices, and residential spaces.

Among the materials, polyurethane (41.0% in 2022) has the largest share in 2021–2023 and it is projected to remain strong until 2030 followed by acrylic, nitrocellulose, unsaturated polyester, and others [17]. In addition to varied finishes including high-gloss, semi-gloss, matte, and satin, polyurethane coatings come in a variety of formulas, including water- and oil-based alternatives [10, 12]. The growing consumer focus on sustainability is driving up demand for such eco-friendly items. In response, a large number of businesses are introducing wood finishing products that are biobased and eco-friendly. According to the regional purposes, Asia Pacific such as India, China, Japan, and South Korea lead to the domination of the wood coating markets by about 46.2% in 2022 [17]. The applications of wood coating (Fig. 5) are varied such as for furniture, joinery, flooring and deckling, siding, and others. Additionally, the technology used for wood coating includes several techniques such as waterborne, solvent-borne, powder coating, radiation cured, and others.

## 5 Impact Categories

Most of the studies agreed that using traditional wood adhesive and coating has brought a lot of health and environmental defects. Therefore, the usage of natural-based wood adhesive and coating is more favorable. Additionally, nanotechnology has been added to the wood adhesive and wood coating to improve the quality of the wood products. In this regard, the life cycle flow diagram is necessary to constrain the study (Fig. 4). It is challenging to locate the life cycle assessment of nanoparticle usage in wood items. The majority of studies would rather highlight the life cycle assessment (LCA) of the wood adhesive or coating than deal with the nanoparticles directly. Furthermore, a broad range of studies has examined the life cycle assessment (LCA) of nanotechnology applications, with a particular focus on wood and other materials like semiconductors, medicine, and so on. To address this challenge, this chapter provides a summary on how wood adhesive and wood coating are being evaluated by the LCA method. In addition to that, the LCA for nanoparticle application is summarized in order to enrich the insight of this study.

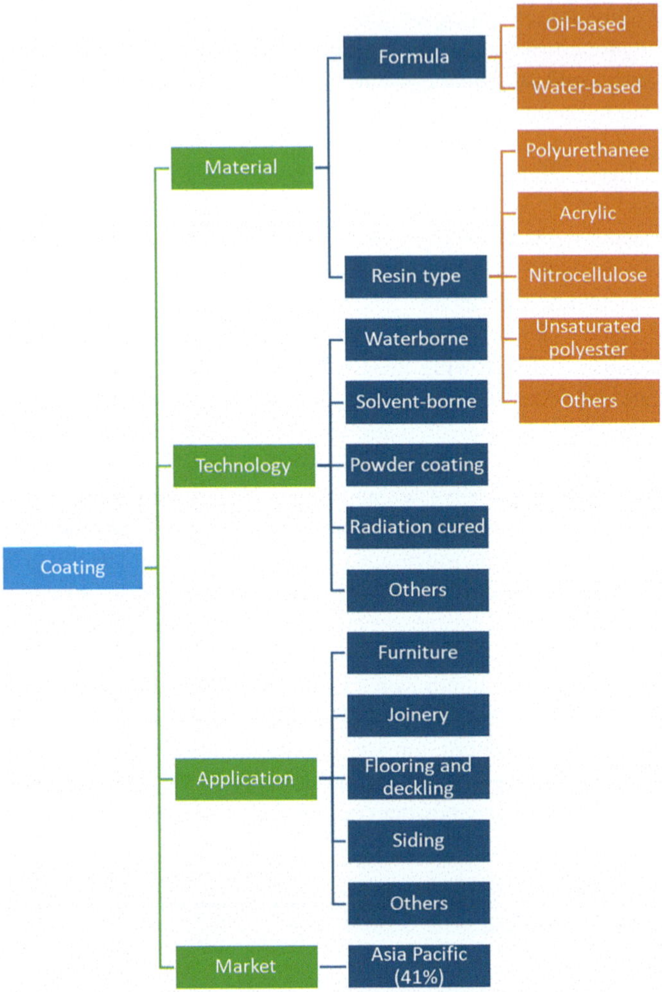


**Fig. 4** Global demand and application of wood coating

Different types of nanoparticles, such as nanotubes, quantum dots, films, plates, and fullerenes, as well as single, fused, aggregated, tubular, and irregular morphologies, are employed in technology [7, 9]. In the matter of application (Fig. 6), nanoparticles are used in very wide sectors such as medicine, energy, food, electronics, environment, and so on [2, 11]. Although there have been many documented benefits of using nanoparticles in various areas, there may be unidentified health and environmental dangers associated with the broad usage of nanoparticles in development and applications.

A typical impact category in LCA studies (Fig. 7) is evaluated on how input influences the output. In the input parameters, the parameters may involve the materials,

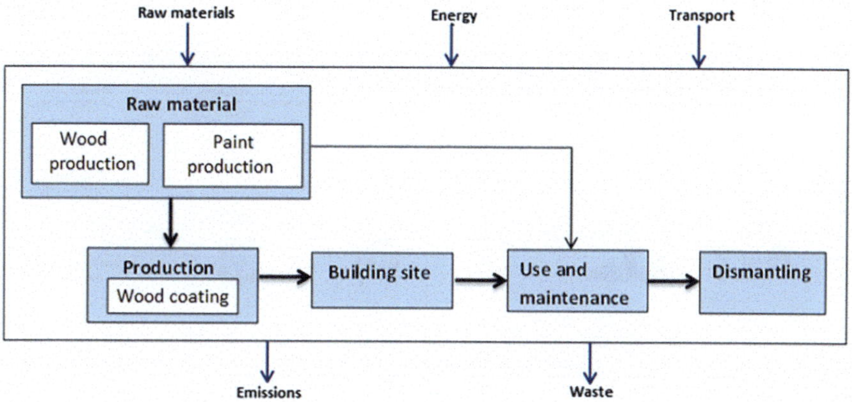


**Fig. 5** Flow diagram of life cycle flow assessment wood production from raw material into dismantling [15]

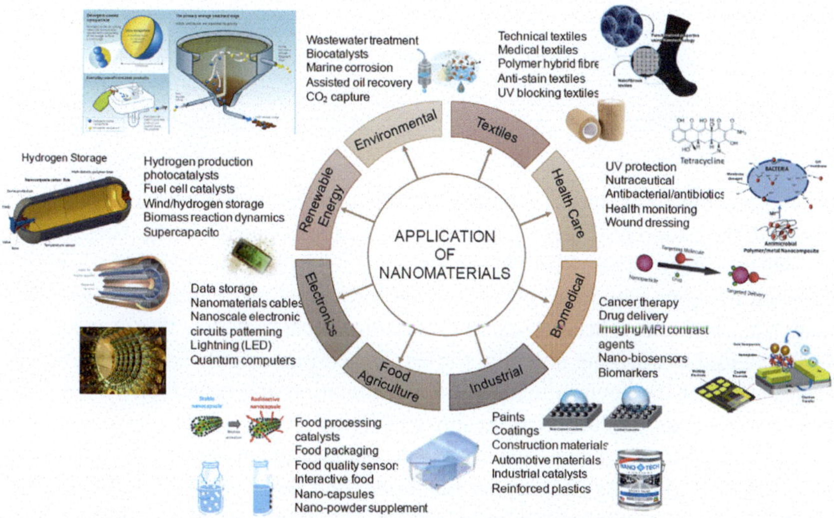


**Fig. 6** Various applications of nanoparticles in human life activities [31]

resources, energy production, utility, and so on [15]. These parameters will influence the output such as the wastewater, solid waste, and emission generation. Additionally, the output of the waste of heat or energy, washout material, or substances may be included as one of the parameters to calculate the impact categories. The impact categories may be defined into several types including environmental (such as $CO_2$ equivalent—$CO_2$eq), human health (health exposure, lethal dose), technology (energy supply), social (society response), or economic effects (budgets).

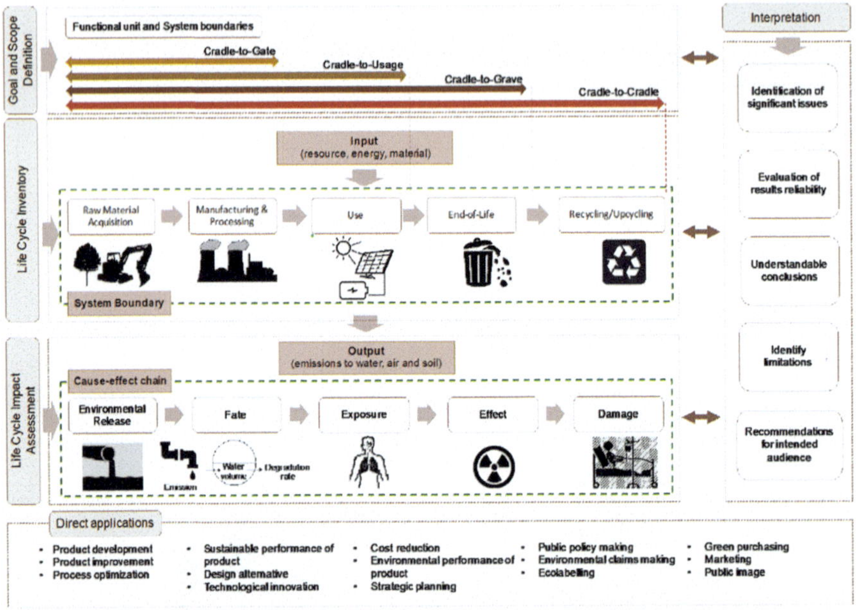


**Fig. 7** A typical framework in LCA studies: components analysis for input and output [31]

A limited number of studies examined the environmental performance of nanoparticles up to the endpoint level of life cycle impact assessment (LCIA). These studies evaluated the effects of nanoparticles on midpoint-level processes like eutrophication, acidification, ozone depletion, photo-oxidant formation, and climate change. Three impact categories have been researched the most: acidification (38%), ecotoxicity (36%), and global warming potential (56%). The enormous effects these key impact categories had on the environment, on bioaccumulation upsetting food systems, and ultimately on human health and the health of all living creatures led to frequent studies of them. According to the reviewed research, the average global warming potential ranges from 0.7 to 3.0 kg $CO_2$ eq per functional unit (FU), while the average emissions from acidification are much more variable, ranging from 1.0 to 20.0 kg $SO_2$ eq per functional unit. In contrast, the range for ecotoxicity, where the FU is expressed in kg, is 3.0–8.0 CTUe/kg.

Likewise, the determination of LCA for the bio-adhesive is influenced by the input and estimated output parameters. For example, takes into account the bio-adhesive derived from starch- and corn-based bio-adhesive LCA study case [4] (Table 1) which is produced from four different technologies, (1) hydrolysis with bio-oil (cassava starch), (2) hydrolysis with N-Methylol Acrylamide (corn starch), (3) pre-treatment with $FeSO_4$ and $H_2O_2$ oxidation (corn starch), and (4) pre-treatment with NaClO and ECH oxidation (corn starch).

According to the evaluation in Table 1, the cassava starch-based technology using bio-oil has less global warming contribution (0.92) compared to the other three

**Table 1** Study case for starch- and corn-based bio-adhesive with different scenarios of synthesis technology

| Impact categories | No. tech | Procedure | Evaluations | Discharge to water |
|---|---|---|---|---|
| • Global warming, GW (kg $CO_2$ eq)<br>• Freshwater Ecotoxicity, FET (g 1,4-DCB)<br>• Human Health, HH, Human toxicity cancer (CTUh) in $\times 10^{-10}$ | 1 | Hydrolysis with bio-oil | • GW: 0.92<br>• FET: 33.49<br>• HH: 7.14 | • $NH_4Cl$<br>• $H_2SO_4$<br>• H2O<br>• $(NH_4)_2SO_4$ |
| | 2 | Hydrolysis with N-Methylol Acrylamid | • GW: 1.67<br>• FET: 79.65<br>• HH: 2.20 | • $NH_4Cl$<br>• $H_2SO_4$<br>• HCL<br>• NaCl |
| | 3 | Pre-treatment with $FeSO_4$ and $H_2O_2$ oxidation | • GW: 1.10<br>• FET: 12.98<br>• HH: 5.80 | • $FeH_3$<br>• $H_2SO_4$ |
| | 4 | Pre-treatment with NaClO and ECH oxidation | • GW: 1.82<br>• FET: 1409<br>• HH: 3.06 | • NaCl<br>• HCL<br>• $H_2SO_4$ |

technologies. The Freshwater ecotoxicity shows that cornstarch-based bio-adhesive with NaClO and ECH pretreatment (FET: 1409) have very high levels. This result suggests that involving the NaClO and ECH in the process and discharging the unused material into the environment may influence the freshwater quality, for example, eutrophication or enrich the excessive nutrients in the water body [3, 46]. Furthermore, the human health cancer-related evaluation shows that cassava-starch-based bio-adhesive has a higher level than other technologies. It may be due to the chemical discharges from this technology being more varied than other technologies therefore the complexity and potential other chemical reactions of cancer may occur [42]. This kind of evaluation is also applied in the wood coating. The input parameter may influence the output including the products and the discharge to the environment. The impact categories is affected by the chosen technology, variety of the chemicals, and the quantity of the generation of the products.

In the manner of the nanoparticle, several studies attempted to evaluate using several types of software (Table 2). The greenhouse gas, air pollution, electricity generation, economic input, and output, as well as energy consumption, are the most categories evaluated in nanoparticle technology. However, the LCA of nanoparticle considered has many lacking on how the material is treated. Nanoparticle production or synthesis is a distinct material with unique mechanical, optical, and physical properties, therefore, LCA evaluation must include the advanced technology and inventory to synchronize the condition of the material. So far, the LCA evaluation shows the results of the nanoparticle-related study in typical treatment with common material. Therefore, the enrichment of the LCA inventory is considered to be performed for further analysis.

**Table 2** Nanoparticle application in LCA study using several types of method analyses

| Materials | Impact categories | Unit | Software | References |
|---|---|---|---|---|
| Nanocrystaline | Greenhouse gases (GHC), air pollution emission, electricity generation | 1 kWh electricity | Ecoindicator 95, Environmental Priority Strategies (EPS), Eco Sweden, Eco Netherlands and Environmental Design of Industrial products (EDIP)/SimaPro | [18] |
| Nanoclay polymer composite | Projected fuel savings, Carbon Dioxide ($CO_2$) reduction, economic inputs and outputs, GHG emissions, toxic releases | 16.9 million light-duty vehicles, 210 million vehicles on the road | Economic Input–Output Life Cycle Assessment (EIO-LCA) through Economic Input–Output (EIO) model/GaBi 4 | [26] |
| Nanoscale platinum-group metal particles | Economic inputs and outputs, economic purchases, emissions of conventional pollutants and greenhouse gases, RCRA hazardous waste, toxic releases | Projected motor vehicles in the US between 2005 and 2030 | EIO-LCA through EIO model/GaBi 4 | [27] |
| Various oxide nanoparticles | Energy consumption, $CO_2$ emissions | $CO_2$eq | Ecoinvent—non stated | [32] |
| Nanoclay polypropylene layered silicate nanocomposite agricultural film | GHG emissions | Coverage of 650 $m^3$, $CO_2$eq | SimaPro Derived from the latter energy and material data from the pilot plant | [35] |

## 6 Limitations

As nanoparticle impacts can happen at any stage of the life cycle, the end-of-life system boundary is crucial in mitigating all potential impacts because, in contrast to the manufacturing and production stages, nanoparticles released during the usage and end-of-life stages are unlikely to be in perfect condition. The limitations include the fact that most authors assumed that nanoparticle products are managed at the end of their useful lives in a manner similar to that of conventional products because they had little understanding of the flows of end-of-life nanoparticles and the possibility of emissions from different waste management procedures (such as recycling and reuse). In the case of LCA, the researchers' ability to accurately assess the environment is hampered by the lack of data. Low data transparency results from certain firms withholding the energy and materials used in the creation of nanoparticles.

A lack of correlation between the mathematical models might result from uncertainties arising from the selection of models and frameworks during the modeling process. In this instance, addressing the probabilistic distributions, assessing different distribution impacts quantitatively, and prioritizing the model's output uncertainties should all help to increase the availability of data in models that have been made public. Only basic LCA models that concentrate on the primary variables and environmental hotspots (environmental uncertainty) can be created using the results. Because parameters are so important in LCA, some people might have chosen non-representative and non-inconclusive parameters. To make it more realistic of emission values and possible effects, it can be enhanced by addressing the technological scale-up, processing, performance, and destiny models in the environment and toxicity assessment.

## 7 Conclusions

The life cycle assessment (LCA) of wood-containing nanoparticles offers a comprehensive perspective of the environmental consequences linked to this novel material innovation. Wood's mechanical qualities—such as strength, durability, and resistance to biological deterioration—are greatly enhanced by nanoparticles, which might result in longer-lasting goods and possibly lower replacement costs. The prolongation of the product lifespan has the potential to mitigate the environmental impact of raw material extraction and product manufacturing, while also aiding in the conservation of resources. Furthermore, the improved functionality of wood infused with nanoparticles can result in a decrease in the need for chemical treatments, thereby reducing the negative effects on the environment during the usage phase. Nevertheless, the risks to human health and the environment posed by the manufacture, use, and disposal of nanoparticles outweigh the advantages. Nanoparticles' environmental impact is a problem because their manufacturing procedures frequently require high energy consumption and the possible emission of hazardous materials. The breakdown or burning of wood treated with nanoparticles at the disposal stage may cause the release of nanoparticles into the environment, endangering ecosystems and public health. Therefore, even though adding nanoparticles to wood products has potential benefits, it is imperative to take a comprehensive approach that includes using sustainable manufacturing techniques, conducting thorough risk assessments for health and the environment, and enacting sensible regulations to reduce any potential negative effects over the course of the product's lifecycle. This well-rounded viewpoint guarantees that public health and ecological integrity are not jeopardized by technology progress.

## References


1. Adhikari S, Ozarska B (2018) Minimizing environmental impacts of timber products through the production process "From sawmill to final products." Environ Syst Res 7(1):6. https://doi.org/10.1186/s40068-018-0109-x
2. Akhtari M, Ganjipour M (2013) Effect of Nano-silver and Nano-copper and Nano-zinc oxide on Paulownia wood exposed to white-rot fungus. In: Paper presented at the 44th annual meeting of the international research group on wood protection, Çesme, Turkiye
3. Akinnawo SO (2023) Eutrophication: Causes, consequences, physical, chemical and biological techniques for mitigation strategies. Environ Chall 12:100733. https://doi.org/10.1016/j.envc.2023.100733
4. Arias A, Feijoo G, Moreira MT (2021) Evaluation of starch as an environmental-friendly bioresource for the development of wood bioadhesives. Molecules 26(15). https://doi.org/10.3390/molecules26154526
5. Arias A, Feijoo G, Moreira MT (2022) New environmental approach based on a combination of planetary boundaries and life cycle assessment in the wood-based bioadhesive market. ACS Sustain Chem & Eng 10(34):11257–11272. https://doi.org/10.1021/acssuschemeng.2c03058
6. Bak M, Takács D, Rákosa R, Németh ZI, Németh R (2023) One-step process for the fabrication of hydrophobic and dimensional stable wood using functionalized silica nanoparticles. Forests 14(3). https://doi.org/10.3390/f14030651
7. Bastianini F, Hidalgo AIC, Hook DZ, Smith JA, Cumming D, Dunbar A (2024) Using Ag nanoparticles in the electron transport layer of perovskite solar cells to improve efficiency. Sol Energy 268:112318. https://doi.org/10.1016/j.solener.2024.112318
8. Bian Y, Bai M, Cao J, Li J (2023) A strong soybean meal adhesive enhanced by aluminum hydroxide nanoparticles via a low-cost and simple organic-inorganic hybrid strategy. Int J Adhes Adhes 125:103442. https://doi.org/10.1016/j.ijadhadh.2023.103442
9. Borges CC, Tonoli GHD, Cruz TM, Duarte PJ, Junqueira TA (2018) Nanoparticles-based wood preservatives: The next generation of wood protection? Cerne 24(4):397–407. https://doi.org/10.1590/01047760201824042531
10. Bulian F, Graystone JA (2009) Chapter 4—raw materials for wood coatings (2)—solvents, additives and colorants. In: Bulian F, Graystone JA (eds) Wood coatings. Elsevier, Amsterdam, pp 95–135. https://doi.org/10.1016/B978-0-444-52840-7.00004-7
11. Cai X, Riedl B, Wan H, Zhang S (2010) Montmorillonite nanoparticle distribution and morphology in melamine-urea-formaldehyde resin-impregnated wood nanocomposites. Wood Fiber Sci:285–291
12. Chang H, Tu K, Wang X, Liu J (2015) Fabrication of mechanically durable superhydrophobic wood surfaces using polydimethylsiloxane and silica nanoparticles. RSC Adv 5(39):30647–30653. https://doi.org/10.1039/C5RA03070F
13. Dunky M (2003) Adhesives in the wood industry. Handbook of adhesive technology. New York, ABD, pp 872–941
14. Froese M (2023) The demand for bio-based adhesives is growing significantly. Design World
15. Fufa S, Jelle B, Bohne RA, Lausselet C (2013) LCA of nano-coated wooden claddings utilizing accelerated ageing test results
16. Furasova A, Krassas M, Tountas M, Khmelevskaia D, Logunov L, Zhirihin D, Kymakis E, Makarov S (2024) Photovoltaic parameters improvement via size control of monodisperse Mie-resonant nanoparticles in perovskite solar cells. Chem Eng J 493:152771. https://doi.org/10.1016/j.cej.2024.152771
17. Grand RR (2023) Wood coatings market size, share & trends analysis report by resin type (polyurethane, acrylic, nitrocellulose, unsaturated polyester), by technology, by application, by region, and segment forecasts, 2023–2030. Grand Review Research
18. Greijer H, Karlson L, Lindquist S-E, Anders H (2001) Environmental aspects of electricity generation from a nanocrystalline dye sensitized solar cell system. Renew Ene 23(1):27–39. https://doi.org/10.1016/S0960-1481(00)00111-7

19. Guild FJ, Tsang WL, Taylor AC (2022) Silica nano-particle filled polymers: Debonding and microstructure. Compos Sci Technol 218:109202. https://doi.org/10.1016/j.compscitech.2021.109202
20. Huang J, Yang L, Wang W, Yu B, Cao C, Bi Z, Wang J, Shen J, Song L (2024) Biomimetic, highly adhesive, supramolecular nano-coatings for fire-safe polymeric foams. Prog Org Coat 186:107995. https://doi.org/10.1016/j.porgcoat.2023.107995
21. Hussin MH, Abd Latif NH, Hamidon TS, Idris NN, Hashim R, Appaturi JN, Brosse N, Ziegler-Devin I, Chrusiel L, Fatriasari W, Syamani FA, Iswanto AH, Hua LS, Al Edrus SSAO, Lum WC, Antov P, Savov V, Rahandi Lubis MA, Kristak L, Reh R, Sedliačik J (2022) Latest advancements in high-performance bio-based wood adhesives: A critical review. J Market Res 21:3909–3946. https://doi.org/10.1016/j.jmrt.2022.10.156
22. Irshad MA, Abdullah LM, Nasim I, Nawaz R, Zahoor AF, Al-Mutairi AA, Al-Hussain SA, Irfan A, Zaki MEA (2024) Efficient chromium removal from leather industrial wastewater in batch experimental study: Green synthesis and characterization of zinc oxide nanoparticles using Ficus benghalensis extracts. Ecotoxicol Environ Saf 281:116616. https://doi.org/10.1016/j.ecoenv.2024.116616
23. Kaboorani A, Auclair N, Riedl B, Landry V (2017) Mechanical properties of UV-cured cellulose nanocrystal (CNC) nanocomposite coating for wood furniture. Prog Org Coat 104:91–96. https://doi.org/10.1016/j.porgcoat.2016.11.031
24. Kandelbauer A, Widsten P (2009) Antibacterial melamine resin surfaces for wood-based furniture and flooring. Prog Org Coat 65(3):305–313. https://doi.org/10.1016/j.porgcoat.2008.12.003
25. Kumar A, Nguyen TA, Sharma S, Bhat R (2022) Chapter 1—Nanotechnology in paper and wood engineering: An introduction. In: Bhat R, Kumar A, Nguyen TA, Sharma S (eds) Nanotechnology in paper and wood engineering. Elsevier, pp 3–13. https://doi.org/10.1016/B978-0-323-85835-9.00015-5
26. Lloyd SM, Lave LB (2003) Life cycle economic and environmental implications of using nanocomposites in automobiles. Environ Sci Technol 37(15):3458–3466. https://doi.org/10.1021/es026023q
27. Lloyd SM, Lave LB, Matthews HS (2005) Life cycle benefits of using nanotechnology to stabilize platinum-group metal particles in automotive catalysts. Environ Sci Technol 39(5):1384–1392. https://doi.org/10.1021/es049325w
28. Montanari C, Olsén P, Berglund LA (2021) Sustainable wood nanotechnologies for wood composites processed by in-situ polymerization. Front Chem 9
29. Nakamura N, Ohta S (2024) Precise control methods of the physicochemical properties of nanoparticles for personalized medicine. Curr Opin Biotechnol 87:103108. https://doi.org/10.1016/j.copbio.2024.103108
30. Ni K, Du G, Liu C, Wu Y, Yang H, Yin C, Su H, Ran X, Gao W, Yang L (2023) Cross-linked entanglement of aldehyde and amine-functionalized nanocellulose reinforced with biomineralization to produce an all-bio-based adhesive. Chem Eng J 465:142888. https://doi.org/10.1016/j.cej.2023.142888
31. Nizam NUM, Hanafiah MM, Woon KS (2021) A content review of life cycle assessment of nanomaterials: Current practices, challenges, and future prospects. Nanomaterials 11(12). https://doi.org/10.3390/nano11123324
32. Osterwalder N, Capello C, Hungerbühler K, Stark WJ (2006) Energy consumption during nanoparticle production: How economic is dry synthesis? J Nanopart Res 8(1):1–9. https://doi.org/10.1007/s11051-005-8384-7
33. Packham DE (2009) Adhesive technology and sustainability. Int J Adhes Adhes 29(3):248–252. https://doi.org/10.1016/j.ijadhadh.2008.06.002
34. Qian L, Long L, Xu JF (2014) Surface modification of nano-titanium and its effect on the antibacterial property of waterborne wood coating. Key Eng Mater 609–610:88–93. https://doi.org/10.4028/www.scientific.net/KEM.609-610.88
35. Roes AL, Marsili E, Nieuwlaar E, Patel MK (2007) Environmental and cost assessment of a polypropylene nanocomposite. J Polym Environ 15(3):212–226. https://doi.org/10.1007/s10924-007-0064-5

36. Saw S, Mahto S, Chandra N (2024) Morphological, structural, and electronic properties of green synthesized ZnO nanoparticles by experimental and DFT+U method—A review. Phys Lett A 518:129697. https://doi.org/10.1016/j.physleta.2024.129697
37. Song X, Guo W, Zhu Z, Han G, Cheng W (2024) Preparation of uniform lignin/titanium dioxide nanoparticles by confined assembly: A multifunctional nanofiller for a waterborne polyurethane wood coating. Int J Biol Macromol 258:128827. https://doi.org/10.1016/j.ijbiomac.2023.128827
38. Temizel-Sekeryan S, Hicks AL (2020) Global environmental impacts of silver nanoparticle production methods supported by life cycle assessment. Resour Conserv Recycl 156:104676. https://doi.org/10.1016/j.resconrec.2019.104676
39. Wang Y, Wu X, Wang Y, Tian Y, Mu H, Li J (2021) Hydrophobic and UV-resistant properties of environmentally friendly nano-ZnO-coated wood. 75(2):138–147. https://doi.org/10.1515/hf-2019-0312
40. Xiong J, Chen F, Sun L, Yu X, Zhao J, Hu Y, Wang Y (2019) Characterization of VOC emissions from composite wood furniture: Parameter determination and simplified model. Build Environ 161:106237. https://doi.org/10.1016/j.buildenv.2019.106237
41. Yang R, Gu Y, Qin J, Liu Q, Liu Q (2023) Potential role of Chinese medicine nanoparticles to treat coronary artery disease. Heliyon 9(9):e19766. https://doi.org/10.1016/j.heliyon.2023.e19766
42. Yasasve M, Manjusha M, Manojj D, Hariharan NM, Sai Preethi P, Asaithambi P, Karmegam N, Saravanan M (2022) Unravelling the emerging carcinogenic contaminants from industrial waste water for prospective remediation by electrocoagulation—A review. Chemosphere 307:136017. https://doi.org/10.1016/j.chemosphere.2022.136017
43. Yue H, Xu C, Yao J, He M, Yin G, Cui Y, Yang C, Guo J (2022) Characterization and properties of plywood bioadhesive derived from cottonseed protein and sawdust cellulose. Cellulose 29(10):5869–5881. https://doi.org/10.1007/s10570-022-04611-9
44. Zhang B, Pizzi A, Petrissans M, Petrissans A, Baptiste C (2023) Self-cross-linked tannin-aminated tannin surface coatings for particleboard. J Renew Mater 11(12):4097–4121
45. Zhang B, Pizzi AP, Pétrissans M, Petrissans A, Colin B (2023b) A melamine-dialdehyde starch wood particleboard surface finish without formaldehyde. J Renew Mater 11. https://doi.org/10.32604/jrm.2023.028888
46. Zhou Y, Wang L, Zhou Y, Mao X-z (2020) Eutrophication control strategies for highly anthropogenic influenced coastal waters. Sci Total Environ 705:135760. https://doi.org/10.1016/j.scitotenv.2019.135760

# Economic Aspects of Nanomaterials Application in Bioadhesives and Wood-Based Composites

**Nikolay Neykov, Viktor Savov, and Aureliu-Florin Halalisan**

**Abstract** The constant drive to increase productivity leads to the search for multiple solutions to improve the economic efficiency of industrial products. At the same time, industries in developed societies are under constant pressure to reduce carbon emissions. In some countries, the target is a 40% reduction by 2030. The European Union placed the ambitious goal of a 50% reduction by 2030 and a carbon–neutral economy by 2050. All these targets pose significant challenges to industrial producers, and they have to improve the economic scale and the boundary of the production capabilities. Nanomaterials and nanotechnologies can significantly improve the properties of wood-based panels and binders for that production. Incorporating nanomaterials in wood-based composites improves their waterproofness and thermal and electrical conductivity. It also can lead to reduced free formaldehyde content in the boards, acceleration of the hot pressing cycle (reduction of press factor), etc. However, the question of the economic efficiency of using different nanomaterials in the production of wood-based panels remains controversial. Nanotechnologies and materials produced by the nano-industries are considered a contemporary way to shift the production possibilities curves of industrial sectors and national economies far ahead. These productions stay between the sustainable use of natural resources and the requirement for economic efficiency, and nanomaterials can provide additional opportunities for improvement. Despite the great number of publications on wood-based composites (WBC) and adhesives for their production, there are few in the field of economic aspects. The problem deserves particular attention due to the boundaries of applying nanomaterials and bioadhesives to large-scale manufacturing. The current study focuses on some market-derived features of WBCs produced with bioadhesives and nanomaterials. It tries to reveal the product and the market

N. Neykov (✉)
Faculty of Ecology and Landscape Architecture, University of Forestry, Sofia, Bulgaria
e-mail: nneykov@ltu.bg

V. Savov
Faculty of Forest Industry, University of Forestry, Sofia, Bulgaria
e-mail: victor_savov@ltu.bg

A.-F. Halalisan
Department of Forest Engineering, Transilvania University of Brasov, Brasov, Romania
e-mail: aureliu.halalisan@unitbv.ro



P. Antov et al. (eds.), *Nanomaterials Additives in Bioadhesives for Wood Composites*, Smart Nanomaterials Technology, https://doi.org/10.1007/978-981-96-3400-2_11

basement of industrial application of these materials. For this purpose in the study, SWOT analysis is used to summarize some of the main features of the bioadhesives, the nanomaterials embedded in them and the final products–WBCs. The analysis enhances these products' economic understanding from their pure mechanical and chemical properties. The results revealed that WBs produced with bioadhesives and, like an option, nanomaterials are very promising in their strengths. Despite the market pressure for efficiency, these products have a great possibility to improve the product mix of the producers.

**Keywords** Nanomaterials · Bioadhesives · Wood-based composites · Economics

## 1 Introduction

Nanotechnology involves producing and using materials that have a dimension of a unit of 100 nm or less [47]. This dimension refers to the orientation of molecules and atoms to control and design the properties of nanomaterials [10]. Nanomaterials can be summarised as inorganic-based, organic-based, carbon-based and composite-based nanomaterials [16]. Composite nanomaterials combine metal–metal oxide-based, organic and carbon or metal–organic-based materials [16, 25, 33, 41]. Organic nanomaterials include organic materials such as cyclodextrin, micelles, dendrimers and liposomes [16]. The compounds in WBC in the form of nanoparticles are silicon dioxide, titanium dioxide, halloysite, aluminium oxide, zinc oxide, clay minerals (bentonite, montmorillonite), and nanocellulose carbon nanotubes [23]. WBs are manufactured by combining fragmented wood and wood fibre into a cohesive mat of wooden composites (NESHAP). Particleboards are among the wood composites commonly used [30] in many industries.

Research on the issues of wood composites and the non-wood materials used for their production is extensive. It includes numerous articles related to bioadhesives and the properties of the adhesives used for their production. The problems of phenolic resins are among the most researched. Antov et al's. [5] study discusses recent advances in using nanomaterials, such as nanocellulose, nanolignin, and nanoclay, in synthesising sustainable, bio-based wood adhesives for the production of wood-based composites. Savov et al. [37] summarised the most recent advances and technological developments in the fabrication, design, characteristics, and applications of eco-friendly wood and wood-based composites. Savov et al. [38] investigated the effect of the adhesive system on the properties of fiberboard panels bonded with hydrolysis lignin and phenol–formaldehyde. Reh et al. [36] performed novel research on the problems dedicated to eco-friendly wood-based composites. Bekhta et al. [9] examined laboratory-produced particleboards magnesium and sodium lignosulfonates (LS), used pure and in mixtures with urea formaldehyde (UF) resin. The results define them as eco-friendly composites. In previous research, some current authors [4, 6] determined that bio-based adhesives provide an eco-friendly and sustainable alternative to the conventional adhesive systems used in the wood-based panel industry. Jasmani et al. [17] profoundly analysed the application of nanomaterials in wood-based products. They propose a detailed description of the properties that nanomaterials give to wood and wood-based products as wood-based composites (WBs).

## 2 Methodology

The present study made a summary and analysis of various studies related to the use of nanomaterials and the technologies that use them. From the available research on the issues of nanotechnology, the green economy and the application of various resins in the production of wood composites, assumptions were made about the economic benefits and challenges that manufacturers of chipboard, MDF and other composite wood products face when using nanomaterials of biological origin. All the research has been done on the related topics, and the hypothesis placed in the current study is summarized in a SWOT analysis. The strengths of nanomaterials are an integral part of the bioadhesives and properties of wood composites. That forces all the strengths and, together with them, the possibilities to be considered in the complex whole of the product—wood composite. The weaknesses are summarized in focus on the unknowns surrounding the new materials and the costs that would arise from their application. The possibilities are described as a synergistic effect of the three main components of wood-based composites—bioadhesives, the need for these composites and nanomaterials.

## 3 Result and Discussion

### *3.1 Strengths of Nanomaterials Application in Bioadhesives and Wood Composites*

#### 3.1.1 Innovativeness

Nanotechnology is among the leading instruments driving the world's economic development in the twenty-first century [12]. The use of nanomaterials in producing wood composites and, more precisely, in the adhesives and resins used will bring this sector into the scope of research and scientific development. According to Shaik et al. [40], economic impact assessments of 2D nanomaterial-based heat transfer fluids will be among the top research areas. Nanomaterials, through bioadhesives, construct the physical form of wood composites. Wood composites themselves require cleaner raw materials [8].

Furthermore, the European Forest Strategy (EUFS), released in July 2021, defines four indicators: sustainability, innovation, investments, and promotion. That opens the niche for nanomaterials in the adhesives used in wood composites production. The introduction of nanomaterials in the production lines of wood composites leads to the following types of innovations [51].

- Process Innovations. Hybrid biocomposites made of bioresources materials led to technological advances in manufacturing [27]. The application of nanomaterials changes the processes of pressing, moistening [3], drying, etc. For example, the nano polymers encapsulate and hold in a chemical until the physical environment changes [11]. According to EUFS, bio-adhesives as part of the production processes in wood composites improve the sustainability of manufacturing. Savov et al. [37] summarized the leading technological innovative alternatives for binding improvements in wood composites like nanosilver—mainly for coating [28, 45]; starch [14, 44]. Some authors [36] directly propose magnesium and sodium lignosulfonates [7] as an additive in the mixture of urea–formaldehyde resin to reduce the toxicity of wood composites manufacturing. Lugovitskaya [24] examines the combination of nanoparticles and other lignosulfonate-containing nanostructures, which can improve the glueing and pressing processes in some wood composite manufacturing. Implementation of the bio-adhesives the nanomaterials as part of bettering their properties will lead to improving the efficiency of machinery and drying systems using environmentally friendly chemicals [2, 52].
- Product innovations. As some authors stated, actions are necessary to trigger creation, focused on new wood-based composites with higher performance and sustainability. Analysing the product innovations that can be introduced through bioadhesives must consider the benefit or added value that wood composites will provide consumers. Wood composites are positioned in three main market segments—traditional customers looking for the functionality of wood and innovative and environmentally concerned consumers [31]. Many researchers [4, 18, 48] focus on using wood composites in the circular economy as an alternative for recycling. The application of nanomaterials would strengthen the properties of bioadhesives in the new environmentally oriented wood composites.
- ***Market innovations***. As Reh et al. [36] state in their research contemporary external environment of industrial enterprises poses new challenges in the production methods and materials. Bioadhesives can gain competitive advantages over the manufacturer that implement them in the products by understanding market innovation as changing the perception of how the product is seen in the marketplace. Due to the derivative nature of the market demand for nanomaterials and bioadhesives for wood composites, they should also be present in the characteristics of the potential product. The production of wood composites with embedded bioadhesives should form a new marketing mix.

### 3.1.2 Product Value and Enhancements

Levchenko et al. [22], in their research, focused on the waste, derived nanomaterials and mentioned that these materials could be the core for manufacturing products for environmental remediation and energy storage. In this way, applying nanomaterials and bioadhesives to wood composites will create a new value. Jasmani et al. [17] stated that the core product of wood-based composites like panels could be improved significantly using nanomaterials. Inorganic nanoparticles like zinc oxide or silver nanoparticles can add effective biocide and antibacterial properties to wood-based composites. It is visible from the opinions of the researchers in the field that nanomaterials and bioadhesives contain promising capabilities to enhance the wood-based composites as a sophisticated structure of sub-products (the five levels of the full product). The product levels of the wood-based composites can be explained as follows:

- Core Benefit: The core benefit is the need or wants that the customer satisfies when they buy the wood-based composites. Regardless of the adhesive mixture used to produce wood composites, they meet a corresponding requirement for materials for the furniture industry, construction, mechanical engineering, etc. Bioadhesives and the nanomaterials added to them are not particularly important for business, and Bioadhesives, as a separate product, satisfy the need to produce wood composites. A fundamental need may arise when the composite manufacturer looks specifically for bioadhesives.
- Basic Product: At the second level, woo-based composites appear as an actual product and include the essential features the product must have. Here biocomposites and nanomaterials are vital, forming the structure of the items that satisfy the core benefit. Here any additional requirement to the composites can be included in the structure of the WBC for them to have the form of the different product, for example, not particleboards with urea formaldehyde glue, but with some bioadhesive along with nanomaterials included.
- Expected Product: Here is the power of bioadhesives and nanomaterials. The customer will expect to receive such if they want environmentaly friendly products. Customers can receive their desired outcome by including bioadhesives and nanomaterials in WBC.
- Augmented Product: It presents features that will exceed customer expectations. That is where manufacturers start distinguishing your product from competitors. Here the manufacturers and the other distinguishing features of the WBCs can provide product lines with particle boards with silver nanoparticles, waterproof panels, etc. Nanomaterials require an enhanced supply chain and business ecosystem.
- Potential Product: With this type of product introduction, novel materials make product transformations possible. The well-developed innovation ecosystem, including universities and laboratories and WBC manufacturers of WBCs, will create a strategic competitive advantage.

The following main nanomaterials in the production of wood composites lead to improved properties of the expected augmented and potential product [17].

- Nanocopper oxide and alkane surfactant improve the waterproofness of the plywood.

- Nanoparticles added into melamine-urea formaldehyde. Nano-ZnO, an eco-friendly biocide, can be recommended for use in interiors with temporarily increased humidity [35].
- Nanocellulose as the reinforcement material. Including functionalised nanocellulose as reinforcing material improves the mechanical properties of epoxy-based cellulose nanocomposites [17]. Reinforcing filler gave even better mechanical properties than nano-silica [21]. Nanocellulose crystals improve the thermomechanical performance of reinforcing fillers in structural materials and coatings [49].
- Silver nanoparticles (Ag-Np): Nosál and Reinprecht [29] examined the influence of silver nanoparticles on the antibacterial and anti-mould efficiency of silver nanoparticles present in melamine-laminated particleboard surfaces. They proved that these nanoparticles reduced the activity of the S. aureus bacterium.

In the research on wood composites and their adhesives, the main emphasis is on the properties of the base product. In fact, in market relations, the properties of a product may be excellent. Still, if a wide range of market needs are not covered, composites may remain an exotic product line in a manufacturer's mix. As proposed by Schwartz [39], the typical or universal structure of the product's value is presented in Fig. 1.

The structure presented above presents the full range of benefits a product can deliver to consumers. When considering wood composites, the benefits can vary significantly according to the composite type. Nanomaterials and bioadhesives can add value to many of the value elements. As a constructive part of the WBC, they form the consumers' perception along with the help of promotion in the following directions:

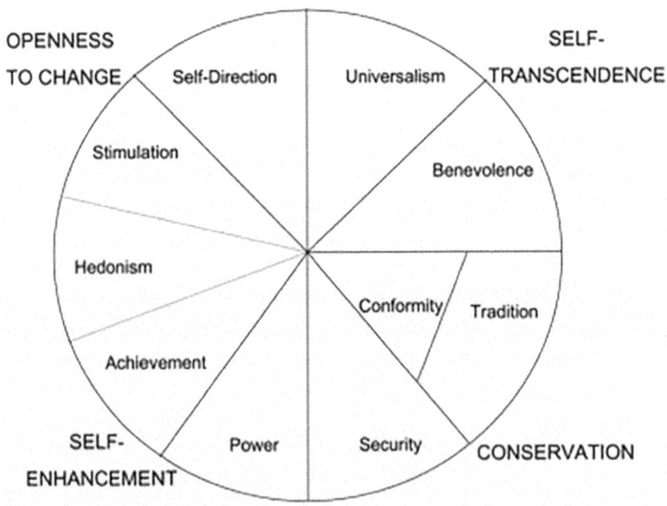


**Fig. 1** The universal value structure [19]

- Stimulation, Achievement, Self-direction. As previously mentioned, the innovator consumer group is among the most active ones purchasing wood composites with bioadhesives. Although the bio component of the WBC is probably not a niche [31], the younger generations are more likely to have active market behaviour [20] to buy products made of WBC with bioadhesives.
- Security. The probably most important factor is that consumers prefer WBC with bioadhesives. Nanomaterials in the adhesives can reduce formaldehyde emissions [23]. Such characteristics can easily position newly developed wood composites in parallel segments of the market's early and late majority [26].

One of the essential advantages that nanomaterials provide is developing the basic product [15] of wood composites. In their research, Stern et al. [43] proposed a model for innovations in the forest bioeconomy from the viewpoint of a value-added pyramid. According to their model, the bottom of the pyramid consists of incremental improvements in traditional pulp, paper, wood products, and wood-based bioenergy. Here the development progress led to new marketing channels and environmental impact reduction in forestry. The WBCs are in the middle of the pyramid. They require a reduction in environmental impact in manufacturing. They induce services and branding developments. That means that bioadhesives provide an opportunity for augmented and potential product development. Nanomaterials are at the top of the pyramid. They give from one side a possibility to gain competitive advantages through new business models and from another to provide further services complying with the existing expected, augmented and potential product levels.

## *3.2 Opportunities*

### 3.2.1 Perspective and Modernity

When enterprises search for new product development ideas, the mainstream of the research and innovation can look at the tendencies inside. The application of nanomaterials is among the most improving area of study. According to Shaik et al. [40], the number of papers dedicated to some nanomaterials applications has risen more than nine times in recent years. Prospects in the application of nanomaterials in bioadhesives for wood composites are determined by three main factors: the development in the production and trade of wood composites, the development of the application of bioadhesives and the development of the nanomaterials market. More and more companies specifically use wood and natural fibre plastic granulates in consumer goods like watch cases, toys, combs or trays–the unique look gives the impression of a high-value product and is well received by customers. Partanen and Carus [32] investigated wood-plastic composites' applicability and market development. They define the time to implement these materials in large-scale production. The production of WBs is constantly increasing in the EU. Figure 2 presents the

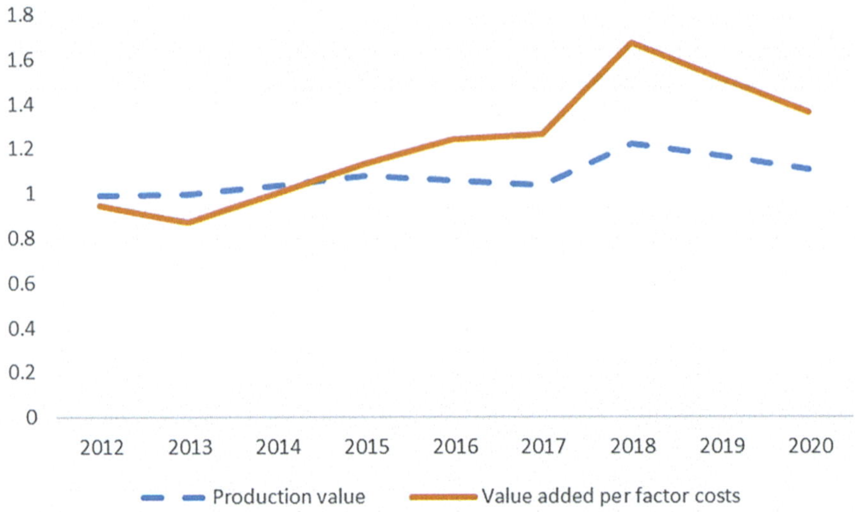


**Fig. 2** Index of the production and value-added of wood composites in the European Union for the period 2012–2020 (SBS-Eurostat)

index of the output and value-added per factor costs in the European Union for the years 2012–2020.

The graph shows that the production of wood composites is constantly increasing. At the same time, added value is outpacing production growth. Such trends show the great possibilities that bioadhesives and nanomaterials offer for product differentiation. Competition, apart from price, should be based on non-price methods. The slow growth of the product index, calculated based on the value of production at current prices, favours the intensive development of added value. These obstacles face the manufacturers' pressure between gaining efficiency and adding value to the customers. In any case, it is clear that nanotechnology's potential is real [12].

### 3.2.2 Innovation Business Ecosystem Improvement

The application of bioadhesives and, in addition, nanomaterials in them requires the complexity of the supply chain of wood composites, including new supplier materials and services. The positive in applying new adhesive compounds are deepening the research institutions-business relationship. As Shayakhmetov et al. [42] point out, the development of the innovation ecosystem in the university is the way to the practical commercialization of intellectual activity in the field of nanotechnology. According to Vaseashta [46], in their research in nanomaterials-based CBRNE Sensors, developing integrated solution pathways using an ecosystem of technological innovations provides tactical superiority. Introducing nanomaterials places multi-disciplinary challenges to universities, research institutions, entrepreneurs, equipment manufacturers and biotech [34].

## 3.3 Weaknesses

The slow growth of the product index (Fig. 2), calculated based on the value of production at current prices, favours the intensive development of added value. These obstacles face the manufacturers' pressure between gaining efficiency and adding value to the customers. Nanomaterials and bioadhesives can push production costs forward due to the research, innovation and investment processes. As Yashawantha et al. [50] state, in the case of 2D nanomaterials, the industrial readiness of value chains and industrial development policies are among the significant concerns most likely to affect these materials' industrial implementation. Shaik et al. [40] point out that nanoparticles could aggregate in the lungs, alimentary tract, liver, heart, spleen, kidneys, and cardiac muscle. Nanocopper used in the bioadhesives for wood composites includes Cu particles can be lethal, according to concentration.

## 3.4 Threats

Wilson [47] reveals that they introduce considerable uncertainty along with the great opportunities that nanomaterials give to industry and humanity. She states that the existing regulations cannot answer to improving nanotechnologies. As part of the international efforts to regulate the use of nanomaterials, the legislation will get more and more restrictive on their industrial application. Delgado [12] summarizes the main issues that implementation of nanomaterials can emerge:

- Wider integration of risk management and ethics regarding nanotechnology. As stated in the weaknesses of the application of nanomaterials, it is possible to increase the internal control for enterprises that produce wood composites using nanomaterials. That will lead to increased fixed and variable costs for composite production.
- Creation of an acceptable regulatory agenda. Delgado defines this as a future result to be achieved, but for the manufacturers, this is a threat of regulation-derived extra expenses and market restrictions.
- Encouraging the stimulation and spread of public dialogue. Public dialogue is a good thing and can have a positive PR effect on public opinion about one or another composite, but it carries considerable risk. Developing the dialogue in a direction that would harm the composites with incorporated nanomaterials is possible, and such a circumstance jeopardized the great perspective of a given composite.

## 4 Conclusion

The present study examines the complex relationship between wood-based composites, the adhesive compounds used for their production, and the nanomaterials that enhance the properties of the adhesive agents. The study systematized the main characteristics of nanomaterials and bioadhesives that define them as products that will be applied in the future of the forest industry. Thanks to the SWOT analysis, it was found that nanomaterials have a great market potential when they are used in wood-based composites. They offer both innovativeness and opportunities to position products in new market segments. Nanomaterials develop all levels of the product—wood-based composites. Through satisfaction of sophisticated combination of environmentaly oriented consumer needs and the need of something new, nanomaterials in bioadhesives transform the traditional WBCs and abe able to rtestart their life cicle.

Along with the strengths and opportunities nanomaterials have someweaknesses and can cause threats to manufacturers of WBCs that apply them to production lines. Nanomaterials require significant costs for D&D. They can increase the risk and can cause unpredictable damages to workers if are not being implemented in a right way. Theer are also many different issues that can appear arised from the society and authorities. In breaf nanomaterials included in bioadhesives as part of WBCs have their future in the big scale manufacturing of oarticle and fiber boards, plywood and other composites.

## References

1. Ahamed M, Al Saalhi MS, Siddiqui MKJ (2010) Silver nanoparticle applications and human health. Clin Chim Acta 411:1841–1848
2. Adhikari S, Ozarska B (2018) Minimizing environmental impacts of timber products through the production process "From Sawmill to Final Products." Environ Syst Res 7(1):1–15
3. Al-Hamad EG, Ragab AM, Elattar MM, Sadek DM (2022) Experimental comparison of fibers and nanomaterials in compression test and cost of high strength concrete in Egypt. Civ Environ Eng 18(2):736–749
4. Antov P, Jivkov V, Savov V, Simeonova R, Yavorov N (2020) Structural application of eco-friendly composites from recycled wood fibres bonded with magnesium lignosulfonate. Appl Sciences 10(21):7526. https://doi.org/10.3390/app10217526
5. Antov P, Lee SH, Lubis MA, Yadav SM (2022) Potential of nanomaterials in bio-based wood adhesives: An overview. Emerg Nanomater: Oppor Chall For Sect:25–63
6. Antov P, Savov V, Neykov N (2020) Sustainable bio-based adhesives for eco-friendly wood composites. A review. Wood Res 65(1):51–62
7. Balea PG, Lunguleasa A, Zeleniuc O, Coşereanu C (2022) Three adhesive recipes based on magnesium lignosulfonate, used to manufacture particleboards with low formaldehyde emissions and good mechanical properties. Forests 13(5):737. https://doi.org/10.3390/f13050737
8. Barbour J. (2004) Wood formation and propertiesл wood quality. In: Burley J (ed) Encyclopedia of forest sciences. Elsevier, pp 1840–1846. https://doi.org/10.1016/B0-12-145160-7/00039-9
9. Bekhta P, Noshchenko G, Réh R, Kristak L, Sedliačik J, Antov P, Savov V (2021) Properties of eco-friendly particleboards bonded with lignosulfonate-urea-formaldehyde adhesives and PMDI as a crosslinker. Materials 14(17):4875

10. Bhagyalakshmi P, Rajan K, Kumar KS (2018) Experimental study on solar energy storage using phase change materials in spherical shell storage system, Orient J Chem 34(1):507. https://doi.org/10.13005/ojc/340156
11. Clausen CA (2014) Innovations in wood protection in the age of nanotechnology. In: 8 th Latin American biodeterioration and biodegradation symposium—Porto Alegre, RS, Brazil—April 7th to 10th, 2013. https://citeseerx.ist.psu.edu/document?repid=rep1&type=pdf&doi=f16235841f092b85072b6d2c4da80d20451bf50f. Accessed 21 Feb 2023
12. Delgado GC (2010) Economics and governance of nanomaterials: Potential and risks. Technol Soc 32(2):137–144
13. EUFS New EU Forest Strategy for 2030. https://ec.europa.eu/environment/strategy/forest-strategy_it. Accessed 19 Feb 2023
14. Ghozali M, Meliana Y, Chalid M (2022) Novel in situ modification for thermoplastic starch preparation based on Arenga pinnata Palm Starch. Polymers 14:4813. https://doi.org/10.3390/polym14224813
15. Gozali L, Irawan AP, Tunjungsari HK, De Candra C, Lim V, Nowescophor, Ali A (2022) The development of several marketing models in performance measurement in industrial and business management systems. In: Proceedings of the international conference on industrial engineering and operations management, Istambul
16. İnada AA, Arman S, Safaei B (2022) A novel review on the efficiency of nanomaterials for solar energy storage systems. J Energy Storage 55 (Part C):105661. https://doi.org/10.1016/j.est.2022.105661
17. Jasmani L, Rusli R, Khadiran T, Jalil R, Adnan S (2020) Application of nanotechnology in wood-based products industry: A review. Nanoscale Res Lett 15(1):207. https://doi.org/10.1186/s11671-020-03438-2
18. Jivkov V, Simeonova R, Antov P, Marinova A, Petrova B, Kristak L (2021) Structural application of lightweight panels made ofwaste cardboard and beech veneer. Materials 14:5064. https://doi.org/10.3390/ma14175064
19. Kajonius PJ, Persson BN, Jonason PK (2015) Hedonism, achievement, and power: Universal values that characterize the Dark Triad. Personality Individ Differ 77:173–178
20. Kanchanapibul M, Lacka E, Wang X, Chan HK (2014) An empirical investigation of green purchase behaviour among the young generation. J Clean Prod 66:528–536
21. Lanna A, Suklueng M, Kasagepongsan C, Suchat S (2020) Performance of novel engineered materials from epoxy resin with modified epoxidized natural rubber and nanocellulose or nanosilica. Adv Polym Technol. https://doi.org/10.1155/2020/2123836
22. Levchenko I, Mandhakini M, Prasad K, Bazaka O, Ivanova EP, Jacob MV, Bazaka K (2022) Functional nanomaterials from waste and low-value natural products: A technological approach level. Adv Mater Technol 7(11):2101471
23. Lykidis C (2023) Formaldehyde emissions from wood-based composites: effects of nanomaterials. In: Taghiyari HR, Morrell JJ, Husen A (eds) Emerging nanomaterials. Springer, Cham. https://doi.org/10.1007/978-3-031-17378-3_10
24. Lugovitskaya TN (2022) Sulfite lignin nanoparticles and nanovesicles as biologically active crop growth stimulants. ACS Appl Nano Mater 5(6):8048–8058
25. Majhi KC, Yadav M (2021) Synthesis of inorganic nanomaterials using carbohydrates. Green Sustain Process Chem Environ Eng Sci: Green Inorg Synth:109–135. https://doi.org/10.1016/B978-0-12-821887-7.00003-3
26. Mattila M, Karjaluoto H, Pento T (2003) Internet banking adoption among mature customers: early majority or laggards? J Serv Mark 17(5):514–528. https://doi.org/10.1108/0887604031048629410486294
27. Mohanty AK, Vivekanandhan S, Pin JM, Misra M (2018) Composites from renewable and sustainable resources: Challenges and innovations. Science 362(6414):536–542
28. Motas JG, Quadrini F, Nedelcu D (2020) Silver nano-coating of liquid wood for nanocomposite manufacturing. Procedia Manuf 47:974–979. https://doi.org/10.1016/j.promfg.2020.04.299
29. Nosál E, Reinprecht L (2019) Antibacterial and anti-mold efficiency of silver nanoparticles present in melamine-laminated particleboard surfaces. BioResources 14(2):3914–3924

30. Onat SM, Kelleci O (2022) Effects of silane treatment on physical and mechanical properties of particleboards prepared with urea formaldehyde. Drvna Industrija 73(4):439–444
31. Osburg VS, Strack M, Toporowski W (2016) Consumer acceptance of wood-polymer composites: A conjoint analytical approach with a focus on innovative and environmentally concerned consumers. J Clean Prod 110:180–190
32. Partanen A, Carus M (2016) Wood and natural fiber composites current trend in consumer goods and automotive parts. Reinf Plast 60(3):170–173
33. Prasanna YS, Deshmukh SS (2021) Significance of nanomaterials in solar energy storage applications. Mater Today: Proc 38:2633–2638
34. Ragauskas AJ (2016) Reporting on the cellulosic nanomaterials business ecosystem. Tappi J 15(6). https://www.osti.gov/servlets/purl/1327729. Accessed on 18 Feb 2023
35. Reinprecht L, Iždinský J, Vidholdová Z (2018) Biological resistance and application properties of particleboards containing nano-zinc oxide. Adv Mater Sci Eng. https://doi.org/10.1155/2018/2680121
36. Reh R, Kristak L, Antov P (2022) Advanced eco-friendly wood-based composites. Materials 15:8651. https://doi.org/10.3390/ma15238651
37. Savov V, Antov P, Zhou Y, Bekhta P (2023) Eco-friendly wood composites: Design, characterization and applications. Polymers 15(4):892
38. Savov V, Valchev I, Antov P, Yordanov I, Popski Z (2022) Effect of the adhesive system on the properties of fiberboard panelsbonded with hydrolysis lignin and phenol-formaldehyde resin. Polymers 14:1768
39. Schwartz SH (1992) Universals in the content and structure of values: Theoretical advances and empirical tests in 20 countries. In: Advances in experimental social psychology, vol 25, pp 1–65
40. Shaik AH, Shaik S, Goyal S, Chandan MR, Veza I, Buradi A, Alarifi IM (2022) A review on environmental and economic impact of 2d nanomaterials-based heat transfer fluids. J Nanomater. https://doi.org/10.1155/2022/3443360
41. Shi L, Hu Y, Bai Y, He Y (2020) Dynamic tuning of magnetic phase change composites for solar-thermal conversion and energy storage. Appl Energy 263:114570. https://doi.org/10.1016/j.apenergy.2020.11457
42. Shayakhmetov U, Larkina AA, Khalikov RM, Sinitsin DA, Nedoseko IV (2021) Methodological tools for university transfer of high-demand nanotechnologies to the regional building industry. Nanotekhnologii v Stroitel'stve 13(1):12–17
43. Stern T, Ranacher L, Mair C, Berghäll S, Lähtinen K, Forsblom M, Toppinen A (2018) Perceptions on the importance of forest sector innovations: Biofuels, biomaterials, or niche products? Forests 9(5):255. https://doi.org/10.3390/f9050255
44. Sulaiman NS, Hashim R, Amini MHM, Sulaiman O, Hiziroglu S (2013) Evaluation of the properties of particleboard made using oil palm starch modified with epichlorohydrin. BioResources 8(1):283–301
45. Taghiyari HR, Ilies DC, Antov P, Vasile G, Majidinajafabadi R, Lee SH (2022) Effects of nanosilver and heat treatment on the pull-off strength of sealer-clear finish in solid wood species. Polymers 14:5516. https://doi.org/10.3390/polym14245516
46. Vaseashta A (2013) Ecosystem of innovations in nanomaterials based CBRNE sensors and threat mitigation. In: Vaseashta A, Khudaverdyan S (eds) Advanced sensors for safety and security. NATO science for peace and security series B: Physics and biophysics. Springer, Dordrecht. https://doi.org/10.1007/978-94-007-7003-4_1
47. Wilson RF (2006) Nanotechnology: the challenge of regulating known unknowns. J Law, Med & Ethics 34(4):704–713
48. Wronka A, Robles E, Kowaluk G (2021) Upcycling and recycling potential of selected lignocellulosicwaste biomass. Materials 14:7772. https://doi.org/10.3390/ma14247772
49. Xu S, Girouard N, Schueneman G, Shofner ML, Meredith JC (2013) Mechanical and thermal properties of waterborne epoxy composites containing cellulose nanocrystals. Polymer 54(24):6589–6598

50. Yashawantha KM, Asif A, Ravindra Babu G, Ramis MK (2021) Rheological behavior and thermal conductivity of graphite–ethylene glycol nanofluid. J Test Eval 49(4):20190255. https://doi.org/10.1155/2022/3443360
51. Yin CJX (2019) Connotation and types of innovation. The Routledge companion to innovation management. Routledge, pp 26–54
52. Zanuttini R, Negro F (2021) Wood-based composites: innovation towards a sustainable future. Forests 12(12):1717. https://doi.org/10.3390/f12121717

The manufacturer's authorised representative in the EU is Springer Nature Customer Service Centre GmbH, Europaplatz 3, 69115 Heidelberg, Germany. If you have any concerns regarding our products, please contact ProductSafety@springernature.com

Printed and bound by CPI Group (UK) Ltd, Croydon, CR0 4YY
30/09/2026
02225253-0006